# High-Temperature Processing of Food Products

# High-Temperature Processing of Food Products

## Unit Operations and Processing Equipment in the Food Industry

*Edited by*

*Seid Mahdi Jafari*

Faculty of Food Science and Technology, Gorgan University of Agricultural Sciences and Natural Resources, Gorgan, Iran

ELSEVIER

WP
WOODHEAD PUBLISHING
An imprint of Elsevier

Woodhead Publishing is an imprint of Elsevier
50 Hampshire Street, 5th Floor, Cambridge, MA 02139, United States
The Boulevard, Langford Lane, Kidlington, OX5 1GB, United Kingdom



ISBN: 978-0-12-818618-3 (print)

ISBN: 978-0-12-819707-3 (online)

For information on all Woodhead publications
visit our website at https://www.elsevier.com/books-and-journals

*Publisher:* Nikki P. Levy
*Acquisitions Editor:* Nina Bandeira
*Editorial Project Manager:* Andrea R. Dulberger
*Production Project Manager:* Fizza Fathima
*Cover Designer:* Mark Rogers

Typeset by STRAIVE, India

# Contents

## Section B Baking and roasting operations



## Section C Thawing operations

# Contributors

**Özgül Altay** Department of Food Engineering, Faculty of Engineering, Ege University, İzmir, Turkey

**Tawheed Amin** Division of Food Science and Technology, SKUAST-Kashmir, Srinagar, India

**Elham Ansarifar** Social Determinants of Health Research Center, Department of Public Health, School of Health, Birjand University of Medical Sciences, Birjand, Iran

**Vahid Baeghbali** Department of Food Hygiene and Quality Control, School of Nutrition and Food Sciences, Shiraz University of Medical Sciences, Shiraz, Iran

**Noemi Baldino** Department of Information, Modeling, Electronics and System Engineering (D.I.M.E.S.), University of Calabria, Rende, CS, Italy

**Omar Bashir** Division of Food Science and Technology, SKUAST-Kashmir, Srinagar, India

**Angelos-Panagiotis Bizymis** Laboratory of Food Chemistry and Technology, School of Chemical Engineering, National Technical University of Athens, Athens, Greece

**Pranjal P. Das** Department of Chemical Engineering, Indian Institute of Technology Guwahati, Guwahati, Assam, India

**Prangan Duarah** Centre for the Environment, Indian Institute of Technology Guwahati, Guwahati, Assam, India

**Georgia Frakolaki** Laboratory of Food Chemistry and Technology, School of Chemical Engineering, National Technical University of Athens, Athens, Greece

**Domenico Gabriele** Department of Information, Modeling, Electronics and System Engineering (D.I.M.E.S.), University of Calabria, Rende, CS, Italy

**Gousia Gani** Division of Food Science and Technology, SKUAST-Kashmir, Srinagar, India

**Virginia Giannou** Laboratory of Food Chemistry and Technology, School of Chemical Engineering, National Technical University of Athens, Athens, Greece

**Sara Hedayati** Nutrition Research Center, School of Nutrition and Food Sciences, Shiraz University of Medical Sciences, Shiraz, Iran

**Işıl İlter** Department of Food Engineering, Faculty of Engineering, Ege University, İzmir; Department of Food Engineering, Faculty of Engineering, Manisa Celal Bayar University, Manisa, Turkey

**Md. Kaviul Islam** Department of Mechanical Engineering, Iowa State University, Union Drive, Ames, IA, United States; School of Science and Engineering, Canadian University of Bangladesh, Rangs Rl Square Pragati Sharani (Bir Uttam Rafiqul Islam Avenue), Dhaka, Bangladesh

**Abida Jabeen** Division of Food Science and Technology, SKUAST-Kashmir, Srinagar, India

**Seid Mahdi Jafari** Faculty of Food Science and Technology, Gorgan University of Agricultural Sciences and Natural Resources, Gorgan, Iran; Nutrition and Bromatology Group, Analytical and Food Chemistry Department, Faculty of Food Science and Technology, University of Vigo, Ourense, Spain

**Nusrat Jan** Division of Food Science and Technology, SKUAST-Kashmir, Srinagar, India

**Mohammad U.H. Joardder** Department of Mechanical Engineering, Rajshahi University of Engineering & Technology, Rajshahi, Bangladesh; School of Mechanical, Medical and Process Engineering, Queensland University of Technology, Brisbane, QLD, Australia

**Figen Kaymak Ertekin** Department of Food Engineering, Faculty of Engineering, Ege University, İzmir, Turkey

**Tryfon Kekes** Laboratory of Food Chemistry and Technology, School of Chemical Engineering, National Technical University of Athens, Athens, Greece

**Özgün Köprüalan** Department of Food Engineering, Faculty of Engineering, Ege University, İzmir, Turkey

**Chi-Ching Lee** Faculty of Engineering and Natural Sciences, Department of Food Engineering, Istanbul Sabahattin Zaim University, Istanbul, Turkey

**Francesca Romana Lupi** Department of Information, Modeling, Electronics and System Engineering (D.I.M.E.S.), University of Calabria, Rende, CS, Italy

**Taha Mukhtar** Division of Food Science and Technology, SKUAST-Kashmir, Srinagar, India

**Marin Neio Demirci** Faculty of Engineering and Natural Sciences, Department of Food Engineering, Istanbul Sabahattin Zaim University, Istanbul, Turkey

**Mihir K. Purkait** Department of Chemical Engineering, Indian Institute of Technology Guwahati, Guwahati, Assam, India

**Nantawan Therdthai** Department of Product Development, Faculty of Agro-Industry, Kasetsart University, Bangkok, Thailand

**Constantina Tzia** Laboratory of Food Chemistry and Technology, School of Chemical Engineering, National Technical University of Athens, Athens, Greece

**Nazrana Rafique Wani** Division of Food Science and Technology, SKUAST-Kashmir, Srinagar, India

# An overview of high-temperature food processes

# 1

*Işıl İlter*[a,b], *Özgül Altay*[a], *Özgün Köprüalan*[a], *Figen Kaymak Ertekin*[a], *and Seid Mahdi Jafari*[c,d]

[a]Department of Food Engineering, Faculty of Engineering, Ege University, İzmir, Turkey, [b]Department of Food Engineering, Faculty of Engineering, Manisa Celal Bayar University, Manisa, Turkey, [c]Faculty of Food Science and Technology, Gorgan University of Agricultural Sciences and Natural Resources, Gorgan, Iran, [d]Nutrition and Bromatology Group, Analytical and Food Chemistry Department, Faculty of Food Science and Technology, University of Vigo, Ourense, Spain

## 1.1 Introduction

High-temperature (HT) food processing has been widely used since ancient times, not only for increasing the flavor properties but also for the protection of foods by destroying enzymes and microorganisms. In addition, heat treatments release natural phytochemicals with functional and bioactive properties and ensure food safety, and the application settings are quite simple and clear (Nayak et al., 2015; van Boekel et al., 2010). However, HT applications in foods can result in nutrient losses as well as the production of hazardous chemicals that affect flavor, texture, and color (Mogol and Gökmen, 2016). HT food processing methods can be classified into heat processes using steam and water such as blanching (boiling), pasteurization, sterilization, cooking in water/steam; heat processes using hot air such as baking, roasting, and oven cooking methods; and heat processes using hot oil like frying and heat processes by direct and radiated energy such as dielectric, Ohmic, and infrared (IR) processing.

Cooking (see Chapters 2 and 3), especially for meat and vegetable foods, is the most widely used HT processing method. Chemical changes in the structure of foods occur during the cooking process, making them more palatable. In addition to the purpose of enhancing flavor, it is often used to pasteurize meat, for example, to inactivate microorganisms that cause spoilage and to make the product safe by killing vegetative pathogens. The heat cooking methods involve a wide variety of processes, i.e., boiling, steaming, cooking in ovens with dry heat and moist heat, vacuum cooking, and *sous-vide* cooking (Palermo et al., 2014). These cooking processes can improve the sensory and textural properties of the food material as well as its functional features. Natural phytochemical characteristics and biological properties are significantly influenced by different cooking techniques (Bernhardt and Schlich, 2006; Hidalgo and Zamora, 2017). Besides, the Maillard and carbonyl-amine reactions through thermal processes produce new bioactive compounds (Zhao et al., 2019).

**High-Temperature Processing of Food Products. https://doi.org/10.1016/B978-0-12-818618-3.00002-1**

Baking (see Chapters 4 and 6) and roasting (see Chapters 5 and 6) are HT processes in which the food undergoes simultaneous physical, chemical, and biological changes. The purpose of applying these processes in foods is to improve the sensory, flavor, and aroma properties (Dobraszczyk et al., 2006). During the baking and roasting processes, heat and mass transfers from the oven cavity to the product surface occur simultaneously with convective, radiative, and conductive heat transfer or combinations thereof. Inside and on the surface of the product, internal heat diffusion, internal moisture diffusion (liquid or steam), internal and surface evaporation, and convective moisture removal mechanisms occur (Isleroglu et al., 2012; Sakin et al., 2007a). Baking is a cooking method applied to flour-based foods and fruits, whereas roasting is a method applied to meat, nuts, seeds, and vegetables. The application of heat during baking eliminates enzymes and microorganisms and, to some extent, decreases the water activity of the food, therefore preserving food. The shelf life of baked foods is further extended by refrigeration or packaging. On the other hand, roasting is a common dry heating processing method having different aims from baking to meeting specific requirements (Bhattacharya, 2014). The most typical characteristics of baked/roasted foods are their strong aromas formed during the processing. The intense heat applied during baking/roasting causes Maillard browning reactions between sugars and amino acids on the surface layers of foods (Toledo and Brody, 2007), in which the reaction appears to be favored by HT and low moisture content. The characteristic caramel-brown color associated with baked foods is also due to Maillard reaction, and these further reactions result in toxic substances, i.e., furfural, hydroxymethylfurfural (HMF), and acrylamide.

Frying (see Chapters 9 and 10) is a complex process for cooking foods by immersing them in oil at temperatures ranging from 170°C to 190°C, which are over the boiling point of water, and it is widely used in domestic and industrial food processing (Farkas, 1994). Immersion frying or deep-fat frying includes simultaneous heat and mass transfer, producing a counter flow of water vapor and oil at the surface of the food. Fried foods are made from a wide range of components, including meats, vegetables, and grains. While the method and products are popular, frying and fried meals have several disadvantages. The end product's oil quality and oil content are the primary sources of product disadvantages. Furthermore, frying causes significant microstructural changes as well as physicochemical changes in major nutritional components (Singh and Debnath, 2011). HT used for frying increase the deterioration of the oil and cause the production of free fatty acids, which change the viscosity, taste, and color of the oil and increase foaming. In addition, the frequent consumption of fried foods is associated with an unhealthy diet, resulting in a higher risk of some diseases (Pankaj and Keener, 2017).

Thawing (see Chapters 7 and 8) is a heat transfer process used for frozen food products both on a domestic and industrial scale that involves phase transition. Heat conduction inside a product, as well as convection and radiation between the object and the surrounding environment, are the primary mechanisms for thawing (Fadiji et al., 2021). Thawing is a complex process and requires an understanding of the product's physical properties across a large temperature range, as well as heat and water transport mechanisms on the product and its surface (Lind, 1991). In an industrial setting, the major aim of optimal thawing is to reduce thawing time while reducing quality losses and microbiological growth (Bedane et al., 2018). Traditional thawing procedures, such as air thawing, water thawing, and vacuum thawing, are slow and cause product quality attributes to deteriorate (Cai et al., 2019). HTs on the product surface are common, resulting in quality losses and the potential for microbiological growth (Bedane et al., 2018). Appropriate thawing methods, which may be completed in less time, decrease the loss of sensory attributes, prevent microbiological growth and maintain the desired quality and safety of the food product. In recent years, new technologies have been developed to reduce dripping and quality deterioration during the thawing process.

This chapter provides an overview of HT food processing methods on general principles and heat and mass transfer mechanisms of the methods including cooking, baking and roasting, frying, and thawing. Also, this chapter reviewed the affecting factors of each method and their effects on food properties. Furthermore, the equipment, which can be well adopted at household and commercial levels, is discussed and recent advances and possible future applications are suggested. More details about these processes are covered in the following chapters of the current book.

## 1.2 Cooking operations

### *1.2.1 General principles of cooking processes*

The human digestive system has adapted to the ingestion of cooked meals. Since plant meals with high fiber content, meat with a hard-to-chew texture, and many other raw foods are inefficiently digested by the human digestive system (Wrangham and Conklin-Brittain, 2003), cooking is an essential process to be applied to the raw foods. Cooking is a process that raises the temperature of food, resulting in a series of simultaneous processes that affect the flavor, texture and nutritional quality, and safety of foods. Different cooking techniques indicate that the temperature of food is raised in different ways. Heat treatment is one of the procedures that are mostly used for food processing because of the favorable effects on eating quality, besides it preserves foods by inhibiting enzymes and microorganisms. Moisture is transferred from the product to the cooking environment simultaneously with the application of heat. Meanwhile, a mass transfer takes place in the form of diffusion of moisture from the inside of the product to the surface and evaporation from the surface of the product (Bengtsson et al., 1976). The impact of simultaneous heat and mass transfer on the changes that occur in food products after cooking is highly significant. Although the main heat transfer mechanism varies with cooking methods, its effect that produces also varies as well.

### *1.2.2 Heat and mass transfer during cooking*

The use of heat treatments in culinary operations has several advantages. It is easy to use and regulate, produces shelf-stable food products, and has beneficial impacts on various food components (e.g., inhibition of antinutritional trypsin in some legumes, improved digestibility of proteins, gelatinization of starches, and release of bound niacin). On the other side, heat affects the components of foods that are responsible for their flavor, color, taste, and texture, and thus weakens them in terms of quality and nutritional value. During the cooking process, while heat is transferred from the environment to the food, moisture is counter currently transferred from the product to the environment. This process is known as simultaneous heat and mass transfer (Skjöldebrand and Hallström, 1980). Savoye (1992) described the four important transfer mechanisms that occur in the cooking process: (a) heat transfer to food, (b) heat transfer within food, (c) mass transfer within food, (d) mass transfer between food and cooking medium. While heat transfer occurs at the boundaries between the environment and the food material by conduction, convection, radiation, or a combination of these mechanisms, it also takes place inside the foodstuff (Gupta, 2001; Therdthai and Zhou, 2003).

Cooking methods can be classified in a variety of ways. The transfer medium (air, water, oil, etc.) or mechanism (heated substances, electromagnetic waves, etc.) of the heat transmitted to the food can be selected in this category. Cooking methods would be summarized in this part based on their general characteristics, heat transfer systems, and process parameters. The main cooking methods are boiling (blanching), steam cooking, vacuum cooking, *sous-vide* cooking, grilling (broiling), and microwave cooking. Although baking, roasting, and frying are accepted as cooking methods, they have been examined in other titles in this chapter.

### 1.2.3 Cooking methods and equipment

Enzymatic activity in food items, particularly in fruits and vegetables, is inhibited and stopped with **boiling or blanching** methods and they are prepared for subsequent processes such as freezing, canning, or direct consumption. In this method, convective heating is applied with steam or hot water. In this respect, it cannot be considered as a single preservation method, but as a pretreatment applied especially before processes such as sterilization, drying, and freezing. To inactivate the enzymes, food is rapidly heated to a determined temperature, held at this temperature for a required time, and then rapidly cooled to ambient temperature. In practice, the time–temperature combinations used for blanching are evaluated separately for each raw material to reach a certain temperature in the thermal center of the foodstuff (Oboh, 2005). Factors affecting the boiling time are the characteristics of the food (product shape size, type, variety, degree of maturity, thermal conductivity) and the heat source (scalding temperature and heating method, convective heat transfer coefficient) (Fellows, 2016).

**Steam cooking** is the heating process by using steam at atmospheric pressure occurring at 100°C. For steam cooking at low pressures, the cooking temperature is $<100°C$, at high pressures, the cooking temperature is $>100°C$, which is also called pressure cooking. In this method, the steam heats and moistens the food and the cooking process is ensured. Steam power is used to cook food in a shorter time without losing its nutritional value (Rosenthal, 2015). After steaming, the second stage can be combined with the radiant or hot air cooking stage to give the food material the typical roasted look and flavor. HTs result in faster processing times and may retain nutrients damaged by long processing.

**Vacuum cooking** is a method applied by heating foods under a certain pressure at temperatures <100°C. Working at low temperatures prevents the food from losing its nutritional content and prevents the formation of harmful ingredients that may occur because of HTs. Heat treatment methods performed under vacuum prevent color loss of foods due to low temperature and minimized oxygen content and preserve flavor (Andrés-Bello et al., 2009). With the vacuum applied to the heating medium, the oxygen content of the environment also decreases, and this limits the oxidation reactions that may occur in the food. In addition, the low oxygen content of the environment prevents enzymatic and nonenzymatic degradation reactions, thus reducing the factors affecting the quality of the product. However, the biggest disadvantage of this method is that it takes a long time to cook (Martínez-Hernández et al., 2013).

***Sous-vide* cooking** is one of the alternative cooking methods that has started to attract attention in recent years and means "**heated under vacuum**." *Sous-vide* foods are defined as foods that are cooked under controlled conditions (at a certain temperature/time) after the food is vacuumed in packages by adding flavoring materials (salt, spices, sauce, etc.) to the raw material (García-Linares et al., 2004; Koç et al., 2017). This packaging technique stands out, especially with the effects of microbiological safety, prevention of moisture and aroma loss (Ilic et al., 2022; Kathuria et al., 2022). Since the product is cooked in the package during the preparation phase, it can be directly consumed after a short heating process as an advantage of sous vide technology. Thus, products that serve ready-made food technology, whose importance has increased exponentially in recent years, are emerging (Martínez-Hernández et al., 2013). *Sous-vide* technology serves this purpose by ensuring that the food can be preserved for a long time without losing its freshness, and can be easily heated and consumed when desired. Furthermore, it also offers reliable, durable and quality products delivered to the consumer under cold chain conditions.

**Grilling or broiling** is a fast and dry-heat cooking method in which food is cooked directly over a dry heat source such as a charcoal or gas grill or in a grill pan. In grill cooking, radiant heat transfer is very high and primary, and although not all heat transfer is radiant, high surface temperatures can be achieved by rapid onset browning reactions and charring (Marcus, 2013a). Electrical equipment or flame sources can be used in this method. Flames are used to give the surface of the food product a barbecue appearance and to give it the popular barbecue taste (Bejerholm et al., 2014).

**Microwave cooking method** uses a combination of conduction and radiation to produce heat. Water molecules vibrate and produce heat energy in the microwave cooking process due to the influence of radiation in the oven. The heat is then dissipated into the food and beverages by conduction and convection, depending on whether the food is solid, liquid, or both. Microwave cooking has a wide range of uses in the food industry. Food ingredients can be dried, sterilized, pasteurized, thawed, and cooked using microwave technology (Gupta and Leong, 2007; Liburdi et al., 2019). Microwave cooking provides several advantages, including a fast heating rate, a short cooking time, the convenience of use, and minimal maintenance costs (Salazar-González et al., 2012; Zhang et al., 2006). In addition, as compared to other traditional cooking techniques, microwave heating has a smaller impact on the nutritional quality and flavor of foods (Vadivambal and Jayas, 2010).

According to production capacity, cooking equipment is designed for high or low needs. The amount of energy required to heat the product, the cost of the final product, the profitability of the operation, the processing method, and the applications are all factors to consider when designing the cooking equipment. The fuel to be used (electricity, gas, liquid fuel, etc.), the safety of the process, the risks of food contamination, the flexibility of use, and the heat transfer all contribute to the cooking method and the equipment. These variables have a massive effect on investment and operating costs (Fellows, 2017). Table 1.1 shows cooking equipment, principles and usage areas.

### 1.2.4 Factors affecting the cooking process and their effects on food properties

Cooking techniques, cooking environment properties (temperature, pressure, temperature distribution, relative humidity (% RH), air velocity, heating power, heat flux provided, heating rate, heat transfer created coefficient) and initial food product characteristics (temperature, moisture content, flow/consistency characteristics) can be counted as the factors that affect the cooking process. It is very difficult to summarize because the factors mentioned are very diverse. The effects of factors may be different for each cooking method and each different food product. Because cooking is a complex process that involves simultaneous heat and mass transfer, many biological and chemical reactions take place simultaneously. Many studies in the literature have discussed the factors affecting the cooking methods, processing factors, and their effects on different food products. Table 1.2 contains these evaluations.

The main purpose of all cooking techniques is to provide the organoleptic quality of foods in terms of certain flavors, aromas, colors, or textures and to ensure that products are microbiologically safe. HT used during the cooking process causes protein coagulation and denaturation, while in some cases it also causes partial or complete gelatinization of starch granules and sugar caramelization. All these changes allow consumers to get the texture, aroma, and flavor they want in their food. The heat that food receives during the cooking process causes some changes in the sensory and nutritional qualities of food (Rosenthal, 2015). In general, the time-temperature combination used for all cooking methods should provide adequate microbial safety, ensure that products are suitable for digestion, provides the desired appearance and flavor, and prevents excessive textural softening, formation of undesirable ingredients, and loss of nutrients. Whatever is cooked, foods continue their metabolic activities, albeit slightly, until the cooking process. An increase in temperature during cooking may result in the degradation of the structural integrity of the food product and the ending of metabolic activity (Mehmood and Zeb, 2020). The quality of the product is changed as the changes that occur during cooking depend both on the temperature and time applied.

A wide variety of foods, each with different quality characteristics, undergo similar textural, nutritional, and chemical changes during cooking processes. The temperature and time factor to be applied to each cooking method may differ

**Table 1.1** Cooking equipment, principles and usage areas.

| Cooking equipment | Properties and usage area | Reference |
|---|---|---|
| Jacketed kettles (or boiling pans) | It's used in a variety of products, including creams, jams, sauces, and stews.<br>Steam, hot water, or hot oil are used to heat the apparatus, which has double or triple walls constructed of hemispherical stainless steel. Some models have a chilling procedure after cooking, as well as a mixing mechanism and scrapers to keep the food from clinging to the walls | Fellows (2017) |
| Bratt pans | These pans are multifunctional multipurpose cookers that may be used for braising, boiling, steaming, poaching, stewing, and dry heating.<br>Bratt pans are deep, rectangular cookware with stable lids. For cooking at temperatures >100°C, use a pressure Bratt pan with a snap-close lid | Fellows (2017) |
| Steamers | This type of cooker can also be used for cooking, roasting, caramelizing, sautéing, and then chilling products for vacuum cooking, cooling, and mixing. Manufacturers claim to have 50% more heat transfer area than hemispherical water heaters or traditional jacketed stoves, resulting in shorter processing times. Direct injection or infusion of steam can also be used to heat foods. When steam condenses, the latent heat of condensation is transferred to the product.<br>Steamers come in a variety of shapes and sizes, including big, enclosed cabinets for cooking rice, noodles, or meat. A PLC controls the cooking time and temperature, as well as displays temperature curves that may be stored and printed to fulfill HACCP criteria | Mashiane et al. (2021) |
| Vacuum Cooking | In the food industry, the vacuum cooking method is generally preferred in the production of a wide variety of food products such as sugary foods (jam, molasses, etc.), tomato paste, meat, vegetables, and fruits.<br>Vacuum cooking devices consist of 3 basic pieces of equipment. This equipment; a vacuum cooking vessel, condenser, and vacuum pump.<br>In these setups, pressure, mixer speed, and working time can be controlled by the PLC control system. There is a probe or sensor that provides temperature measurement in the boiler | Tomruk et al. (2016) |
| *Sous-vide* Cooking | The *sous-vide* technique is used for industrial meat or food producing services or retail sales and for consumers.<br>In this technique, by using steam ovens or hot water tubs during cooking, homogeneous cooking is provided because of better penetration of heat into the food with the effect of steam or water. There is no need for operations such as turning or mixing the cooked meat or vegetables. In each process, the food product is prepared, vacuum packed and cooked | Koç et al. (2017) |

**Table 1.2** Different cooking methods and factors affecting the cooking process for different food products.

| Cooking method | Product | Factors | Results | Reference |
|---|---|---|---|---|
| Boiling<br>Steaming<br>Microwave cooking<br>Stir-frying | Pumpkin and pumpkin leaves | Temperature<br>Time<br>Power | The total phenolic component loss was mostly seen in boiling and microwave cooking processes.<br>The method providing the highest antioxidant activity and preserving the antioxidant components was steaming. | Mashiane et al. (2021) |
| Boiling<br>Pressure boiling<br>Steaming<br>Pressure steaming | Black Beans | Time<br>Pressure | Decreases in total phenolic content and anthocyanin contents were determined in all cooking treatments. The steaming process allowed the preservation of these components compared to the boiling process. | Xu and Chang (2009) |
| Boiling<br>Steaming<br>Sous vide<br>Vacuum cooking | Artichoke | Temperature<br>Time<br>Pressure | Steam cooking and *sous-vide* methods were the cooking methods providing the highest total phenolic content and total antioxidant capacity results.<br>It was determined that the vacuum cooking method improved the color and textural properties. | Rinaldi et al. (2020) |
| Vacuum cooking | Beef meat | Temperature<br>Time | Higher cooking temperature and longer time led to better textural quality and cooked color perception due to protein denaturation | Devseren et al. (2020) |
| Vacuum cooking<br>Atmospheric cooking | Strawberry jam | Temperature<br>Pressure<br>Time | The amount of HMF increased with increasing time in both cooking methods.<br>The products obtained by vacuum cooking have lower HMF values; their colors are lighter and brighter. | Tomruk et al. (2016) |

| | | | | |
|---|---|---|---|---|
| *Sous-vide*<br>Vacuum cooking<br>Atmospheric cooking | Green peas, carrot | Temperature<br>Time | The cook-vide approach protected antioxidant properties, total phenolic content, and vitamin C content better than other methods. Both *sous-vide* and cook-vide procedures preserved greater nutrients in green peas and carrots. | Koç et al. (2017) |
| Boiling<br>Grilling<br>*Sous-vide* | Wild boar ham | Temperature<br>Time | Boiling led to significantly higher cooking loss than hams cooked by the other two methods. In sensory evaluation, methods other than boiling were appreciated because they provided juiciness in chewing. | Ilic et al. (2022) |
| *Sous-vide*<br>Grilling | Lamp meat | Temperature<br>Time | It is ensured that the sous vide method was applied at a lower temperature but in a longer time, the surface color was lighter, the water loss was low, and oxidation was prevented. | Ortuño et al. (2021) |

even if they are for the same feature. If a portion of food is cooked based on a quality characteristic (e.g., color or texture) and the time–temperature treatment differs from a specified procedure, other quality parameters may be sub-optimal (Martínez-Hernández et al., 2013). Thermally generated chemical reactions and nutrient loss in the cooking material are two basic causes of nutrient loss during cooking. Many nutrients are sensitive to the application of heat, their concentrations decreasing with temperature and time. Each nutrient has a different rate of degradation ( James and James, 2014). For example, ascorbic acid (vitamin C) and folic acid have very low potency and a very high degradation rate. Among the essential amino acids, lysine is the least stable. Vitamin losses mostly depend on blanching, thermal destruction, and to a lesser extent, oxidation. Conditions and processes with lower degradation rates are being developed to reduce these losses. Fast cooking methods cause less nutrient loss if applied at HTs for a short time or at low temperatures for a long time. The reason for the nutritional losses may be the thermal decomposition effect as well as the reactions between them. For example, amino groups of proteins participate in this reaction in Maillard reaction (Shahidi et al., 2014). As a result, changes in the structure of the protein and the dissolution of polysaccharides indirectly cause changes in textural properties. The application of HT causes relaxation of the tertiary.

Heating for longer durations or at HTs generates permanent changes in the tertiary structure, known as denaturation, which can result in functional property changes that are either beneficial or undesirable (Kathuria et al., 2022). Many plant foods include starch, which becomes gelatinized when heated with water during cooking. As the starch granules absorb water, they swell, and their crystalline regions deteriorate. When gelatinized starch has both a softening and a thickening action, it is used to give many food items their viscosity (Rinaldi et al., 2020). Finally, denaturation of proteins and gelatinization of starch heat-set the liquid matrix. This situation cannot occur correctly if proper cooking temperatures and times are not applied. Color, which is one of the physical properties of food products and affects the appeal of the food to the consumer, also undergoes significant changes in cooking processes.

Two types of color change occur because of cooking after HT treatment: alteration of natural pigments in raw materials and browning reactions (Rinaldi et al., 2020; Rosenthal, 2015). Metalloporphyrin pigment is found in both animal (HEM group carrying molecules) and plant origin (chlorophyll molecule) foods. The myoglobin pigment in the muscles turns from red-purple to brown as a result of changes in the iron atom in the center of the amino acids in its structure as a result of heat application. Chlorophyll, one of the color pigments found in plants, is also sensitive to heat and pH changes and turns from bright green to grayish green. The demineralization reaction of the central magnesium atom is responsible for this change. Many natural pigments, such as chlorophyll, are sensitive to pH and temperature differences and cause color change to be seen. Anthocyanin pigment is red, colorless or blue in appearance at different pH values. In addition to the changes in chemical structures, losses occur due to the passage of water-soluble pigments into the water during cooking Browning reactions are more likely to occur when there is a lack of accessible water. Caramelization happens when sugars are cooked, especially when acids or alkalis are present, and it is significant in both sugar and flour confectionery. Maillard browning is the most well-known of these processes, which includes the interaction of carboxyl groups with amino groups (Marcus, 2013b). The reaction produces a variety of brown chemicals as well as distinct odors.

### 1.2.5 Trends in cooking technology

Conventional cooking methods can cause some parts of the food product to be overcooked or undercooked, which may adversely influence the quality. Because it takes a considerable time to deliver sufficient heat to the food product center to reach a safe final temperature. In recent years, it has been seen that consumer demand has increased for high-quality products, which are produced by using alternative technologies, and especially electrical processes have gained importance as alternative minimal processing technology and as potential methods (Bozkurt and Icier, 2010). Developing alternative technologies and products is also essential to move the food industry, and other industries, toward sustainable processing and reduce commercial energy use and thereby responsibly preserve local and global environments (Proctor, 2011). Several novel thermal and nonthermal technologies designed to achieve microbial safety have minimal effect on the nutrients and quality of food. Recently, either individual or combination applications of new and innovative technologies and effects to meet products quality are of great interest (Cummins and Lyng, 2016). Ultrasound treatments, ohmic cooking and IR radiation heating have been considered as alternative novel technologies to current methods to improve product quality and safety, increase energy and processing efficiency, and reduce chemical usage.

## 1.3 Baking and roasting operations

### 1.3.1 General principles of baking and roasting processes

Baking is the process of cooking food in an oven using hot air with or without the addition of water vapor (Isleroglu et al., 2012). This process can be thought of as simultaneous heat and mass transfer that causes physical and biological changes, and chemical reactions in final food products (Purlis, 2012). Sluimer and Krist-Spit (1989) stated that the main heat transfer mechanism in baking is evaporation and condensation. In the baking process, the energy required to be given to the food system is quite high for the water to evaporate and for the baking to be fully ensured. Since studies on optimizing energy consumption and baking conditions for ovens are limited, ovens are often operated by trial and error (Broyart and Trystram, 2002).

Roasting is a dry cooking process to cook or gelatinize food ingredients (without using water or oil), neutralize nonnutritive factors, and make food more palatable. Roasting and toasting are similar cooking methods, and it is possible to separate these two methods according to the foods to which they are applied. For example, roasting is applied to meat, coffee, and nuts, while toasting is applied to cereals (Fellows, 2009a). In addition, it is possible to distinguish between roasting and toasting processes with the combinations of temperature and time. In general, roasting is a procedure that involves a lower temperature for a longer period (up to a few minutes), whereas toasting is a technique that involves a much higher temperature for a short period (e.g., $<1.0\,\text{min}$). Both of these food processing techniques involve simultaneous heat and mass transfer (Fast and Caldwell, 2000).

### 1.3.2 Heat and mass transfer during baking and roasting

Although the terms "roasting" and "baking" are sometimes interchanged, the heat and mass transport equations for roasting and baking are similar (Sakin-Yilmazer et al., 2012). In conventional ovens, heat is transferred from the oven walls to the product surface by radiation, from the heating medium by convection, and from the carrier/plate surface to the product by conduction (Marcotte, 2007). The heat transfer between the heating medium and the food takes place on the surface of the food, which is accepted as a boundary condition (Bird et al., 2002). Moisture diffusion from the product to the product surface and evaporation from the product surface happen simultaneously with heat transfer during the baking process (Bengtsson et al., 1976). When food is placed in a hot oven, the hot air having low humidity forms a moisture vapor pressure gradient that causes moisture evaporation on the food surface, and moisture moves from the inside of the food to the surface. The type of food, heat transfer rate and airflow rate in the oven are the factors affecting moisture transfer. When the moisture transfer on the surface of the food is greater than that inside the food, the surface dries up, the surface temperature is equalized to the air temperature in the oven and crust formation is observed (Marcotte, 2007). Heat-induced changes on the food surface and moisture retention in the interior of some products (cakes, bread, meats, etc.) are desirable quality characteristics (Sakin et al., 2007a). Table 1.3 shows the different types of mass and heat transfer that occur in different parts of food during baking and roasting.

**Table 1.3** Heat and mass transfer types during the baking and roasting process.

| Parts of food | Type of heat transfer | Type of mass transfer |
|---|---|---|
| Boundary-Layer | • Conduction<br>• Convection<br>• Radiation | • Vapor diffusion |
| Interior | • Conduction | • Capillary flow |
| Crust | • Conduction<br>• Vapor movement (Convection) | • Vapor diffusion |
| Evaporation part | • Vapor diffusion<br>• Capillary flow<br>• Surface diffusion | • Conduction<br>• Movement of vapor and liquid water |

### 1.3.3 Baking and roasting methods and equipment

The baked products become light, porous, easily digestible, and delicious with the effect of heat treatment (Manhiça et al., 2012). Different types of ovens are widely used for baking such as direct (forced convection), indirect, microwave, radiation, and IR systems which are the most common industrial ovens (Paton et al., 2013). Roasting which is a dry heat treatment is required continuous mixing to ensure a homogeneous process. Common roasting methods are pan roasting, oven roasting, and sand roasting. Baking and roasting methods, equipment, and principles of the methods are reviewed in Table 1.4.

### 1.3.4 Factors affecting baking and roasting process and their effects on food properties

Baking and roasting processes aim to improve the sensory qualities of foods, increase flavor, and expand the range of flavors, and textures found in foods. Shell formation reduces thermal conductivity and heat penetration rate. The temperature of the air, and therefore, the rate of heat transfer is lowered for products that require a relatively low moisture content, such as biscuits and crackers, to allow moisture to escape from the meal without forming a top crust. The size of the food is another important factor in the baking process. The distance the heat travels to reach the center of the food determines the baking time and this depends on the size of the food (Mizukoshi, 1997). Temperature distribution, % RH of the medium, baking technique (convective oven, steam oven, microwave oven, IR oven, etc.), air velocity and temperature in the oven, air velocity, heating power, heat flux supplied, heating speed and the heat transfer coefficient, the initial properties of the food product (temperature, moisture content and flow/consistency properties), the position of the food product in the oven, and the oven load are other factors that affect the properties of baked foods (Kemerli, 2011).

A group of aroma compounds known as carbonyls is formed depending on factors such as moisture content, temperature, and baking/roasting time of the process (Toledo and Brody, 2007). These compounds often form large complex molecules with a yellow or brown color and a characteristic roasted aroma. The Maillard reaction produces flavor depending on the moisture content, temperature, and processing time, as well as the amount of reducing sugars and amino acids in the food. Unlike low-temperature roasting, HT roasting quickly produces dry surface crust, resulting in an intense Maillard reaction and flavor development. Using reducing sugars, the Maillard reaction also leads to the creation of polymerized chemical compounds such as HMF and acrylamide. In baking and roasting processes, the production and concentration of acrylamide are influenced by the food matrix as well as process factors such as temperature and time, and moisture content of the food. Acrylamide is a neurotoxic component that is classified as a carcinogen in 1994 by the International Agency for Research on Cancer (IARC) (Gökmen et al., 2008; Isleroglu et al., 2012). Studies have concentrated on steam-assisted ovens for baking and roasting foods to reduce harmful compounds like acrylamide and HMF while developing bulk density and textural qualities of food products (Isleroglu et al., 2012).

**Table 1.4** Baking methods, type of equipment, and principles.

| Methods | Type of equipment | Principles | References |
|---|---|---|---|
| ***Baking*** | | | |
| Traditional Baking | Traditional semidirect Furnaces (wood or gas) | The traditional furnace is distinguished by two distinct chambers: the combustion chamber and the baking chamber. Water vapor is sprayed into the oven, which has a considerable impact on the color of the crust. Due to its design and the way dough is fed into it, the temperature profile in the oven is nonuniform. | Cappelli et al. (2021); Manhiça et al. (2012) |
| Industrial continuous and batch baking | Continuous and batch ovens | Batch ovens employ natural convection and conduction, whereas continuous ovens use forced convection and radiation. The heat conductivity of the oven cavity is critical in most batch ovens since they are static. Compared to continuous ovens, batch ovens have greater labor expenses and less homogeneous baking conditions. | Pinelli and Suman (2017); Walker (2004) |
| Air-forced convection baking | Air-forced convection rotary ovens, impingement ovens | Since rotary ovens have several shelves where bread and bakery products can be placed, they can produce in high volumes by taking up little space. In addition, the racks can rotate in this oven type and this ensures even cooking. In impingement ovens, as the product passes through the furnace on a conveyor, nozzles direct high velocity, vertical jets of heated air to the top and bottom of the product. Impingement heating not only provides advantages in baking speed, product yield and oven space but also saves fuel by offering baking at lower temperatures. | Sihler (1973) |

| | | | |
|---|---|---|---|
| Microwave baking | Microwave ovens, IR ovens | The advantage of microwave ovens is that the radiation passes quickly through the dry shell and is usually absorbed by the moist part inside the food. They are extremely costly to run and difficult to control, in addition to having quality issues owing to inconsistent heating. Generators of microwaves (magnetron tubes) convert electrical energy into thermal heat with only 50% efficiency. Microwave technology is similar to radio frequency heating, although it operates at a significantly lower frequency. Compared to conventional microwaves, the energy penetrates the product more deeply and there are fewer hot spots. | Cappelli et al. (2021); Lucas (2014); Walker (2004) |
| Steam baking | Steam ovens | Traditional steamers and pressure cookers steadily build up steam as the temperature increases. In contrast, a steam generator is located outside the cooking area in a steam oven. The internal capacity of the oven increases due to this design, the temperature can be measured more accurately and kept at a constant value and when used, a direct and strong heat transfer is ensured as the steam wraps the food from all directions. As a result, the food warms up fast and the baking time is reduced while the energy consumption is reduced. | Cappelli et al. (2021); Isleroglu et al. (2015); Sakin-Yilmazer et al. (2013); Sakin et al. (2007b) |

*Continued*

**Table 1.4** Continued

| Methods | Type of equipment | Principles | References |
|---|---|---|---|
| ***Roasting*** | | | |
| Pan roasting | Pan (stainless steel, coated enamelware, cast iron, or clay pots) | In pan roasting, surface contact is utilized to direct heat from a flame or a hot plate. Stirring ensures that heat energy is spread uniformly and fast, allowing the roasting process to go more speedily. The core temperature remains lower than the surface temperature in this method, causing the outside surface to overcook while the center remains unaffected. | Ozel et al. (2014) |
| Oven roasting | Roasting ovens | Roasting ovens are mostly used for color and flavor development in coffee, cacao, and nuts. Many types of conduction and convection roasting ovens use gas or electricity as a heat source. Convection ovens are more popular because they heat consistently and provide faster roasting than conventional ovens. In these ovens, heat transfer mechanisms are effective through conduction on the surfaces in direct contact with the product, convection from the hot air in the oven environment to the product, and radiation from the heat source used. Direct-heat ovens provide the advantages of rapid roasting periods, high thermal efficiency, good control over roast settings, and quick start-up. The absence of heating zones and the difficulty of automatic loading and unloading are two disadvantages of continuous furnaces. Impact furnaces, in which hot gas flows at a higher velocity perpendicular to the product surface, have been invented recently. | Berk (2018); Fellows (2017) |

| | | | |
|---|---|---|---|
| Sand roasting | Pans and sand | Sand roasting is a common and quick technique of wheat processing that street food vendors use to create value-added foods. In this method, which is more commonly used in ready-to-eat food production, chickpea, barley, rice, flaxseed and peanuts are roasted in hot air at 150°C to 350°C with sand. Because of HT, grain moisture is converted to superheated vapor, which promotes puffing by creating pressure within the grain. Internal pressure is determined by the temperature of the sand, the temperature profile, the moisture content of the grain, and the roasting duration. The puffiness and crispiness of the roasted items have improved, as well as the color, aroma, taste, and texture. The disadvantages of this system are that it is difficult to control the temperature, there is no homogeneous temperature distribution and it is not a hygienic system with the pollution factors from the sand. | Jogihalli et al. (2017); Kaur et al. (2017); Kora, 2019; Swarnakar et al. (2019) |

### 1.3.5 Trends in baking and roasting technology

In recent years, innovative research in baking aims to increase energy efficiency and baked product quality while improving oven design and operating conditions. New roasting techniques and equipment have emerged because of the increasing food quality of roasting and concerns about food cleanliness (Sruthi et al., 2021). The novel baking and roasting processes are summarized in Table 1.5 with different advantages. By combining these advantages, it is possible to develop hybrid baking and roasting systems.

## 1.4 Thawing operations

### 1.4.1 General principles of the thawing process

Freezing has been practiced for thousands of years as a technique of food preservation method (Evans, 2008). In general, the freezing and thawing procedures have a strong influence on the quality of frozen foods. When it comes to limiting cellular damage and drip loss after thawing, the rate of freezing and the production of microscopic ice crystals are crucial (Hanenian and Mittal, 2004). Since freezing is generally followed by thawing, chemical, physical, and microbiological changes damage foods during thawing. Therefore, effective thawing methods are critical for food processing technologies (Fennema et al., 1973; Kalichevsky et al., 1995). To ensure food quality, rapid thawing at moderate temperatures is preferred to avoid major temperature increases and excessive dehydration. The primary objective of the thawing process is to minimize the time required to thaw and the amount of damage that causes to the food quality such as drip loss, protein structural changes, microbial growth, and textural changes (Uyar et al., 2015).

Thawing of frozen foods is an important process in the food industry, and it is beneficial both economically and microbiologically to do it quickly and carefully. The great quality that may be reached with regulated freezing and frozen storage might be harmed by uncontrolled thawing (de Alwis and Fryer, 1990). Foods are typically thawed around 25°C to 40°C at home, whereas commercial thawing takes place at 50°C to 80°C. This increases the risk of spoilage and contamination by harmful microorganisms by prolonging the thawing process. Foods should be certainly thawed slightly above freezing point in the industry to maintain a hard texture for subsequent processing. Some foods are cooked immediately, which means they are quickly heated to a temperature that inhibits the growth of microorganisms. Foods that have not been cooked, such as ice cream or cream, should be consumed as soon as possible after thawing (Fellows, 2009b).

### 1.4.2 Heat transfer during thawing

Except for the differences in thermal characteristics of water and ice, thawing is the polar opposite of freezing (Delgado and Sun, 2001). The thawing period will be longer than the freezing time because water has a larger heat capacity and poorer thermal conductivity than ice. The thawing process is complicated, and a mathematical model

**Table 1.5** Innovative technologies in food baking and roasting.

| Methods | Principles |
|---|---|
| ***Baking*** | |
| Progress in microwave baking | Low airflow in a microwave oven results in very low heat and moisture transfer rates. As a result, current research has concentrated on impingement technology. This method involves hitting the surface of the food with high-velocity air jets, which increases heat rates and mass transfer. Another research examined the impact of combining a halogen light and a microwave oven with other baking methods in terms of energy consumption and product quality. Their findings revealed that bread baked in a microwave oven had a larger specific volume, but bread baked under a halogen light had a lower specific volume and resulted in more weight reduction. The combination had two benefits: microwave baking saves time, and the halogen lamp promotes crust browning (Cappelli et al., 2021). |
| Combining microwave with IR heating | IR-microwave combinations are finding usage area in bakeries (Walker, 2004; Yolacaner et al., 2017) Sumnu et al. (2005), in this study, a microwave oven was utilized in combination with a halogen light that emits near-IR radiation. The latest results show that the hybrid system combines the browning and crispiness benefits of IR heating with the timesaving benefit of microwave heating, with a baking time reduction of up to 75%. Combined ovens, on the other hand, are shown to be effective in reducing the problem of lack of browning and crust development (Yolacaner et al., 2017). |
| Steam-assisted baking | Steam-assisted baking is a hybrid method that combines the various advantages of convection and steam cooking to produce a higher quality product. This approach has several advantages, including shorter baking times, less dangerous chemical production, such as acrylamide generation, and nutrient protection. Isleroglu et al. (2012) compared a domestic convection oven at various baking temperatures by baking a cookie in a steam-assisted oven. They concluded that increasing the temperature of the surface increased the quantity of acrylamide present. They did notice, however, that the surface temperature of cookies baked in the steam-assisted oven was lower (by roughly 20°C) than those baked in natural and forced convection ovens at all baking temperatures. As a result, steam-assisted baking yields cookies of comparable quality (spread ratio, bulk density, and textural quality), but with a significantly shorter total baking time and a significantly lower amount of fat. |

*Continued*

**Table 1.5** Continued

| Methods | Principles |
|---|---|
| Vacuum baking | The efficiency of vacuum baking in enhancing bread qualities and lowering baking time was explored by Ruttarattanamongkol et al. (2011). Their findings revealed that combining vacuum and traditional baking yielded bread with identical attributes to commercially manufactured bread in a substantially shorter amount of time. The result was a little drier product with a noticeably softer crumb texture. |
| ***Roasting*** | |
| Microwave Roasting | The application time of microwave roasting is shorter than the classical hot air roasting method. Low roasting times also ensure a higher quality of the product (Goszkiewicz et al., 2020; Gunel et al., 2020). In the microwave roasting process, the core temperature is higher than the surface temperature, while in the traditional roasting process, even if the outer surface of the product is over-roasted, the heat cannot reach the core completely. |
| Forced convection continuous tumble roasting | Superheated steam is created from the moisture in the sample, resulting in quicker and more even heat transfer. The samples transported through the chamber by a perforated rotor linking to a screw conveyor (Flinn, 2012), are suspended with hot air in a rotating cylinder so that its surface is heated equally. Therefore, heat is transferred throughout the whole surface area of each particle, resulting in a consistently roasted product. Minimal heat loss and efficient energy consumption are achieved by thermal insulation and continual recirculation of warm air (Schoeman et al., 2016). |
| Revtech roasting | The screw feeder in this system feeds low humidity food into a heated spiral tube. Vibration caused by the spinning motors attached on opposing sides at a set frequency moves and mixes the product in the tube. Vibrations are controlled by changing the motor angle, and product flow is controlled by modifying the screw speed. Since the product is in direct contact with this heated tube, high thermal conductivity is ensured, and the heat is distributed on the food surface in a homogeneous and controlled manner. The temperature distribution of the pipe is affected by factors such as the thermal characteristics of the raw material, convection effects at the inlet and outlet of the pipe, and residence times (Keppler et al., 2018). |

**Table 1.5** Continued

| Methods | Principles |
|---|---|
| Superheated steam roasting | Superheated steam roasting is a contemporary cooking method in which sensible heat is given to water, causing an increase in water temperature over the boiling point at a particular pressure, resulting in the generation of superheated steam (Suleman et al., 2020). The roasting process begins when the product in the roasting chamber comes into an interface with the hot steam. Superheated steam has better thermal qualities than air at the same temperature, which improves quality of roasted product. |
| IR-hot air roasting | The advantages of this modern roasting method are that the equipment provides higher final product quality, higher diffusion coefficient and heat transfer efficiency, and faster operation (Ashtiani et al., 2017). Overheating, oxidation, low yield, and high energy costs are all known to occur in traditional hot air ovens when using IR for roasting (Rahimi et al., 2018). One of the disadvantages of this method of roasting is the lower penetrating strength of IR radiation. Furthermore, extended exposure causes biological materials to fracture, and insensitivity to the reflecting characteristics of the coating is another limitation of this technique (Wen et al., 2020). |

to explain it involves knowledge of the physical characteristics of the product across a wide temperature range, except for heat and water transport processes in the product and its surface (Lind, 1991). The thawing process is a heat transfer process that involves phase changes. Thawing is accomplished primarily by heat conduction inside a product, as well as convection and radiation between the object and the surrounding environment (Fadiji et al., 2021). In the thawing process, the temperature change is characterized as an unsteady-state heat transfer process that is a function of time and location. The heat conduction equation may be used to determine the temperature distribution. The temperature dependence of the characteristics of the product is considerable at the temperature area where the phase change occurs. Consequently, finding analytical solutions to the general equation is impossible (Góral et al., 2016).

The partial differential equation for one-dimensional heat conduction in a solid, subject to the acceptable initial and boundary conditions, physically explains the thawing of infinite slabs, infinite cylinders, and food spheres. To avoid chemical and biological decomposition, foods are usually frozen at temperatures much below their melting temperatures. The beginning state of the process is defined by the frozen storage temperature. Convective heat transfer at the food surface is both the physically most realistic and the most prevalent boundary condition for defrosting food. To finish the resolution definition, the process must have an endpoint. When the thermodynamic center temperature of any food hits 0°C, thawing is complete.

The center temperature of the food was 0°C was chosen as the finishing point of the process. Because of the nonlinearity produced by both the temperature-variable thermal characteristics of foods and the nature of the boundary conditions, there are no accurate solutions for the transfer mechanism in the thawing process (Cleland et al., 1986).

### 1.4.3 Thawing methods and equipment

Thawing of frozen foods has generally been accomplished by cold water thawing, room temperature thawing, air thawing, and refrigerator thawing. The thawing process is particularly slower with these traditional methods due to the lower heat conductivity of frozen foods, which reduces food quality (Cai et al., 2019). The risk of microbiological contamination in frozen food products is directly related to thawing time, which refers to the time that takes to reach 1°C throughout the product (Cai et al., 2020a; Liao et al., 2020). Thawing methods, equipment, and principles of the methods are reviewed in Table 1.6.

### 1.4.4 Factors affecting the thawing process and their effects on food properties

The heat transfer from the thawing medium to the food surface and the heat conduction from the food surface to the center of the food determine the thawing rate. Although thawing in "warm" air takes two to three times as long as freezing in cold air, the heat transfer coefficient and temperature difference might be similar. This can be attributed that water has a poorer thermal conductivity than ice (0.5 vs. 2.4 W/m °C); therefore, as the layer of water grows thicker, heat transfer to the core becomes more difficult (Kennedy, 2000).

Furthermore, the freezing media should be as cold as possible during freezing, while the thawing medium should be kept at a constant temperature during thawing to avoid microbiological growth or problems with appearance or discoloration depending on the rise in surface temperature. In the food industry, air thawing is a frequently applied procedure. Air temperatures are programmed and, when the food surface temperature reaches a definite level, generally about 7°C to 10°C, the air temperature is decreased to below that level (Archer et al., 1995). Controlling RH is critical in an industrial air thawing of unpackaged items to avoid weight loss. Heat transfer by condensation of steam is used in vacuum thawing. Because of the high heat transfer coefficient of steam, unpackaged foods can be defrosted faster than in air or water (Kennedy, 2000).

To avoid microbial growth, chemical deterioration, and excessive water loss due to dripping or dehydration, thawing time should be kept at a minimum (Taher and Farid, 2001). In general, the freezing and thawing procedures have a strong influence on the quality of frozen foods. To avoid tissue damage and drip loss during thawing, the rate of freezing and the production of small ice crystals are of great importance. Chemical and physical changes, as well as microbial growth, occur throughout the thawing

**Table 1.6** Thawing methods, types of equipments, and principles.

| Methods | Types of equipment | Principles | Reference |
|---|---|---|---|
| Cold Water Thawing | Cold Water Tank | This method is quicker than thawing in the refrigerator, but it necessitates more caution. During the thawing process, food should be kept in a hermetic box or packing material such as a plastic bag. Microorganisms in the water might contaminate the food if the packaging material leaks. While it may be used to thaw whole fish, it is not suitable for thawing other foods. | Özoğul (2019) |
| Refrigerator Thawing | Refrigerator | One of the most common methods is to thaw the frozen foods in the refrigerator at around 4°C with air movement until it reaches refrigerator temperature. Although thawing in the refrigerator is safe microbiologically, it has some disadvantages such as a long thawing period and significant drip loss | Anderson et al. (2004) |
| Air Thawing | Air Thawing Machine | A wide range of foods is thawed using continuous air thawing machines. During the thawing process, saturated air is utilized to avoid food dehydration. This procedure is normally performed at room temperature. The inability to produce homogenous thawing in the product, as well as the possibility of microbiological and quality losses, are the most significant disadvantages | Brown et al. (2006) |

process (Fennema et al., 1973; Kalichevsky et al., 1995). It is desirable to prevent quick thawing at low temperatures and increase dehydration of food to ensure food quality. The longer the product thawing period, the higher the microbial growth on the product surface. Traditional thawing has the disadvantages of lower nutritional quality and high energy expenditure due to the leaching of soluble proteins in high protein meals (Roberts et al., 1998). The freezing and thawing processes have the greatest impact on the water content of foods. As the water freezes, the concentration of solutes (proteins, carbohydrates, lipids, vitamins, and minerals) in foods increases (Owen and Lawrie, 1998).

In terms of quality, the thawing process in many food products is critical. Microbial deterioration, drip loss, color and texture changes, protein denaturation, and fat and protein oxidation are examples of quality losses that occur during the thawing process in foods. Because of the long processing periods, quality losses are more likely in conventional thawing procedures. It has been stated that quick thawing has a good influence on food quality when compared to the traditional slow thawing procedure. Negative impacts such as microorganism growth in the product, production of dripping loss, surface oxidation, and color changes in the product can be detected due to the extended thawing period in the refrigerator (Bozkır et al., 2014). Li et al. (2020) found that conventional thawing caused more drip loss, worse textural qualities, and color loss than quick thawing in Atlantic salmon. Microwave thawing of mandarin and strawberry samples decreased drip loss compared with traditional thawing, according to Devseren et al. (2018). By using a high voltage electric field, Mousakhani-Ganjeh et al. (2015) were able to decrease the color and texture degradation, and drip loss found in conventional thawing in tuna fish. Consequently, food quality losses can be decreased by using novel thawing procedures to speed up the thawing time.

### 1.4.5 Trends in thawing technology

In recent years, new technologies have been developed to reduce thawing time, dripping, and quality deterioration during the thawing process. Innovative thawing methods, principles, and advantages-disadvantages are summarized in Table 1.7. Various thawing methods can be combined as hybrid methods to give additional advantages. Food thawing technology is anticipated to be further enhanced in the next years, with existing systems being improved and new technologies being implemented.

**Table 1.7** Trends in thawing technology and principles.

| Methods | Principles |
|---|---|
| Microwave thawing | Microwave thawing has several advantages over other technologies, including quick thawing, simple control, high heat efficiency, and energy-saving. Microwave thawing is now widely used at home and in the food industry because of these benefits (Cai et al., 2020b). However, in industrial usage, the most significant problem is uneven heating caused by unequal microwave field distribution and large temperature changes (leakage heating) inside the food product ( Jiang et al., 2017). One of the primary reasons for leakage heating is the high absorption of microwaves by liquid water over ice. As a result of leakage heating, significant-quality losses such as excessive water loss and thermal deterioration occur (Decareau, 1988). |
| Radio frequency | The principles of radio frequency heat generation are similar to those of microwave heat generation, although the electromagnetic wave frequencies are different (Yang et al., 2019). Polar molecules undergo dipole rotation during radio frequency thawing as a result of field shift, and electromagnetic wave energy is transformed into heat in the food product, resulting in thawing (Cai et al., 2019). |

**Table 1.7** Continued

| Methods | Principles |
|---|---|
| | Nonuniform heating produced by the inherent unequal distribution of the electromagnetic field between the product and the surrounding space (mainly air) is a major difficulty for RF heating, especially for food goods with high moisture content (Guo et al., 2019; Huang et al., 2016). |
| Ohmic thawing | Ohmic thawing is a novel technique in which frozen food containing negative electrons is introduced into a high-voltage electrostatic field and uniformly heated (Li and Sun, 2002). In the same temperature range, ohmic thawing treatment takes much less time than conventional procedures, resulting in thawed foods with better-quality features, integrity, taste, and nutrient retention (Icier et al., 2010; Seyhun et al., 2013). |
| High-pressure thawing | Frozen foods are thawed at comparatively low temperatures using high pressure in a high-pressure thawing technique (Li and Sun, 2002). The melting point of ice and specific heat capacity drops as pressure rises, whereas thermal conductivity rises (Cai et al., 2019). The freezing point of water lowers to 21°C when the pressure is raised to 210 MPa, and there is a negative impact at this pressure level (Cai et al., 2019; Wu et al., 2017). Although high-pressure–aided thawing can minimize thawing time and drip loss, it has certain drawbacks, including being costly and causing pressure-induced protein denaturation and meat discoloration (Wu et al., 2016, 2017). |
| Ultrasonic thawing | The vibration energy provided by ultrasonic waves is turned into heat energy inside the food product using ultrasound procedures. Although ultrasonic approaches lower the risk of water loss, microbial activity, and protein denaturation, they have some drawbacks, including high energy consumption, limited penetration, and localized heating (Cai et al., 2018; Wu et al., 2017). |
| IR thawing | Electromagnetic energy collides and enters the product during far IR heating, where it is transformed to heat (Ashtiani et al., 2017). Although far IR thawing can enhance frozen product quality, there is still a danger of overheating owing to insufficient IR wave penetration depth, particularly for thicker items (Cai et al., 2020b). |
| High-voltage electric thawing | The high-voltage electric field thawing process generates corona wind (Wu et al., 2017). The heat transmission in frozen items is accelerated by the corona wind's turbulence and eddies (Mousakhani-Ganjeh et al., 2015). This approach improves the thawing rate, inhibits microbial growth, and decreases energy consumption; nevertheless, the system's downsides include its huge equipment size and expensive cost (Cai et al., 2019; He et al., 2013). |

# 1.5 Frying operations

## 1.5.1 General principles of the frying process

Frying, one of the methods that provide fast and easy cooking of food, is widely applied both at home and in the food industry. Frying involves immersing food in oil at a temperature > the boiling point of water to modify the eating quality of the food (Mir-Bel et al., 2013). Thermal destruction of microorganisms and enzymes, as well as a reduction in water activity on the product surface, offer a protective effect. The moisture content after frying determines the shelf life of fried foods; items that stay wet have a limited shelf life due to moisture and oil content during storage. Many chemical and physical changes occur as a result of simultaneous heat and mass transfer, including moisture loss, oil absorption, crust formation, starch gelatinization, color change with Maillard reaction, hydrolysis, oxidation, and polymerization. Therefore, food gains its unique sensory properties (color, texture, and flavor) desired by consumers (Saguy and Dana, 2003). The consumption of snack foods, which are generally obtained by frying, is increasing rapidly around the world because they are practical and ready-to-eat foods desired in modern life (Dana and Saguy, 2006). However, many studies reported that excess fat consumption has negative effects on health and causes an increase in diseases such as coronary heart disease, cancer, diabetes, and hypertension (Krokida et al., 2001). Moreover, the formation of potentially toxic components such as acrylamide can occur during frying (Granda and Moreira, 2005). Despite consumer awareness of the relationship between nutrition and health has increased, frying remains one of the main cooking methods (Saguy and Dana, 2003).

## 1.5.2 Fried products/frying oils

The food industry includes a wide variety of fried foods. Thin products, such as chips, are often dried to less than 5% moisture content and have a high oil content. Oil oxidation, rather than microbial growth, is a significant problem due to the high oil content during storage. Thick foods having a high water content (almost 50% wb) and a low fat level such as French fries are cooked in an oven or go through a second frying step before being consumed. When a second frying stage is employed, the fat level of the finished product is often higher than that of fresh fried. Fried foods, such as donuts, which are also quite popular, have a high oil content ranging from 15% to 20% (wb), as high as deep-fried meals such as fish and chicken. The contrast between the crunchy, oily outer layer and the soft, cooked center is the most distinctive aspect of these products (Dobraszczyk et al., 2006).

Foods can be fried in different types of fats and oils, such as vegetable oils, fats, animal fats, or a combination of them. The frying oil has several functions as a heat transfer medium and provides texture and taste to fried foods. Fluidity, mild taste, low propensity to form smoking or fuming, low tendency to gum and oxidative stability of oil are the most significant factors to consider when selecting frying oils (Bouchon, 2009). Saturated fatty acids increased the stability of frying oils, although they are nutritionally undesirable (Sanibal and Mancini-Filho, 2004). In comparison to monoenoic unsaturated fatty acids or saturated fatty acids, polyunsaturated fatty acids have lower thermo-oxidative stability (Kita et al., 2005).

Many countries currently employ alternative vegetable oils, refined seed oils, partly hydrogenated seed oils, palm oil/seed oil or soybean oil blends, and palm olein or super olein as frying oils. Prefried French fries, frozen chips, and other fried foods are frequently made with palm oil or partly hydrogenated palm oil. Due to its great oxidative stability and good performance, palm olein, as a frying oil, is preferred by large snack food manufacturers in various countries. The catering and fast food industry frequently use special longevity oils that include more unsaturated fat products or hydrogenated oils (Kochhar, 2000). Besides, olive oil is a good choice for frying, because of its polyunsaturated fatty acids and the presence of phenolic antioxidants that promote oxidation resistance. Olive oil, on the other hand, is not preferred since it is costly to utilize in the industry (Dobraszczyk et al., 2006). Despite their high content of saturated fatty acids, animal fats are preferred due to their distinct flavor and inexpensive cost (Rosseli, 1998). The significant parameters to consider in the selection of frying oil are thermal and oxidative degradation at HTs in the presence of air and moisture (Kita et al., 2005).

### *1.5.3 Heat and mass transfer during frying*

Frying is a simultaneous heat and mass transfer process in which the water in the food evaporates as a result of the heat transferred from HT oil to the food, and the oil diffuses counter-currently into the voids of the food (Tarmizi and Niranjan, 2013). While the heat transfer proceeds by convection in the oil, it takes place by conduction between the food surface and the oil and inside the food. Oil temperature and heating rate determine the effect of heat in the frying process. In deep frying, the food is surrounded by oil and thus uniform heat conduction occurs from all surfaces of the food (Sumnu and Şahin, 2009). Water, which is important for mass transfer, also plays a critical role in heat transfer. While the water in the food turns into vapor, it receives evaporation latent heat from the oil, so burning does not occur on the surface of the food. The heat on the surface is transmitted to the center of the food and it is cooked (Blumenthal and Stier, 1991). In the first stage, the surface temperature of the product reaches the boiling point temperature of water from the initial temperature in a very short time. The fast evaporation of water on the surface is the second stage, which includes the loss of moisture on the surface and hence the formation of crust. The product begins to lose moisture in the third stage. The temperature in the center of the food reaches the boiling point of water during this stage, which lasts the longest. The last stage of the vapor bubbles is defined by the observable moisture loss from the product reaching its endpoints (Dueik and Bouchon, 2011; Farinu and Baik, 2006). Large pores formed during the removal of water from the food are filled with oil particles. As the temperature rises in the center of the food, starch gelatinization, and protein denaturation occur. After a short time, because of drying on the surface of the food, inward crusting occurs. In the next stage, the surface temperature of the food rises to the temperature of the oil. During the frying process, heat and mass transfer occurs in two independent zones (shell and interior) separated by a moving boundary and at temperatures over the boiling point of water. During the frying process, several chemical reactions and physical changes occur, resulting in a unique crispy texture and flavor (Dobarganes et al., 2000; Farkas et al., 1996; Krokida et al., 2001).

### 1.5.4 Frying methods and equipment

The frying process allows the food to be cooked in a short time by gaining the desired color, texture, and flavor because of immersing the food in edible oil at 150°C to 200°C or adding a thin layer of oil between the food and the heated surface (Farkas et al., 1996). Since frying provides the preferred sensory properties quickly and economically, it is one of the most frequently used cooking methods at home and in fast-food restaurants (Moreira et al., 1995; Varela et al., 1988). Under suitable conditions, fried foods have the desired flavor, golden crust, and crunchy texture (Rossell, 2001; Warner and Fehr, 2008).

The frying process is generally applied in two different ways; deep and shallow (contact) frying. In shallow frying, there is a thin layer of oil between the heating surface and the product. Homogeneous frying cannot take place because only one surface of the food is in contact with oil. In the deep-frying process, the food surface is completely covered by oil and the heat transfer is equal at every point, so frying becomes uniform (Yılmaz et al., 2002). Deep frying is more preferred than shallow frying because of the higher quality of the product in terms of desired color, texture, and flavor properties.

While shallow frying is performed only with batch systems, deep frying operations can be applied in batch or continuous systems. Batch frying systems are generally preferred in small companies that produce ready meals, restaurants, and fast-food production places (Dunford, 2016). In these systems, the frying process is carried out by dipping the food, which is placed in the basket, into the heated oil in a fixed system with a manual or mechanical control and keeping the food in the oil until the frying process is completed. The temperature and amount of oil are controlled electronically. Continuous frying systems are used in enterprises that produce large amounts of fried foods. Here, the raw food is fed continuously from one end of the frying device by the conveyor, and fried food is removed from the other end of the device when the frying process is completed. Some designs provide uniform distribution of the product, regulation of the flow, and immersion of the product in the oil with the paddles they contain (Tian, 2013).

Batch fryers can have one or more chambers with oil capacities ranging from 5 to 25L. The baskets, made of high-quality stainless steel, are usually oil-immersed and may have a manual or automatic basket lift system. To heat the oil in batch fryers, electricity, gas, or fuel can be used (Dobraszczyk et al., 2006). Continuous fryers are used to prepare large quantities of food at speeds up to 25,000kg per hour and fully automated machines that include a frying pan with the oil temperature controller and a conveyor belt moving the food through the oil (Moreira, 2006). The oil can be heated directly in the fry pot with a gas burner battery or an electric heater, or it can be heated in an external heat exchanger that is continuously provided with oil. To provide optimal temperature control and product quality, some continuous fryers are built with multiple independently adjustable heating zones distributed throughout the fryer (Dobraszczyk et al., 2006; Moreira et al., 1999). Deep fryers may produce a lot of pressure (9–32 psi). Due to their uniform color distribution and excellent texture, deep oil fryers were created specifically for frying certain foods like chicken. Although pressure fryers can reduce frying time in half, the presence of steam in the fryer can increase the oxidation of frying oil (Moreira et al., 1999).

### 1.5.5 Factors affecting frying process and oil absorption and their effects on food properties

The deep-frying method is often used for the cooking of potato chips, frozen French fries, tortilla/corn chips, donuts, chicken, and fish. Generally, a coating is applied on fish, chicken, and vegetables before they are fried to protect them from HTs (Moreira et al., 1999). The frying procedure is usually performed under atmospheric conditions at temperatures ranging from 170°C to 190°C. The quality of the food, the kind of oil, the surface/volume ratio, the rate at which air is introduced to the oil, the temperature, the heating type, the immersion period, and the material of the fry pot affect the changes in the food and oil. The long using of the oil causes unpleasant properties in the food. Long-term exposure to HTs and ambient air can result in highly oxidized chemicals that are potentially toxic compounds (Del Re and Jorge, 2006). A variety of factors, including oil composition and quality, frying time and temperature, product composition, moisture content, shape, porosity, prefrying, surface treatments, initial interfacial tension, and crust size have been related to fat absorption in fried foods (Moreira et al., 1997). Oil absorption is a complicated mechanism that is still not fully understood. Factors affecting this phenomenon include the original product structure, multiple interchanges between the product and the heating medium, and variations in product and oil qualities. Table 1.8 shows the factors that affect oil absorption.

The frying process causes physicochemical changes as well as microstructural changes in the food components that affect the quality of the fried product (Dueik and Bouchon, 2011; Farinu and Baik, 2006). The changes that occur in food with the effect of heat during frying can be listed as starch gelatinization, protein denaturation, color and aroma development, decrease in moisture content, increase in temperature, crust formation, shrinkage, or swelling of the product (Farkas et al., 1996).

During the frying process, the structure of the food is divided into two parts: the shell (outer layer) and the inner part (core). The desired properties of fried foods are that the crust should be dry, crispy, and oily, and the interior part should be moist and cooked. While the crust part, where moisture loss occurs rapidly, gains a crispy structure, oil absorption takes place. On the inside, more moderate changes are taking place. Starch granules can be gelatinized at around 60°C to 70°C, swell with the effect of intracellular water and coat the inside of the cell. In the 60–80°C range, the cell wall separating the cells breaks down and the cells disperse. Thus, an elemental tissue is formed and proteins can be denatured (Dueik and Bouchon, 2011).

**Table 1.8** Factors affecting oil absorption in the frying process of foods.

| Factors | Mechanism | Reference |
|---|---|---|
| Water escape | The evaporation rate of water from food during the frying process is proportional to the temperature difference between the boiling points of oil and water Because the crust developed on the surface of the food has limited thermal conductivity, it influences heat and mass transport and slows dehydration. In general, the more water that is removed from the surface, the more oil that is absorbed. Water vapor enhances surface porosity, allowing capillary channels to emerge when dehydration occurs at temperatures >100°C. Characterization of product microstructure is thus critical in identifying macroscopic transfers such as oil absorption | Vitrac et al. (2002); Moreira et al. (1997); Pinthus et al. (1995) |
| Capillary pressure | The oil absorption process is affected by capillary forces. A small pore's capacity to draw liquid upward is known as capillarity. There is a pressure difference between the two sides of the contact because of this. Furthermore, the higher the oil absorption, the lower the contact angle between the oil and the product surface. The higher the liquid's surface tension, the more oil it absorbs | Moreira et al. (1999) |
| Vapor condensation and vacuum effect | Intensive drying happens during frying at a temperature > water boiling. The solid matrix of foods acts as a barrier to the formation of water vapor bubbles, resulting in a pressure gradient in the food. As a result, water escapes during frying when the product still has a high free water content that is sensitive to evaporation rate, and the resulting excessive pressure in the material interferes with oil absorption. When the product is taken from the fryer, the core temperature lowers, the steam condenses, and the product's pressure drops dramatically. As a result of the large difference in internal and exterior pressures, a "vacuum effect" occurs, causing surface oil to permeate the product | Gamble and Rice (1987); Moreira and Barrufet (1996) |

When the temperature in the outer layers of the product rises >100°C, numerous changes occur at the cellular and intracellular levels, forming the crust. These chemical and physical changes can be counted as the emergence of intracellular substances due to product shearing, gelatinization and subsequent dehydration of starch, protein denaturation, evaporation of water, and rapid dehydration and fat absorption in the cells on the shell side (Bouchon and Aguilera, 2001). The problem frequently encountered in frying under these conditions is excessive browning or burning of the product before it is fully cooked. In addition, with the effect of HT and oxygen, deterioration reactions occur in the oil in a short time, and it causes a decrease in the nutritional components of the food and the formation of toxic compounds such as acrylamide, which has a negative effect on health (Choe and Min, 2007; Moreira et al., 1999).

### 1.5.6 Trends in frying technology

Consumer awareness of the importance of nutrition in human health and the risks of eating high-energy, low-nutrient foods has prompted further research toward the production of healthy and safe fried foods. Recently, experiments aimed at reducing the oil content of fried foods and improving the frying medium's stability have been conducted for this purpose. Alternative frying methods or combinations of them, such as vacuum frying, ultrasound-assisted frying, microwave frying, and IR frying have been popular in recent years. Table 1.9 gives detailed information regarding novel frying technologies.

**Table 1.9** Novel technologies for food frying.

| Methods | Principles | Advantages |
|---|---|---|
| Vacuum Frying | The vacuum frying method involves cooking without using any air. The boiling point of water and oil is decreased while frying is carried out under a vacuum and without reaching HTs like atmospheric frying (Devseren et al., 2020; Fan et al., 2005). | The absence of air during frying minimizes reactions including lipid oxidation, acrylamide formation, discoloration, and browning, as well as provides nutritious foods (Albertos et al., 2016; Liu et al., 2017). |
| Microwave Frying | Since microwave heating increases the heating rate and thermal efficiency, microwave frying is a novel alternative frying technology that has a high heat and mass transfer efficiency (Wang et al., 2019) | Microwave frying has the advantage of a quicker heating process that reduces processing time while also decreasing the overall oil content in the finished fried product (Devi et al., 2021). |
| | Ultrasound is now widely used as a nonthermal synergistic method to | The advantage of ultrasound assisted frying is that the fried food |

*Continued*

**Table 1.9** Continued

| Methods | Principles | Advantages |
|---|---|---|
| Ultrasound Assisted Frying | increase the production rate. In most cases, ultrasound is used as a pretreatment before frying. The ultrasonic assisted vacuum frying method has been documented in the literature (Devi et al., 2021). | absorbs less oil and has a higher quality (Su et al., 2021). |
| IR Frying | The exposure to heat from an IR source via matter with wavelengths ranging from 0.78 to 1000mm in radiant heating systems is known as IR radiant heating. An IR radiant heating system has better thermal efficiency and heats up faster (Devi et al., 2021). | IR heating in food processing has been shown to minimize processing time and energy costs by achieving less nutrient degradation, increased heat flux, improved thermal efficiency, and much more homogenous heating (Yao et al., 2020). |

Novel frying techniques involve varying degrees of advantages in enhancing heat and mass transfer and reducing energy consumption when compared to conventional frying techniques. Furthermore, literature research shows that trend techniques increase quality aspects such as oil absorption, texture, color, nutritional components, and oil degradation. The solution to the problems related to conventional frying technology has been the use of new technologies separately or together to obtain better fried food.

## 1.6 Conclusion

This paper reviews the theoretical basis of HT unit operations, as well as recent breakthroughs in the study of the impact of HT procedures on the chemical and biological aspects of food. Furthermore, based on traditional procedures, the likely mechanisms driving these alterations are addressed, and further implications and future research objectives are suggested. Technology advancements that result in energy or labor cost reductions or improved product quality are often mentioned. The major HT unit operations including cooking, baking, roasting, frying, and thawing are used in food preservation since ancient times and induce many changes in the food materials as well as value addition. The effect of temperature, time, and moisture content and other processing parameters markedly influences the quality of the heat-induced foods. The availability of cost-effective machinery is still a constraint for their wide use in novel techniques for HT food processing. Manufacturers must invest more capital in novel food processing equipment, but this, along with increased quality assurance, lowers production costs and waste. This improves production efficiency, reduces energy cost, in many cases, low labor cost, and earns more income and market share for higher quality goods. There is a need for more study on the relationship between product qualities, consumer acceptance, structure, and processing techniques for all HT procedures.

## References


Albertos, I., Martin-Diana, A.B., Jaime, I., Diez, A.M., Rico, D., 2016. Protective role of vacuum vs. atmospheric frying on PUFA balance and lipid oxidation. Innovative Food Sci. Emerg. Technol. 36, 336–342. https://doi.org/10.1016/J.IFSET.2016.07.006.

Anderson, B.A., Sun, S., Erdogdu, F., Singh, R.P., 2004. Thawing and freezing of selected meat products in household refrigerators. Int. J. Refrig. 27 (1), 63–72. https://doi.org/10.1016/S0140-7007(03)00093-8.

Andrés-Bello, A., García-Segovia, P., Martínez-Monzó, J., 2009. Effects of vacuum cooking (cook-vide) on the physical-chemical properties of sea bream fillets (*Sparus aurata*). J. Aquat. Food Prod. Technol. 18 (1–2), 79–89. https://doi.org/10.1080/10498850802581773.

Archer, G.P., Kennedy, C.J., Wilson, A.J., 1995. Position paper: towards predictive microbiology in frozen food systems – a framework for understanding microbial population dynamics in frozen structures and in freeze-thaw cycles. Int. J. Food Sci. Technol. 30 (6), 711–723. https://doi.org/10.1111/J.1365-2621.1995.TB01419.X.

Ashtiani, S.H.M., Salarikia, A., Golzarian, M.R., 2017. Analyzing drying characteristics and modeling of thin layers of peppermint leaves under hot-air and infrared treatments. Inf. Process. Agric. 4 (2), 128–139. https://doi.org/10.1016/J.INPA.2017.03.001.

Bedane, T.F., Altin, O., Erol, B., Marra, F., Erdogdu, F., 2018. Thawing of frozen food products in a staggered through-field electrode radio frequency system: a case study for frozen chicken breast meat with effects on drip loss and texture. Innovative Food Sci. Emerg. Technol. 50, 139–147. https://doi.org/10.1016/J.IFSET.2018.09.001.

Bejerholm, C., Tørngren, M.A., Aaslyng, M.D., 2014. Cooking of meat. In: Encyclopedia of Meat Sciences. Vol. 1. Elsevier, pp. 370–376, https://doi.org/10.1016/B978-0-12-384731-7.00187-2.

Bengtsson, N.E., Jakobsson, B., Dagerskog Sık, M., 1976. Cooking of beef by oven roasting: a study of heat and mass transfer. J. Food Sci. 41 (5), 1047–1053. https://doi.org/10.1111/J.1365-2621.1976.TB14386.X.

Berk, Z., 2018. Food Process Engineering and Technology. Elsevier Inc., Academic Press.

Bernhardt, S., Schlich, E., 2006. Impact of different cooking methods on food quality: retention of lipophilic vitamins in fresh and frozen vegetables. J. Food Eng. 77 (2), 327–333. https://doi.org/10.1016/J.JFOODENG.2005.06.040.

Bhattacharya, S., 2014. Roasting and toasting operations in food: process engineering and applications. In: Conventional and Advanced Food Processing Technologies. Wiley, ISBN: 9781118406328, pp. 221–248, https://doi.org/10.1002/9781118406281.CH10.

Bird, R.B., Stewart, W.E., Lightfoot, E.N., 2002. Transport Phenomena, second ed. John Wiley & Sons, Inc.

Blumenthal, M.M., Stier, R.F., 1991. Optimization of deep-fat frying operations. Trends Food Sci. Technol. 2 (C), 144–148. https://doi.org/10.1016/0924-2244(91)90659-7.

Bouchon, P., 2009. Chapter 5: Understanding oil absorption during deep-fat frying. Adv. Food Nutr. Res. 57, 209–234. https://doi.org/10.1016/S1043-4526(09)57005-2.

Bouchon, P., Aguilera, J.M., 2001. Microstructural analysis of frying potatoes. Int. J. Food Sci. Technol. 36 (6), 669–676. https://doi.org/10.1046/J.1365-2621.2001.00499.X.

Bozkır, H., Baysal, T., Rayman Ergün, A., 2014. Gıda Endüstrisinde Uygulanan Yeni Çözündürme Teknikleri. Akademik Gıda 12 (3), 38–44.
Bozkurt, H., Icier, F., 2010. Ohmic cooking of ground beef: effects on quality. J. Food Eng. 96 (4), 481–490. https://doi.org/10.1016/J.JFOODENG.2009.08.030.
Brown, T., Evans, J.A., James, C., James, S.J., Swain, M.V.L., 2006. Thawing of cook-freeze catering packs. J. Food Eng. 74 (1), 70–77. https://doi.org/10.1016/J.JFOODENG.2005.02.016.
Broyart, B., Trystram, G., 2002. Modelling heat and mass transfer during the continuous baking of biscuits. J. Food Eng. 51 (1), 47–57. https://doi.org/10.1016/S0260-8774(01)00036-X.
Cai, L., Cao, M., Cao, A., Regenstein, J., Li, J., Guan, R., 2018. Ultrasound or microwave vacuum thawing of red seabream (*Pagrus major*) fillets. Ultrason. Sonochem. 47, 122–132. https://doi.org/10.1016/J.ULTSONCH.2018.05.001.
Cai, L., Cao, M., Regenstein, J., Cao, A., 2019. Recent advances in food thawing technologies. Compr. Rev. Food Sci. Food Saf. 18 (4), 953–970. https://doi.org/10.1111/1541-4337.12458.
Cai, L., Dai, Y., Cao, M., 2020a. The effects of magnetic nanoparticles combined with microwave or far infrared thawing on the freshness and safety of red seabream (*Pagrus major*) fillets. LWT 128. https://doi.org/10.1016/J.LWT.2020.109456, 109456.
Cai, L., Zhang, W., Cao, A., Cao, M., 2020b. Effects of different thawing methods on the quality of largemouth bass (*Micropterus salmonides*). LWT 120. https://doi.org/10.1016/J.LWT.2019.108908, 108908.
Cappelli, A., Cini, E., Nunes, C., Rocha, D., 2021. Challenges and opportunities in wheat flour, pasta, bread, and bakery product production chains: a systematic review of innovations and improvement strategies to increase sustainability, productivity, and product quality. Sustainability 13 (5). https://doi.org/10.3390/SU13052608, 2608.
Choe, E., Min, D.B., 2007. Chemistry of deep-fat frying oils. J. Food Sci. 72 (5). https://doi.org/10.1111/J.1750-3841.2007.00352.X.
Cleland, D.J., Cleland, A.C., Earle, R.L., Byrne, S.J., 1986. Prediction of thawing times for foods of simple shape. Int. J. Refrig. 9 (4), 220–228. https://doi.org/10.1016/0140-7007(86)90094-0.
Cummins, E.J., Lyng, J.G., 2016. Emerging technologies in meat processing: production, processing and technology. In: Emerging Technologies in Meat Processing: Production, Processing and Technology. Wiley, pp. 1–431, https://doi.org/10.1002/9781118350676.
Dana, D., Saguy, I.S., 2006. Review: mechanism of oil uptake during deep-fat frying and the surfactant effect-theory and myth. Adv. Colloid Interf. Sci. 128–130, 267–272. https://doi.org/10.1016/J.CIS.2006.11.013.
de Alwis, A.A.P., Fryer, P.J., 1990. The use of direct resistance heating in the food industry. J. Food Eng. 11 (1), 3–27. https://doi.org/10.1016/0260-8774(90)90036-8.
Decareau, R.V., 1988. Microwaves in the food processing industry. In: Food Science and Technology. Academic Press Inc., USA, p. 234. https://agris.fao.org/agris-search/search.do?recordID=US880001788.
Del Re, P.V., Jorge, N., 2006. Vegetal oils behavior in discontinued frying of frozen pre-fried products. Food Sci. Technol. 26 (1), 56–63. https://doi.org/10.1590/S0101-20612006000100010.
Delgado, A.E., Sun, D.W., 2001. Heat and mass transfer models for predicting freezing processes – a review. J. Food Eng. 47 (3), 157–174. https://doi.org/10.1016/S0260-8774(00)00112-6.
Devi, S., Zhang, M., Ju, R., Bhandari, B., 2021. Recent development of innovative methods for efficient frying technology. Crit. Rev. Food Sci. Nutr. 61 (22), 3709–3724. https://doi.org/10.1080/10408398.2020.1804319.

Devseren, E., Okut, D., Koç, M., Karataş, H., Kaymak-Ertekin, F., 2020. Vakum Pişirmede İşlem Koşullarının Kırmızı Etin Renk, Tekstür, Mikroyapı ve Duyusal Özellikleri Üzerine Etkisi. Akademik Gıda 18, 347–356. https://doi.org/10.24323/akademik-gida.850826.
Devseren, E., Yılmaz, T., Petmez, B., Okut, D., Koç, M., Kaymak-Ertekin, F., 2018. Farklı Dondurma ve Çözündürme Yöntemleri ile Ön İşlem Uygulamalarının Mandalina ve Çilekte Kalite Özellikleri Üzerine Etkileri. Gıda 43 (3), 370–383. https://doi.org/10.15237/GIDA.GD17108.
Dobarganes, C., Márquez-Ruiz, G., Velasco, J., 2000. Interactions between fat and food during deep-frying. Eur. J. Lipid Sci. Technol. 102, 521–528. https://onlinelibrary.wiley.com/doi/epdf/10.1002/1438-9312%28200009%29102%3A8/9%3C521%3A%3AAID-EJLT521%3E3.0.CO%3B2-A.
Dobraszczyk, B.J., Ainsworth, P., Ibanoglu, S., Bouchon, P., 2006. Baking, extrusion and frying. In: Food Processing Handbook, pp. 237–290, https://doi.org/10.1002/3527607579.CH8.
Dueik, V., Bouchon, P., 2011. Development of healthy low-fat snacks: understanding the mechanisms of quality changes during atmospheric and vacuum frying. Food Rev. Int. 27 (4), 408–432. https://doi.org/10.1080/87559129.2011.563638.
Dunford, N., 2016. Industrial Deep Fat Frying. www.fapc.biz.
Evans, J.A., 2008. Frozen Food Science and Technology. Blackwell Publishing Ltd.
Fadiji, T., Ashtiani, S.H.M., Onwude, D.I., Li, Z., Opara, U.L., 2021. Finite element method for freezing and thawing industrial food processes. Foods 10 (4). https://doi.org/10.3390/FOODS10040869.
Fan, L.P., Zhang, M., Mujumdar, A.S., 2005. Vacuum frying of carrot chips. Dry. Technol. 23 (3), 645–656. https://doi.org/10.1081/DRT-200054159.
Farinu, A., Baik, O.D., 2006. Deep fat frying of foods—transport phenomena. Food Rev. Int. 21 (4), 389–410. https://doi.org/10.1080/87559120500222896.
Farkas, B.E., 1994. Modeling Immersion Frying as a Moving Boundary Problem. University of California. https://books.google.com.tr/books?id=DmHTtgAACAAJ&dq=inauthor:%22Brian+Emerson+Farkas%22&hl=tr&sa=X&redir_esc=y.
Farkas, B.E., Singh, R.P., Rumsey, T.R., 1996. Modeling heat and mass transfer in immersion frying. I, model development. J. Food Eng. 29 (2), 211–226. https://doi.org/10.1016/0260-8774(95)00072-0.
Fast, R.B., Caldwell, E.F., 2000. Breakfast Cereals, and how they Are Made. American Association of Cereal Chemists.
Fellows, P., 2009a. Baking and roasting. In: Food Processing Technology, pp. 538–551, https://doi.org/10.1533/9781845696344.3.538.
Fellows, P., 2009b. Heat processing. In: Food Processing Technology. vol. 2, pp. 339–366, https://doi.org/10.1533/9781845696344.3.339.
Fellows, P., 2016. Food processing. In. Nature 531 (7593), 139. https://doi.org/10.1038/531139a.
Fellows, P., 2017. Industrial cooking. In: Food Processing Technology, pp. 539–562, https://doi.org/10.1016/b978-0-08-100522-4.00010-9.
Fennema, O.R., Powrie, W.D., Marth, E.H., 1973. Low-temperature preservation of foods and living matter. Food Sci. 3 (xi). 598 p. https://books.google.com/books/about/Low_Temperature_Preservation_of_Foods_an.html?hl=tr&id=FUZ-QgAACAAJ.
Flinn, B., 2012. New Roasting Technology.
Gamble, M.H., Rice, P., 1987. Effect of pre-fry drying of oil uptake and distribution in potato crisp manufacture. Int. J. Food Sci. Technol. 22 (5), 535–548. https://doi.org/10.1111/J.1365-2621.1987.TB00519.X.

García-Linares, M.C., Gonzalez-Fandos, E., García-Fernandez, M.C., García-Arias, M.T., 2004. Microbiological and nutritional quality of SousVide or traditionally processed fish: influence of fat content. J. Food Qual. 27 (5), 371–387. https://doi.org/10.1111/J.1745-4557.2004.00676.X.

Gökmen, V., Açar, Ö.Ç., Serpen, A., Morales, F.J., 2008. Effect of leavening agents and sugars on the formation of hydroxymethyl furfural in cookies during baking. Eur. Food Res. Technol. 226 (5), 1031–1037. https://doi.org/10.1007/S00217-007-0628-6/FIGURES/5.

Góral, D., Kluza, F., Spiess, W.E.L., Kozłowicz, K., 2016. Review of thawing time prediction models depending on process conditions and product characteristics. Food Technol. Biotechnol. 54 (1), 3–12. https://doi.org/10.17113/FTB.54.01.16.4108.

Goszkiewicz, A., Kołodziejczyk, E., Ratajczyk, F., 2020. Comparison of microwave and convection method of roasting sunflower seeds and its effect on sensory quality, texture and physicochemical characteristics. Food Struct. 25. https://doi.org/10.1016/J.FOOSTR.2020.100144, 100144.

Granda, C., Moreira, R.G., 2005. Kinetics of acrylamide formation during traditional and vacuum frying of potato chips. J. Food Process Eng. 28 (5). https://doi.org/10.1111/J.1745-4530.2005.034.X.

Gunel, Z., Torun, M., Sahin-Nadeem, H., 2020. Sugar, d-pinitol, volatile composition, and antioxidant activity of carob powder roasted by microwave, hot air, and combined microwave/hot air. J. Food Process. Preserv. 44 (4). https://doi.org/10.1111/JFPP.14371.

Guo, C., Mujumdar, A.S., Zhang, M., 2019. New development in radio frequency heating for fresh food processing: a review. Food Eng. Rev. 11 (1), 29–43. https://doi.org/10.1007/S12393-018-9184-Z.

Gupta, Leong, E.W.W., 2007. Microwaves and Metals. Wiley.

Gupta, T., 2001. Individual heat transfer modes during contact baking of Indian unleavened flat bread (chapati) in a continuous oven. J. Food Eng. 47 (4), 313–319. https://doi.org/10.1016/S0260-8774(00)00135-7.

Hanenian, R., Mittal, G.S., 2004. Effect of freezing and thawing on meat quality. J. Food Agric. Environ., 74–80. https://agris.fao.org/agris-search/search.do?recordID=FI2005026383.

He, X., Liu, R., Nirasawa, S., Zheng, D., Liu, H., 2013. Effect of high voltage electrostatic field treatment on thawing characteristics and post-thawing quality of frozen pork tenderloin meat. J. Food Eng. 115 (2), 245–250. https://doi.org/10.1016/J.JFOODENG.2012.10.023.

Hidalgo, F.J., Zamora, R., 2017. Food processing antioxidants. Adv. Food Nutr. Res. 81, 31–64. https://doi.org/10.1016/BS.AFNR.2016.10.002.

Huang, Z., Marra, F., Wang, S., 2016. A novel strategy for improving radio frequency heating uniformity of dry food products using computational modeling. Innovative Food Sci. Emerg. Technol. 34, 100–111. https://doi.org/10.1016/J.IFSET.2016.01.005.

Icier, F., Izzetoglu, G.T., Bozkurt, H., Ober, A., 2010. Effects of ohmic thawing on histological and textural properties of beef cuts. J. Food Eng. 99 (3), 360–365. https://doi.org/10.1016/J.JFOODENG.2010.03.018.

Ilic, J., Tomasevic, I., Djekic, I., 2022. Influence of boiling, grilling, and sous-vide on mastication, bolus formation, and dynamic sensory perception of wild boar ham. Meat Sci. 188. https://doi.org/10.1016/J.MEATSCI.2022.108805, 108805.

Isleroglu, H., Kemerli, T., Kaymak-Ertekin, F., 2015. Effect of steam-assisted hybrid cooking on textural quality characteristics, cooking loss, and free moisture content of beef. Int. J. Food Prop. 18 (2), 403–414. https://doi.org/10.1080/10942912.2013.833219.

Isleroglu, H., Kemerli, T., Sakin-Yilmazer, M., Guven, G., Ozdestan, O., Uren, A., Kaymak-Ertekin, F., 2012. Effect of steam baking on acrylamide formation and browning kinetics of cookies. J. Food Sci. 77 (10), 257–263. https://doi.org/10.1111/j.1750-3841.2012.02912.x.

James, S.J., James, C., 2014. Heat processing methods. In: Encyclopedia of Meat Sciences. Vol. 2. Elsevier Ltd., https://doi.org/10.1016/B978-0-12-384731-7.00132-X.

Jiang, H., Liu, Z., Wang, S., 2017. Critical reviews in food science and nutrition microwave processing: effects and impacts on food components microwave processing: effects and impacts on food components. Crit. Rev. Food Sci. Nutr. https://doi.org/10.1080/10408398.2017.1319322.

Jogihalli, P., Singh, L., Sharanagat, V.S., 2017. Effect of microwave roasting parameters on functional and antioxidant properties of chickpea (*Cicer arietinum*). LWT Food Sci. Technol. 79, 223–233. https://doi.org/10.1016/J.LWT.2017.01.047.

Kalichevsky, M.T., Knorr, D., Lillford, P.J., 1995. Potential food applications of high-pressure effects on ice-water transitions. Trends Food Sci. Technol. 6 (8), 253–259. https://doi.org/10.1016/S0924-2244(00)89109-8.

Kathuria, D., Dhiman, A.K., Attri, S., 2022. Sous vide, a culinary technique for improving quality of food products: a review. Trends Food Sci. Technol. 119 (May 2021), 57–68. https://doi.org/10.1016/j.tifs.2021.11.031.

Kaur, R., Kaur, M., Gill, B.S., 2017. Phenolic acid composition of flaxseed cultivars by ultra-performance liquid chromatography (UPLC) and their antioxidant activities: effect of sand roasting and microwave heating. J. Food Process. Preserv. 41 (5). https://doi.org/10.1111/JFPP.13181, e13181.

Kemerli, T., 2011. Kurabiye Tipi Hamur Ürünlerinin Konvensiyonel Ve Buhar Destekli Hibrid Fırınlarda Pişirme İşleminin Karşılaştırılması Ve Pişirme İşleminin Eş Zamanlı Isı-Kütle Transferinin Modellenmesi. Ege University.

Kennedy, C., 2000. Managing Frozen Foods., p. 286.

Keppler, S., Bakalis, S., Leadley, C.E., Sahi, S.S., Fryer, P.J., 2018. Evaluation of dry heat treatment of soft wheat flour for the production of high ratio cakes. Food Res. Int. 107, 360–370. https://doi.org/10.1016/J.FOODRES.2018.02.041.

Kita, A., Lisińska, G., Powolny, M., 2005. The influence of frying medium degradation on fat uptake and texture of French fries. J. Sci. Food Agric. 85 (7), 1113–1118. https://doi.org/10.1002/JSFA.2076.

Koç, M., Baysan, U., Devseren, E., Okut, D., Atak, Z., Karataş, H., Kaymak-Ertekin, F., 2017. Effects of different cooking methods on the chemical and physical properties of carrots and green peas. Innovative Food Sci. Emerg. Technol. 42, 109–119. https://doi.org/10.1016/J.IFSET.2017.06.010.

Kochhar, S.P., 2000. Stabilisation of frying oils with natural antioxidative components. Eur. J. Lipid Sci. Technol. 102 (8–9), 552–559. https://doi.org/10.1002/1438-9312(200009)102:8/9<552::AID-EJLT552>3.0.CO;2-V.

Kora, A.J., 2019. Applications of sand roasting and baking in the preparation of traditional Indian snacks: nutritional and antioxidant status. Bull. Natl. Res. Cent. 43 (1), 1–11. https://doi.org/10.1186/S42269-019-0199-2.

Krokida, M.K., Oreopoulou, V., Maroulis, Z.B., Marinos-Kouris, D., 2001. Effect of pre-treatment on viscoelastic behaviour of potato strips. J. Food Eng. 50 (1), 11–17. https://doi.org/10.1016/S0260-8774(00)00189-8.

Li, B., Sun, D.W., 2002. Novel methods for rapid freezing and thawing of foods – a review. J. Food Eng. 54 (3), 175–182. https://doi.org/10.1016/S0260-8774(01)00209-6.

Li, J., Shi, J., Huang, X., Zou, X., Li, Z., Zhang, D., Zhang, W., Xu, Y., 2020. Effects of pulsed electric field on freeze-thaw quality of Atlantic salmon. Innovative Food Sci. Emerg. Technol. 65. https://doi.org/10.1016/J.IFSET.2020.102454, 102454.
Liao, X., Xiang, Q., Cullen, P.J., Su, Y., Chen, S., Ye, X., Liu, D., Ding, T., 2020. Plasma-activated water (PAW) and slightly acidic electrolyzed water (SAEW) as beef thawing media for enhancing microbiological safety. LWT 117. https://doi.org/10.1016/J.LWT.2019.108649, 108649.
Liburdi, K., Benucci, I., Esti, M., 2019. Effect of microwave power and blanching time in relation to different geometric shapes of vegetables. LWT 99, 497–504. https://doi.org/10.1016/J.LWT.2018.10.029.
Lind, I., 1991. Mathematical modelling of the thawing process. J. Food Eng. 14 (1), 1–23. https://doi.org/10.1016/0260-8774(91)90051-S.
Liu, R., Zhang, Y., Wu, L., Xing, Y., Kong, Y., Sun, J., Wei, Y., 2017. Impact of vacuum mixing on protein composition and secondary structure of noodle dough. LWT Food Sci. Technol. 85, 197–203. https://doi.org/10.1016/J.LWT.2017.07.009.
Lucas, T., 2014. Baking. In: Bakery Products Science and Technology, second ed., ISBN: 9781119967156, pp. 335–354, https://doi.org/10.1002/9781118792001.CH19.
Manhiça, F.A., Lucas, C., Richards, T., 2012. Wood consumption and analysis of the bread baking process in wood-fired bakery ovens. Appl. Therm. Eng. 47, 63–72. https://doi.org/10.1016/J.APPLTHERMALENG.2012.03.007.
Marcotte, M., 2007. Heat and mass transfer during baking. In: Transactions on State of the Art in Science and Engineering. vol. 13, pp. 1755–8336, https://doi.org/10.2495/978-1-85312-932-2/08.
Marcus, J.B., 2013a. Culinary arts basics: healthy cooking fundamentals. In: Culinary Nutrition., https://doi.org/10.1016/b978-0-12-391882-6.00003-0.
Marcus, J.B., 2013b. Food science basics: healthy cooking and baking demystified. In: Culinary Nutrition., https://doi.org/10.1016/b978-0-12-391882-6.00002-9.
Martínez-Hernández, G.B., Artés-Hernández, F., Colares-Souza, F., Gómez, P.A., García-Gómez, P., Artés, F., 2013. Innovative cooking techniques for improving the overall quality of a Kailan-hybrid broccoli. Food Bioprocess Technol. 6 (8), 2135–2149. https://doi.org/10.1007/S11947-012-0871-0.
Mashiane, P., Mashitoa, F.M., Slabbert, R.M., Sivakumar, D., 2021. Impact of household cooking techniques on colour, antioxidant and sensory properties of African pumpkin and pumpkin leaves. Int. J. Gastron. Food Sci. 23 (August 2020), 100307. https://doi.org/10.1016/j.ijgfs.2021.100307.
Mehmood, A., Zeb, A., 2020. Effects of different cooking techniques on bioactive contents of leafy vegetables. Int. J. Gastron. Food Sci. 22 (August). https://doi.org/10.1016/j.ijgfs.2020.100246, 100246.
Mir-Bel, J., Oria, R., Salvador, M.L., 2013. Reduction in hydroxymethylfurfural content in 'churros', a Spanish fried dough, by vacuum frying. Int. J. Food Sci. Technol. 48 (10), 2042–2049. https://doi.org/10.1111/IJFS.12182.
Mizukoshi, M., 1997. Baking Mechanism in Cake Production. https://agris.fao.org/agris-search/search.do?recordID=US1997058858.
Mogol, B.A., Gökmen, V., 2016. Thermal process contaminants: acrylamide, chloropropanols and furan. Curr. Opin. Food Sci. 7, 86–92. https://doi.org/10.1016/J.COFS.2016.01.005.
Moreira, R., 2006. The engineering aspects of deep-fat frying. In: Food Science and Technology. CRC Press, https://doi.org/10.1201/B15995-124.
Moreira, R., Palaiu, J., Sun, X., 1995. Simultaneous heat and mass transfer during the deep fat frying of tortilla chips. J. Food Process Eng. 18 (3), 307–320. https://doi.org/10.1111/J.1745-4530.1995.TB00369.X.

Moreira, R.G., Barrufet, M.A., 1996. Spatial distribution of oil after deep-fat frying of tortilla chips from a stochastic model. J. Food Eng. 27 (3), 279–290. https://doi.org/10.1016/0260-8774(95)00010-0.

Moreira, R.G., Castell-Perez, M.E., Barrufet, M.A., 1999. Deep Fat Frying: Fundamentals and Applications. Aspen. https://link.springer.com/book/9780834213210.

Moreira, R.G., Sun, X., Chen, Y., 1997. Factors affecting oil uptake in tortilla chips in deep-fat frying. J. Food Eng. 31 (4), 485–498. https://doi.org/10.1016/S0260-8774(96)00088-X.

Mousakhani-Ganjeh, A., Hamdami, N., Soltanizadeh, N., 2015. Impact of high voltage electric field thawing on the quality of frozen tuna fish (*Thunnus albacares*). J. Food Eng. 156, 39–44. https://doi.org/10.1016/J.JFOODENG.2015.02.004.

Nayak, B., Liu, R.H., Tang, J., 2015. Effect of processing on phenolic antioxidants of fruits, vegetables, and grains—a review. Crit. Rev. Food Sci. Nutr. 55 (7), 887–918. https://doi.org/10.1080/10408398.2011.654142.

Oboh, G., 2005. Effect of blanching on the antioxidant properties of some tropical green leafy vegetables. LWT Food Sci. Technol. 38 (5), 513–517. https://doi.org/10.1016/j.lwt.2004.07.007.

Ortuño, J., Mateo, L., Rodríguez-Estrada, M.T., Bañón, S., 2021. Effects of sous vide vs grilling methods on lamb meat colour and lipid stability during cooking and heated display. Meat Sci. 171. https://doi.org/10.1016/J.MEATSCI.2020.108287.

Owen, J.E., Lawrie, R.A., 1998. The effect of an artificially induced high pH on the susceptibility of minced porcine muscle to undergo oxidative rancidity under frozen storage. Int. J. Food Sci. Technol. 10 (2), 169–180. https://doi.org/10.1111/J.1365-2621.1975.TB00019.X.

Ozel, M.Z., Yanik, D.K., Gogus, F., Hamilton, J.F., Lewis, A.C., 2014. Effect of roasting method and oil reduction on volatiles of roasted *Pistacia terebinthus* using direct thermal desorption-GCxGC-TOF/MS. LWT Food Sci. Technol. 59 (1), 283–288. https://doi.org/10.1016/J.LWT.2014.05.004.

Özoğul, Y., 2019. Innovative technologies in seafood processing. In: Food Science & Technology, first ed. CRC Press, https://doi.org/10.1201/9780429327551/INNOVATIVE-TECHNOLOGIES-SEAFOOD-PROCESSING-YE.

Palermo, M., Pellegrini, N., Fogliano, V., 2014. The effect of cooking on the phytochemical content of vegetables. J. Sci. Food Agric. 94 (6), 1057–1070. https://doi.org/10.1002/JSFA.6478.

Pankaj, S.K., Keener, K.M., 2017. A review and research trends in alternate frying technologies. Curr. Opin. Food Sci. 16, 74–79. https://doi.org/10.1016/J.COFS.2017.09.001.

Paton, J., Khatir, Z., Thompson, H., Kapur, N., Toropov, V., 2013. Thermal energy management in the bread baking industry using a system modelling approach. Appl. Therm. Eng. 53 (2), 340–347. https://doi.org/10.1016/J.APPLTHERMALENG.2012.03.036.

Pinelli, M., Suman, A., 2017. Thermal and fluid dynamic analysis of an air-forced convection rotary bread-baking oven by means of an experimental and numerical approach. Appl. Therm. Eng. 117, 330–342. https://doi.org/10.1016/J.APPLTHERMALENG.2017.02.037.

Pinthus, E.J., Weinberg, P., Saguy, I.S., 1995. Oil uptake in deep fat frying as affected by porosity. J. Food Sci. 60 (4), 767–769. https://doi.org/10.1111/J.1365-2621.1995.TB06224.X.

Proctor, A., 2011. Introduction to green chemistry. In: Proctor, A. (Ed.), Alternatives to Conventional Food Processing. RSC Publishing, Cambridge, https://doi.org/10.1039/9781849730976.

Purlis, E., 2012. Baking process design based on modelling and simulation: towards optimization of bread baking. Food Control 27 (1), 45–52. https://doi.org/10.1016/J.FOODCONT.2012.02.034.

Rahimi, D., Kashaninejad, M., Ziaiifar, A.M., Mahoonak, A.S., 2018. Effect of infrared final cooking on some physico-chemical and engineering properties of partially fried chicken nugget. Innov. Food Sci. Emerg. Technol. 47, 1–8. https://doi.org/10.1016/J.IFSET.2018.01.004.

Rinaldi, M., Littardi, P., Cavazza, A., Santi, S., Grimaldi, M., Rodolfi, M., Ganino, T., Chiavaro, E., 2020. Effect of different atmospheric and subatmospheric cooking techniques on qualitative properties and microstructure of artichoke heads. Food Res. Int. 137 (August). https://doi.org/10.1016/j.foodres.2020.109679, 109679.

Roberts, J.S., Balaban, M.O., Zimmerman, R., Luzuriaga, D., 1998. Design and testing of a prototype ohmic thawing unit. Comput. Electron. Agric. 19 (2), 211–222. https://doi.org/10.1016/S0168-1699(97)00044-6.

Rosenthal, A.J., 2015. Cooking: domestic techniques. In: Encyclopedia of Food and Health, first ed. vol. 3. Elsevier Ltd., https://doi.org/10.1016/B978-0-12-384947-2.00199-9.

Rosseli, J.B., 1998. Industrial frying process. Grasas Aceites 49 (3–4), 282–295. http://grasasyaceites.revistas.csic.es.

Rossell, J.B., 2001. Frying: Improving Quality. CRC Press.

Ruttarattanamongkol, K., Wagner, M.E., Rizvi, S.S.H., 2011. Properties of yeast free bread produced by supercritical fluid extrusion (SCFX) and vacuum baking. Innovative Food Sci. Emerg. Technol. 12 (4), 542–550. https://doi.org/10.1016/J.IFSET.2011.07.006.

Saguy, I.S., Dana, D., 2003. Integrated approach to deep fat frying: engineering, nutrition, health and consumer aspects. J. Food Eng. 56 (2–3), 143–152. https://doi.org/10.1016/S0260-8774(02)00243-1.

Sakin, M., Kaymak-Ertekin, F., Ilicali, C., 2007a. Modeling the moisture transfer during baking of white cake. J. Food Eng. 80 (3), 822–831. https://doi.org/10.1016/j.jfoodeng.2006.07.011.

Sakin, M., Kaymak-Ertekin, F., Ilicali, C., 2007b. Simultaneous heat and mass transfer simulation applied to convective oven cup cake baking. J. Food Eng. 83 (3), 463–474. https://doi.org/10.1016/j.jfoodeng.2007.04.007.

Sakin-Yilmazer, M., Kaymak-Ertekin, F., Ilicali, C., 2012. Modeling of simultaneous heat and mass transfer during convective oven ring cake baking. J. Food Eng. 111 (2), 289–298. https://doi.org/10.1016/j.jfoodeng.2012.02.020.

Sakin-Yilmazer, M., Kemerli, T., Isleroglu, H., Ozdestan, O., Guven, G., Uren, A., Kaymak-Ertekin, F., 2013. Baking kinetics of muffins in convection and steam assisted hybrid ovens (baking kinetics of muffin). J. Food Eng. 119 (3), 483–489. https://doi.org/10.1016/j.jfoodeng.2013.06.019.

Salazar-González, C., Martín-González, M.F.S., López-Malo, A., Sosa-Morales, M.E., 2012. Recent studies related to microwave processing of fluid foods. Food Bioprocess Technol. 5 (1), 31–46. https://doi.org/10.1007/S11947-011-0639-Y/TABLES/3.

Sanibal, E.A.A., Mancini-Filho, J., 2004. Frying oil and fat quality measured by chemical, physical, and test kit analyses. J. Am. Oil Chem. Soc. 81 (9), 847–852. https://doi.org/10.1007/S11746-004-0990-8.

Savoye, I., Trystram, G., Duquenoy, A., Brunet, P., Marchin, F., 1992. Heat and mass transfer dynamic modelling of an indirect biscuit baking tunnel-oven. Part I: modelling principles. J. Food Eng. 16 (3), 173–196. https://doi.org/10.1016/0260-8774(92)90033-3.

Schoeman, L., Du Plessis, A., Manley, M., 2016. Non-destructive characterisation and quantification of the effect of conventional oven and forced convection continuous tumble (FCCT) roasting on the three-dimensional microstructure of whole wheat kernels using X-ray micro-computed tomography (μCT). J. Food Eng. 187, 1–13. https://doi.org/10.1016/J.JFOODENG.2016.04.015.

Seyhun, N., Ramaswamy, H.S., Zhu, S., Sumnu, G., Sahin, S., 2013. Ohmic tempering of frozen potato puree. Food Bioprocess Technol. 6 (11), 3200–3205. https://doi.org/10.1007/S11947-012-1002-7.

Shahidi, F., Samaranayaka, A.G.P., Pegg, R.B., 2014. Maillard reaction and browning. In: Encyclopedia of Meat Sciences. vol. 1. Elsevier Ltd., https://doi.org/10.1016/B978-0-12-384731-7.00130-6.

Sihler, A.L., 1973. Baking and Roasting. Am. Anthropol. 75 (5), 1721–1725. https://doi.org/10.1525/AA.1973.75.5.02A00350.

Singh, R.P., Debnath, S., 2011. Heat and mass transfer in foods during deep-fat frying. In: Batters and Breadings in Food Processing, pp. 185–206, https://doi.org/10.1016/B978-1-891127-71-7.50016-4.

Skjöldebrand, C., Hallström, B., 1980. Convection oven frying heat and mass transport in the product. J. Food Sci. 45 (5), 1347–1353. https://doi.org/10.1111/J.1365-2621.1980.TB06552.X.

Sluimer, P., Krist-Spit, C.E., 1989. Heat Transport in Dough during the Baking of Bread. https://agris.fao.org/agris-search/search.do?recordID=US8907448.

Sruthi, N.U., Premjit, Y., Pandiselvam, R., Kothakota, A., Ramesh, S.V., 2021. An overview of conventional and emerging techniques of roasting: effect on food bioactive signatures. Food Chem. 348. https://doi.org/10.1016/J.FOODCHEM.2021.129088, 129088.

Su, Y., Gao, J., Tang, S., Feng, L., Azam, S.M.R., Zheng, T., 2021. Recent advances in physical fields-based frying techniques for enhanced efficiency and quality attributes. Crit. Rev. Food Sci. Nutr. https://doi.org/10.1080/10408398.2021.1882933.

Suleman, R., Hui, T., Wang, Z., Liu, H., Zhang, D., 2020. Comparative analysis of charcoal grilling, infrared grilling and superheated steam roasting on the colour, textural quality and heterocyclic aromatic amines of lamb patties. Int. J. Food Sci. Technol. 55 (3), 1057–1068. https://doi.org/10.1111/IJFS.14388.

Sumnu, G., Sahin, S., Sevimli, M., 2005. Microwave, infrared and infrared-microwave combination baking of cakes. J. Food Eng. 71 (2), 150–155. https://doi.org/10.1016/J.JFOODENG.2004.10.027.

Sumnu, S.G., Şahin, S., 2009. Advances in deep fat frying. In: Contemporary Food Engineering, first ed. CRC Press. https://doi.org/10.1201/9781420055597.

Swarnakar, A.K., Srivastav, P.P., Das, S.K., 2019. Optimization of preconditioning process of pressure parboiled brown rice (unpolished) for microwave puffing and its comparison with hot sand bed puffing. J. Food Process Eng. 42 (3). https://doi.org/10.1111/JFPE.13007, e13007.

Taher, B.J., Farid, M.M., 2001. Cyclic microwave thawing of frozen meat: experimental and theoretical investigation. Chem. Eng. Process. Process Intensif. 40 (4), 379–389. https://doi.org/10.1016/S0255-2701(01)00118-0.

Tarmizi, A.A.H., Niranjan, K., 2013. Combination of moderate vacuum frying with high vacuum drainage-relationship between process conditions and oil uptake. Food Bioprocess Technol. 6 (10), 2600–2608. https://doi.org/10.1007/S11947-012-0921-7/FIGURES/8.

Therdthai, N., Zhou, W., 2003. Recent advances in the studies of bread baking process and their impacts on the bread baking technology. Food Sci. Technol. Res. 9 (3), 219–226. https://doi.org/10.3136/FSTR.9.219.

Tian, X., 2013. Chemical Factors Affecting Degradation Processes of Vegetable Oils during Frying., https://doi.org/10.7282/T39S1P2C.

Toledo, R.T., Brody, A.L., 2007. Food science, technology, and engineering overview for food product development. In: Developing New Food Products for a Changing Marketplace, second ed, pp. 209–231, https://doi.org/10.1201/9781420004328-8/

FOOD-SCIENCE-TECHNOLOGY-ENGINEERING-OVERVIEW-FOOD-PRODUCT-DEVELOPMENT-ROMEO-TOLEDO-AARON-BRODY.

Tomruk, D., Devseren, E., Koç, M., Ocak, Karatas, H., Kaymak-Ertekin, F., 2016. Developing a household vacuum cooking equipment, testing its performance on strawberry jam production and its comparison with atmospheric cooking. Agron. Res. 14 (Special Issue 2), 1475–1487.

Uyar, R., Bedane, T.F., Erdogdu, F., Koray Palazoglu, T., Farag, K.W., Marra, F., 2015. Radiofrequency thawing of food products – a computational study. J. Food Eng. 146, 163–171. https://doi.org/10.1016/J.JFOODENG.2014.08.018.

Vadivambal, R., Jayas, D.S., 2010. Non-uniform temperature distribution during microwave heating of food materials-a review. Food Bioprocess Technol. 3 (2), 161–171. https://doi.org/10.1007/S11947-008-0136-0/TABLES/3.

van Boekel, M., Fogliano, V., Pellegrini, N., Stanton, C., Scholz, G., Lalljie, S., Somoza, V., Knorr, D., Jasti, P.R., Eisenbrand, G., 2010. A review on the beneficial aspects of food processing. Mol. Nutr. Food Res. 54 (9), 1215–1247. https://doi.org/10.1002/MNFR.200900608.

Varela, G.M., Bender, A.E., Morton, I.D., 1988. Frying of Food: Principles, Changes, New Approaches. VCH.

Vitrac, O., Dufour, D., Trystram, G., Raoult-Wack, A.L., 2002. Characterization of heat and mass transfer during deep-fat frying and its effect on cassava chip quality. J. Food Eng. 53 (2), 161–176. https://doi.org/10.1016/S0260-8774(01)00153-4.

Walker, C.E., 2004. Oven technologies. In: Encyclopedia of Grain Science. Elsevier, pp. 396–407, https://doi.org/10.1016/B0-12-765490-9/00121-X.

Wang, Q., Li, S., Han, X., Ni, Y., Zhao, D., Hao, J., 2019. Quality evaluation and drying kinetics of shitake mushrooms dried by hot air, infrared and intermittent microwave-assisted drying methods. LWT 107, 236–242. https://doi.org/10.1016/J.LWT.2019.03.020.

Warner, K., Fehr, W., 2008. Mid-oleic/ultra low linolenic acid soybean oil: a healthful new alternative to hydrogenated oil for frying. J. Am. Oil Chem. Soc. 85 (10), 945–951. https://doi.org/10.1007/S11746-008-1275-1/TABLES/2.

Wen, Y.X., Chen, L.Y., Li, B.S., Ruan, Z., Pan, Q., 2020. Effect of infrared radiation-hot air (IR-HA) drying on kinetics and quality changes of star anise (*Illicium verum*). Dry. Technol. 39 (1), 90–103. https://doi.org/10.1080/07373937.2019.1696816.

Wrangham, R., Conklin-Brittain, N.L., 2003. Cooking as a biological trait. Comp. Biochem. Physiol. A Mol. Integr. Physiol. 136 (1), 35–46. https://doi.org/10.1016/S1095-6433(03)00020-5.

Wu, H.L., Wang, G.H., Xiang, W.Z., Li, T., He, H., 2016. Stability and antioxidant activity of food-grade phycocyanin isolated from *Spirulina platensis*. Int. J. Food Prop. 19 (10), 2349–2362. https://doi.org/10.1080/10942912.2015.1038564.

Wu, X.F., Zhang, M., Adhikari, B., Sun, J., 2017. Recent developments in novel freezing and thawing technologies applied to foods. Crit. Rev. Food Sci. Nutr. 57 (17), 3620–3631. https://doi.org/10.1080/10408398.2015.1132670.

Xu, B., Chang, S.K.C., 2009. Total phenolic, phenolic acid, anthocyanin, flavan-3-ol, and flavonol profiles and antioxidant properties of pinto and black beans (*Phaseolus vulgaris* L.) as affected by thermal processing. J. Agric. Food Chem. 57 (11), 4754–4764. https://doi.org/10.1021/JF900695S.

Yang, H., Chen, Q., Cao, H., Fan, D., Huang, J., Zhao, J., Yan, B., Zhou, W., Zhang, W., Zhang, H., 2019. Radiofrequency thawing of frozen minced fish based on the dielectric response mechanism. Innovative Food Sci. Emerg. Technol. 52, 80–88. https://doi.org/10.1016/J.IFSET.2018.10.013.

Yao, L., Fan, L., Duan, Z., 2020. Effect of different pretreatments followed by hot-air and far-infrared drying on the bioactive compounds, physicochemical property and microstructure of mango slices. Food Chem. 305. https://doi.org/10.1016/J.FOODCHEM.2019.125477, 125477.

Yılmaz, İ., Şimşek, O., Işıklı, M., 2002. Fatty acid composition and quality characteristics of low-fat cooked sausages made with beef and chicken meat, tomato juice and sunflower oil. Meat Sci. 62 (2), 253–258. https://doi.org/10.1016/S0309-1740(01)00255-8.

Yolacaner, E.T., Sumnu, G., Sahin, S., 2017. Microwave-assisted baking. In: The Microwave Processing of Foods, second ed, pp. 117–141, https://doi.org/10.1016/B978-0-08-100528-6.00006-1.

Zhang, M., Tang, J., Mujumdar, A.S., Wang, S., 2006. Trends in microwave-related drying of fruits and vegetables. Trends Food Sci. Technol. 17 (10), 524–534. https://doi.org/10.1016/J.TIFS.2006.04.011.

Zhao, C., Liu, Y., Lai, S., Cao, H., Guan, Y., San Cheang, W., Liu, B., Zhao, K., Miao, S., Riviere, C., Capanoglu, E., Xiao, J., 2019. Effects of domestic cooking process on the chemical and biological properties of dietary phytochemicals. Trends Food Sci. Technol. 85, 55–66. https://doi.org/10.1016/J.TIFS.2019.01.004.

# Section A

# Cooking operations

# Thermophysical aspects of cooking processes

2

*Mohammad U.H. Joardder*[a,b] *and Md. Kaviul Islam*[c,d]
[a]Department of Mechanical Engineering, Rajshahi University of Engineering & Technology, Rajshahi, Bangladesh, [b]School of Mechanical, Medical and Process Engineering, Queensland University of Technology, Brisbane, QLD, Australia, [c]Department of Mechanical Engineering, Iowa State University, Union Drive, Ames, IA, United States, [d]School of Science and Engineering, Canadian University of Bangladesh, Rangs RI Square Pragati Sharani (Bir Uttam Rafiqul Islam Avenue), Dhaka, Bangladesh

## 2.1 Introduction

Cooking is as ancient as civilization, and it is regarded as both a science and an art. It has been the most used method of food processing across the globe throughout history. It helps to destroy microorganisms, increasing taste and bioavailability of food. Apart from a few foods, including fruits, most raw foods are not readily available for consumption. These foods have less bioaccessibility and bioavailability in their raw form. Cooking increases the bioaccessibility of nutrients during digestion of these foods. Moreover, cooking causes killing and inactivation of microorganisms present in the food, which then enhances its shelf life ( Joardder and Masud, 2019). The actual history of the initiation and progress of cooking cannot be precisely traced back, but it has certainly evolved significantly over time. The history and gradual development of cooking techniques have been presented in Fig. 2.1.

It has commonly been assumed that early humans initiated cooking by tossing a raw piece of something into a fire. This type of cooking can be described as roasting, where foods including meat were placed in an open fire or hung above it using a stick. Before cooking, humans used to eat raw meat and other foods. After the discovery of fire, the cooking of food gained popularity as it made food softer and easy to consume. During the Paleolithic period, the food steaming process was in use by the early human Aurignacian culture in what is now France, who steamed food wrapped in leaves over hot embers. The earliest oven to be discovered dates back 6500 years, and baking became a popular cooking technique. Although baking was first used 6500 years ago, it was a more accessible and very popular cooking method in the Renaissance period. Around 3100 BCE, Egyptian civilization was flourishing, and more types of animals and food-bearing plants were cultivated. A brief history of food preservation including cooking can be found in the work of Joardder and Masud (2019).

During the medieval period, castles, monasteries, and palaces used to have formal kitchen facilities equipped with cookware and utensils; however, ordinary people had no dedicated indoor cooking facilities in the form of a kitchen. Cast iron cooking

**High-Temperature Processing of Food Products.** https://doi.org/10.1016/B978-0-12-818618-3.00009-4

**Fig. 2.1** Development of cooking techniques.

utensils were widely manufactured and became popular in the mid-17th century, which helped to replace the common open fire cooking system. Cooking in pots or utensils ensures more effective utilization of heat energy than that during typical open fire cooking. As science and technology developed significantly, food preservation and preparation became easier through processes such as refrigeration, freezing, heating, and canning. Commercial microwave (MW) ovens were first introduced in 1947. Similarly, the induction heating system is now used in cooking throughout the world due to its unique heating characteristics. Application of electricity-driven cooking systems including electric oven, MW oven, infrared cooker, and induction cooker make cooking far more easy than conventional cooking systems. In the following section, the underlying physics of cooking is discussed in detail.

## 2.2 Underlying physics of cooking

Cooking involves simultaneous heat and mass transfer processes. During cooking, energy is provided from a heat energy source to accelerate the mass transfer to and from the food. Heat energy can be provided by utilizing different forms of heating devices, including the conventional stove, through which heat is transferred from the source to the food by conduction, convection, or radiation (Joardder et al., 2019). Similarly, heat can be directly or indirectly provided to the foods depending on the nature of food and the availability of heating facilities. The overall cooking process encompasses the following energy transfer phenomena, as shown in Fig. 2.2:

**I.** Energy transfer in the form of work (fluid flow)
**II.** Energy transfer in the form of mass
**III.** Energy transfer in the form of heat

In these energy transport phenomena, the tendency of equilibrium is often referred to as the "driving force," which plays a vital role in the kinetics of cooking. In this regard, temperature, concentration, and velocity gradients are the driving forces of heat, mass,

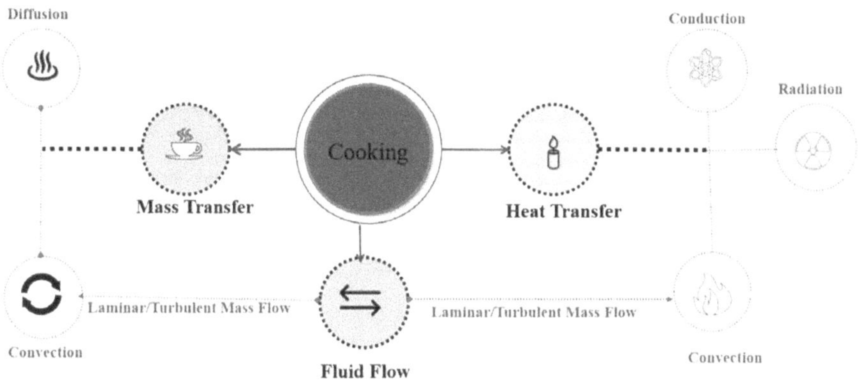


**Fig. 2.2** Different forms of energy transfer during cooking.

and momentum transfer, respectively. The transport phenomena follow laws of conservation including mass, energy, and momentum (Masud et al., 2019). Mass includes water transfers to and from the sample, maintaining different internal and external transport phenomena. Fluid flow, especially of air, is vital when convection is important in transporting heat and mass. Phase-change convection, such as boiling, is important when the cooking medium is oil or water.

## 2.2.1 Heat transfer

Heating is one of the vital transport phenomena that take place in most of the food processing technologies, including cooking. Heating results in the killing of potentially harmful microorganisms and also makes food easier to digest. Heat is applied for a wide range of purposes including reducing enzymatic activity, increasing bioavailability, and increasing the shelf life of foods. In the cooking operation, one or more modes of heat transfer phenomena take place. Heat transfer happens between two regions between which there is a temperature difference. Therefore, the temperature gradient is the driving force of heat transfer phenomena. In general, heat travels from the high-temperature zone to the low-temperature zone. However, radiation travels in both directions as long as the two bodies maintain a temperature that is greater than absolute zero.

Heat transfer within food materials can be categorized as conduction heat transfer. Conduction is a volumetric heat transfer phenomenon. The rate of conduction heat transfer directly depends on thermal diffusivity, food sample thickness, and surface area, along with the temperature gradient. Therefore, food that is cut into small pieces takes less time to cook than its larger counterparts. Convection heat transfer is a surface phenomenon where fluid flow is involved in moving heat from the surface of an object to the surroundings. Boiling is referred to as phase-change convection heat transfer as it transforms liquid to vapor. Convection heat transfer significantly depends on the type of fluid, the velocity of fluid, and the type of surface, among other things. Moreover, boiling significantly depends on the pressure and surface condition of heating elements (utensils). A highly porous cooking surface is more conducive in the initiation of boiling phenomena.

In contrast to convection and conduction, radiation heat transfer is an exceptional heat transfer that does not require any medium. Any object with a temperature of more than absolute zero spontaneously emits thermal energy as a form of electromagnetic wave. Many of the available cooking facilities are now equipped with radiation heat transfer. The infrared cooker is an example of a radiation heat transfer facilitated cooking system.

### 2.2.2 Mass transfer

Most food processing operations, including cooking, drying, crystallization, humidification, distillation, leaching, evaporation, and absorption, among others, are involved in mass transfer. Both diffusion and convection mass transfer phenomena take place during cooking. Moreover, both inward and outward mass transfer mechanisms can be observed during cooking. A portion of water, volatile components, and nutrients comes out from the food materials to the cooking medium. On the other hand, water, oil, and spices enter into the food materials during cooking. The degree of transfer of both kinds depends on the nature of cooking and medium of cooking.

### 2.2.3 Simultaneous heat and mass transfer

In reality, cooking can be treated as a method of heat transfer with concurrent mass transfer phenomena. This means both heat and mass transfer phenomena take place simultaneously during cooking. In addition to this, the turbulence of gaseous and liquid media is also an important phenomenon that controls the cooking process. Food surfaces are mainly subjected to convection and radiation heat transfer, in general; however, conduction also takes place while the food touches the bottom part of the cooking utensil. Eventually, the distribution of temperature and mass transfer highly depends on transport properties including thermal diffusivity, specific heat capacity, as well as convective heat transfer coefficient (Tocci and Mascheroni, 1996; Isleroglu and Kaymak-Ertekin, 2016).

Heat transfer accelerates the rate of mass transfer during cooking. Convection mass transfer takes place at the outer surface of food and facilitates diffusion mass transfer towards the core of the food. Moreover, pressure-driven moisture flow and diffusion occurs in driving mass from the inside of the sample (Tzempelikos et al., 2015). In addition to this, evaporation is the single most important phenomenon that takes place during cooking. Simultaneous heat and mass transfer take place during evaporation. Different combinations of heat and mass transfer phenomena take place in various types of cooking system. Fig. 2.3 depicts the prime heat and mass transfer phenomena of the different main cooking techniques.

## 2.3 The medium of cooking

The overall environment of cooking mainly depends on the medium of cooking as well as the types of heating source, as shown in Fig. 2.4. There are different ways

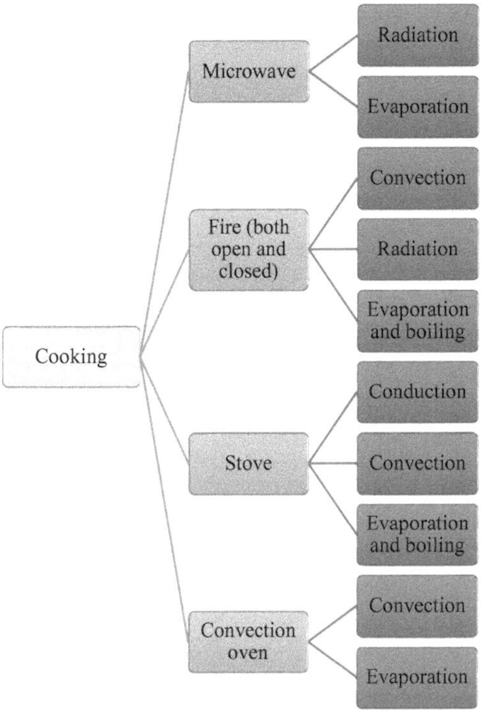


**Fig. 2.3** Simultaneous heat and mass transfer phenomena in different cooking techniques.

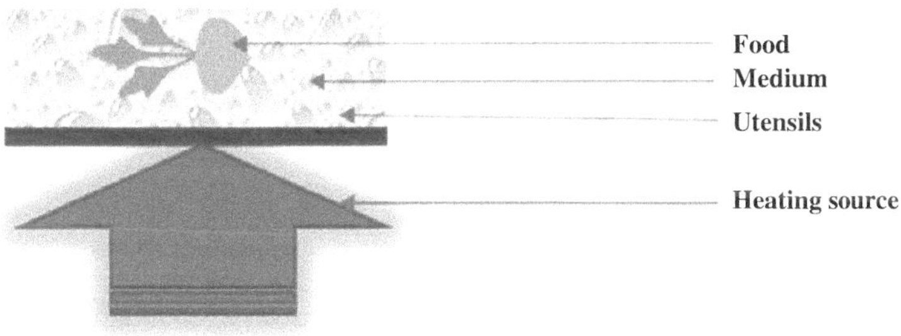


**Fig. 2.4** Distribution of heat from the heating source to food during cooking.

to execute the cooking process depending on the medium to distribute the heat uniformly. Direct contact of heat is not acceptable in most forms of cooking. To distribute the heat from the heating source to food materials, utensils and mediums of cooking are used as presented in Fig. 2.3. Water, air, and oil are the main cooking media, as described below.

## 2.3.1 Water

Water acts as an indirect medium of cooking. It takes heat from the source and then the heated water is used to heat the foods. High specific heat makes water favorable for storing higher heat energy and distributing it to the entire food. Moreover, water is the safest medium in which food can be cooked. It helps to prevent food from dehydrating during cooking where drying of food is not intended. It serves as a medium in the process of steaming and boiling. It is also used as a medium for extraction of flavor from the cooked items in the form of soup. Due to the low boiling temperature of water, a substantial amount of energy is lost during boiling. However, boiling helps in preparing many recipes for better taste.

### 2.3.2 Oil

Oil can be used as an excellent medium of cooking. Oil can be heated to a higher temperature than that of water. Water starts boiling at 100°C, whereas oil can reach up to 200°C without demonstrating any boiling phenomena. Higher temperature helps to reduce the cooking time. The Maillard reaction occurs on the outer surface of fried food. During deep frying,oil conducts more heat because it increases the surface area of food in contact with heat rather than food directly in contact with the hot pan; therefore, better heat transfer is provided by oil frying rather than dry frying.

### 2.3.3 Air

Apart from water and oil, the air is used in cooking of certain types of food. A cooking system that is purely derived by conduction or radiation heat transfer mechanisms does not need a fluid medium to distribute temperature. Moreover, an advanced cooking system such as an air fryer does not require any liquid medium of cooking. In these cooking systems, heat transfer takes place in the presence of flowing air. For instance, air frying is an innovative approach that uses air as the heating medium of food. Some cooking systems do not need any cooking medium. For example, cooking in a MW oven does not need any medium of cooking as it can heat the food directly.

## 2.4 Heating sources

Heating is indispensable in the cooking process. No cooking system can be run without a heating system. However, all heating systems transfer heat to the cooking pan or food using either the convection, conduction, or radiation mode of heat transfer. In the following sections, the common heating strategies during cooking are described briefly.

### 2.4.1 Direct flame

Smoking or roasting, providing heat directly to the food, is one of the oldest practiced food processing techniques for fish and meat (Adeyeye and Oyewole, 2016). In smoking, food is placed in the smoke from the burning of wood (Adeyeye, 2016). Convection and radiation heat transfer, along with dissipation of volatile components of food materials, take place during the smoking process. The high temperature in smoking makes the cooking of food faster, as the heat directly transfers to food without the interface of any utensils or cooking medium. In this process, the fire needs to be maintained appropriately to provide controlled heating; otherwise, over- or under-cooking may take place, eventually resulting in food that is inconsumable. Controlling the flame during burning of wood is far more difficult than during burning of coal. Eventually, wood is replaced by coal in the smoking of food and more uniform temperature distribution is attained.

### 2.4.2 Closed fireplace

In the closed cooking process, food is mainly cooked by radiation emitted from flame or hot refractory materials. The fire or glowing embers are mainly kept inside a closed fireplace. The closed fireplace must have a large enough entrance forcontinuous flow of air, as it is a flame-based system.Otherwise, a closed door results in smoke generation due to incomplete combustion of wood or coal, which is likely to make foods more smoky. Small pieces of wood are suitable for keeping the fire topped up. This type of cooking process is less suitable for grilling highly fatty meat, whereas it is suitable for less fatty meat and fish, and baking vegetables, including potatoes.

### 2.4.3 Conduction through utensils

Stoves are prevalent in both household and commercial cooking. However, the type and size of oven varies according to the amount of food that is to be cooked using the oven. There are various types of cooking stoves, such as wood stoves, biomass cookstoves, gas stoves, and electric stoves. There is a cooktop on the stoves where heat is produced. For food cooking purposes, various-shaped utensils are used to transfer the heat to the food. The food is cooked as heat is continuously transferred to the base of the cooking medium. The difference between open fire cooking and stove cooking is that in the open fire cooking process, heat is directly applied to the food, whereas in stoves, heat is directly applied to the cooking pan then transferred to the food.

### 2.4.4 Convection heating

A fan and exhaust system present in the convection system circulates hot air around the cavity of the oven. Using a fan and exhaust system reduces the moisture quantity in the food outside layer, making it juicy inside and brown and crisp on the outside. When the convection setting is powered off, it operates as a thermal oven. The heat is more evenly distributed in the convection oven than a microwave oven, which makes it suitable for a wide variety of cooking techniques such as grilling, which is very popular in many parts of the globe.

### 2.4.5 Microwave

Electromagnetic waves of a certain wavelength are applied in a microwave oven, which is used to heat and cook food. Microwave (MW) is not considered as thermal

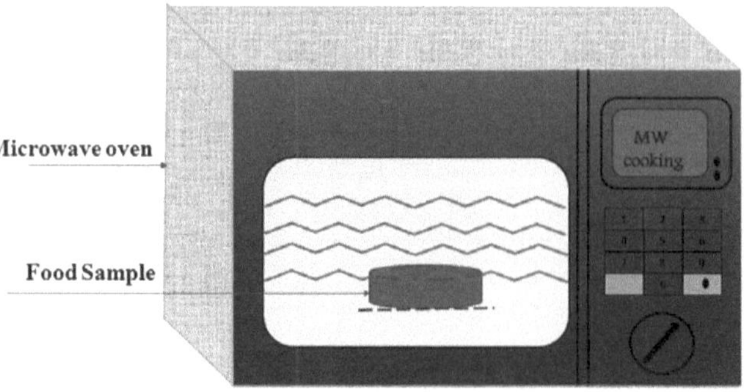


**Fig. 2.5** Electromagnetic radiation applied to foods in a microwave oven.

radiation as the wavelength of MW does not fall within the range of thermal radiation (0.1 to 100 μm). It only heats the food rather than the whole space inside the MW oven by using electromagnetic radiation. A magnetron is used for producing electromagnetic waves, as shown in Fig. 2.5. Nowadays, MW heating is getting popular due to its salient features like reduced cooking time, lower energy consumption, and volumetric heating. Electromagnetic waves with a frequency range from 300 MHz to 300 GHz are classified as microwaves (Sagar and Suresh Kumar, 2010). When MW interacts with the water molecules present in food, this results in movement of water molecules and causes heat generation within the food materials, as shown in Fig. 2.5.

Reduced heating time is required during MW heating in comparison with conductive or convective heating. This is caused due to the internal heating mechanism of MW, whereas other heating mechanisms offer heat transfer from the outside to the interior of the food ( Joardder et al., 2018). Consequently, cooking time in a MW oven is very fast and also does not preheat the oven. Although the MW heating process is very fast, food that requires a high amount of energy to cook, such as raw meat, is not suitable to cook in a microwave oven. However, recently developed combined convective-MW ovens can be used for more versatile cooking. Despite the benefits of MW heating, it is still out of the reach of many people due to its high price.

## 2.5 Performance of common cooking methods

The performance of cooking systems depends on many factors including operating cost, speed of cooking, energy consumption, quality retention, and availability. Selection of cooking methods depends on these factors. It is difficult to grade different cooking systems, as this would depend on multiple variables.

### *2.5.1 Cost*

The cost of the various cooking methods depends on the structure and technology of the method. Both initial and operating costs need to be considered when calculating the total cost. The open fire system is very simple and devoid of complex structures; that is why the cost is very low. As the convection oven has a relatively complicated system consisting of a fan and exhaust system with a robust structure, the cost is higher than the open flame cooking method. Similarly, cooking systems with a wide range of cost are available across the globe. A comparison of cost among various cooking methods is represented in Table 2.1.

### 2.5.2 Speed of cooking

The speed of the cooking methods depends on how fast heat is transferred from the heat source to the food. A microwave oven works by using electromagnetic wave propagation phenomena, and there is little chance of heat energy being wasted in the enclosed space. Hence, cooking time is faster than other cooking methods. Open fire cooking usually takes place in open space, so there is a large possibility of heat loss from the source to the surroundings. Moreover, the cooking speed also depends on the efficiency of the heating mechanism during cooking under a certain cooking method.

### 2.5.3 Availability

Among the cooking techniques, stoves are the most widely available option for daily cooking. Due to simplicity and adaptability to cook a wide variety of foods with this process, it is used in both houses and commercial settings such as restaurants. Convection cooking ovens are relatively expensive and have a complicated mechanism, which is why they are mostly limited to wealthy people. The open fire cooking process is not suitable for confined spaces because of the lack of oxygen supply; so, it is suitable only in open space. Although microwave ovens are widely available across the globe, they are still out of reach of less-privileged people.

### 2.5.4 Energy consumption

Energy consumption is vital in choosing the cooking option. Different cooking systems take different amounts of cooking energy. Types of energy source and energy efficiency make the difference in overall energy consumption of different cooking

**Table 2.1** Cost comparison of various cooking methods.

| Method | Cost | Speed |
|---|---|---|
| Open fire | Very low | Very slow |
| Closed fire | Low | Faster than the open fire method |
| Stove | Moderate | Moderate |
| Convection oven | Higher | Faster |
| Microwave oven | Higher | Fastest |

systems. For example, open flame cooking systems offer very low energy efficiency, whereas MW ovens show higher energy efficiency On the other hand, electricity consumption is much higher in a convection oven than that of a microwave oven. However, overall energy consumption depends on the cooking time and cooking style.

### *2.5.5 Quality*

Significant nutrient degradation can occur during cooking. The main reasons for the quality degradation are heat and oxidation, as well as the volatile nature of many nutrients, which migrate from food during cooking. As much as 90% of nutrients may be lost in severe cooking conditions. Overall degradation of quality depends on the cooking time, intensity of heating, degree of oxidation, and rate of evaporation to the atmosphere. In terms of quality, microwave oven cooking is one of the highest-ranking cooking systems due to its faster heating capacity (Lee et al., 2018).

## 2.6 Concluding remarks

Cooking is an indispensable process for human beings. It makes food more compatible with digestion. Appropriate cooking techniques enhance both the bioavailability and the taste of food. On the other hand, inappropriate cooking techniques result in significant degradation of nutrients of foods; therefore, special care must be taken during cooking. The cooking process involves simultaneous heat and mass transfer. Cooking medium and heating sources are the main factors that control the overall performance of cooking. Water and oil are the main cooking media used across the globe. Biomass, natural gas, and electricity are the common means of producing heat to cook food. Optimum cooking conditions need to be maintained so that the right amounts of energy and time are used to retain maximum food quality.

## References

Adeyeye, S.A.O., 2016. Traditional fish processing in Nigeria: a critical review. Nutr. Food Sci. 46 (3), 321–335. https://doi.org/10.1108/NFS-11-2015-0148.

Adeyeye, S.A.O., Oyewole, O.B., 2016. An overview of traditional fish smoking in Africa. J. Culin. Sci. Technol. 14 (3), 198–215. https://doi.org/10.1080/15428052.2015.1102785.

Isleroglu, H., Kaymak-Ertekin, F., 2016. Modelling of heat and mass transfer during cooking in steam-assisted hybrid oven. J. Food Eng. 181, 50–58. https://doi.org/10.1016/j.jfoodeng.2016.02.027.

Joardder, A., Masud, M., Karim, A., 2018. Relationship between intermittency of drying, microstructural changes, and food quality. In: Intermittent and Nonstationary Drying Technologies. CRC Press, pp. 123–137.

Joardder, M.U.H., Masud, M.H., 2019. Food Preservation in Developing Countries: Challenges and Solutions. Springer International Publishing.

Joardder, M.U.H., Washim, A., Azharul, K., 2019. Heat and Mass Transfer Modelling During Food Drying: Empirical to Multiscale Approaches. Taylor & Francis.

Lee, S., Choi, Y., Jeong, H.S., Lee, J., Sung, J., 2018. Effect of different cooking methods on the content of vitamins and true retention in selected vegetables. Food Sci. Biotechnol. 27 (2), 333–342. https://doi.org/10.1007/s10068-017-0281-1.

Masud, M.H., Ahamed, R., Joardder, M.U.H., Hasan, M., 2019. Mathematical model of heat transfer and feasibility test of improved cooking stoves in Bangladesh. Int. J. Ambient Energy 40 (3), 317–328. https://doi.org/10.1080/01430750.2017.1392354.

Sagar, V.R., Suresh Kumar, P., 2010. Recent advances in drying and dehydration of fruits and vegetables: a review. J. Food Sci. Technol. 47 (1), 15–26. https://doi.org/10.1007/s13197-010-0010-8.

Tocci, A.M., Mascheroni, R.H., 1996. Heat and mass transfer coefficients during the refrigeration. J. Food Eng. 26 (2), 147–160.

Tzempelikos, D.A., Mitrakos, D., Vouros, A.P., Bardakas, A.V., Filios, A.E., Margaris, D.P., 2015. Numerical modeling of heat and mass transfer during convective drying of cylindrical quince slices. J. Food Eng. 156, 10–21. https://doi.org/10.1016/j.jfoodeng.2015.01.017.

## Further reading

Nagarajarao, R.C., 2016. Recent advances in processing and packaging of fishery products: a review. Aquat. Procedia 7, 201–213. https://doi.org/10.1016/j.aqpro.2016.07.028.

# Cooking equipment for the food industry

# 3

*Sara Hedayati*[a], *Vahid Baeghbali*[b], *and Seid Mahdi Jafari*[c,d]
[a]Nutrition Research Center, School of Nutrition and Food Sciences, Shiraz University of Medical Sciences, Shiraz, Iran, [b]Department of Food Hygiene and Quality Control, School of Nutrition and Food Sciences, Shiraz University of Medical Sciences, Shiraz, Iran, [c]Faculty of Food Science and Technology, Gorgan University of Agricultural Sciences and Natural Resources, Gorgan, Iran, [d]Nutrition and Bromatology Group, Analytical and Food Chemistry Department, Faculty of Food Science and Technology, University of Vigo, Ourense, Spain

## 3.1 Introduction

The discovery of fire can be considered as the beginning of food processing as it led to the development of cooking and enabled humans to expand their diet. However, it did not have any scientific background. During the past century, food science was developed and food processing industry was established (Aguilera, 2018). Cooking is a critical step in food processing and the main purpose of industrial cooking is to produce products of high organoleptic quality while ensuring their microbiological safety. At large volume production cooking environments such as centralized kitchens of schools, hospitals, large hotels, food service outlets, home-delivered foods, large staff canteens, mass catering establishments, and industrial production of ready-to-eat meals, industrial cooking equipment is required. These appliances may have a single function or be a multipurpose equipment (Fellows, 2017). Different types of energy such as electricity, gas, liquid or solid fuels can be used in industrial cooking equipment. The source of energy is selected by considering cost, efficiency, safety, flexibility of use, efficiency of heating, and risk of contaminating food. In many countries, electricity is more common due to its ease of installation, operation, and maintenance (Featherstone, 2015).

The materials used in cooking equipment is a critical issue that should be considered in designing these appliances. The surfaces of equipment which are in contact with food should made from noncorrosive, inert, nonabsorbent, nonporous, and smooth materials (Imholte, 1984). Stainless steel is the most desired material in cooking equipment due to its corrosion resistance. Different types of stainless still such as No. 316 and No. 304 are used in cooking equipment. Other metals such as cast iron, monel (an alloy of nickel and copper), aluminum, copper, brass, or bronze can be used in some parts of cooking equipment with some considerations. For instance, cast iron should not be used in equipment which are in direct contact with food. Monel is resistant to corrosion and can be used for equipment which contain salt brines because sodium chloride may cause the corrosion of stainless steel (Katsuyama,

**High-Temperature Processing of Food Products. https://doi.org/10.1016/B978-0-12-818618-3.00004-5**

1993). But if the product contains corn or legumes, monel should not be used in the cooking equipment because cupper of monel leads to discoloration. Aluminum can be used in parts which need high thermal conductivity as it is a great heat conductor but it should not be in contact with acidic foods and not be cleaned by alkaline solutions because it will be corroded. Copper is not recommended in fabricating food equipment because it is corroded in contact with air and forms a layer of soluble compounds which dissolve in food and has negative effects on product quality and cause health concerns. Moreover, metals such as galvanized iron, lead, cadmium and antimony must not be used in cooking equipment.

Cooking equipment should not have open seams, crevices, rough spots and angle corners, because it is not easy to clean the angles and may cause food contamination and spoilage. Different parts of equipment should be dismantled and reassembled quickly and easily and should be resistant to chemical sanitizers. To facilitate cleaning the horizontal parts of equipment should be at least 15 cm (6 in) above the floor and if the equipment occupies a large area the clearance should be 46 cm (18 in) or more (Imholte, 1984) Industrial cooking equipment can be categorized into batch and continuous systems. In batch system, the product is loaded into the equipment, heated, and unloaded as a single batch. Whereas in continuous systems the product is placed on a conveyor which passes through cooking zone followed by cooling. Generally, batch cooking equipment is used for smaller volumes of production such as startups or industries with low production rates while, continuous systems are used for large-scale production. Transition from batch to continuous processing is very interesting for manufacturers because continuous processing production goes on without interruptions, control over processing conditions is more accurate, heating is fast and even, is labor saving and improves consistency and quality of food products. Moreover, less hours are spent for cleaning or maintenance in continuous heating.

Cooking can be performed with dry or moist heat. The dry cooking methods have been discussed in Chapters 6 and 10, thus in this chapter we focus on cooking with moist heat. In this method, the food product is heated in hot water, under vacuum, atmospheric, or pressurized steam.

## 3.2 Different types of cooking equipment

### 3.2.1 *Boiling pans*

Boiling pan also known as jacketed kettle or steam kettle is one of the common equipment in professional kitchens and food processing plants. Boiling pan is used for cooking different food products such as jams, sauces, custards, creams and foods which are cooked in water. This equipment consists of a double jacket tank for indirect heating of food with a capacity of 10 to 1000 L. Cooking kettles must not have sharp corners at the junctures of bottom and sidewalls to food safety and prevent microbial contamination. Generally, cooking kettles have long curves at the juncture. All components of boiling pans are made of stainless-steel; however, the grade of stainless-steel may vary in different components. For instance, the components that are in direct contact with food are made of acid resistant stainless-steel of 1.4404 (AISI 316L) or 1.4301 (AISI 304) grade, while the casing is made of stainless-steel of 1.4509 (AISI 441) grade.

The inner jacket pressure may vary from 0.1 to 1.5 bar. Thus a safety valve should be considered in the system to avoids overpressure of the steam in the double jacket. To increase the heating efficiency and prevent adhering of food to the walls, boiling pan are equipped with stirrer and scrapers with adjustable speed and direction. The diameter of tanks should be large and have shallow depth to facilitates stirring. Boiling pans can be free-standing or tilting. However, the tilting ones are more common because they facilitate the pouring and cleaning operations. Some systems are equipped with motorized tilting and the vessel can be tilted beyond 90 degrees. The tilting axis should be in the middle of the pan to keep the pouring lip high and allow a tall vessel to be filled (Fellows, 2017). The upper well rim must be jacketed and isolated to prevent the risk of injury for user. In terms of heating energy source, the boiling pans are categorized into electric boiling pans, gas boiling pans, direct steam boiling pans, and coal-fired boiling pans.

Boiling pans are suitable for boiling, steaming, poaching, braising or simmering different food products. Thus the temperature control is very critical in these appliances. The temperature of boiling pan can be precisely controlled by either mechanical or digital systems. Thermal insulation of the pan should be considered to keep heat inside the well, save energy and keep operating ambient temperature low. Modern boiling pans have data monitoring ability to control temperature, adjust tilting and pouring speeds, upload recipes and download HACCP data (Fig. 3.1).

### 3.2.2 Bratt pan

A Bratt pan, also known as tilt skillet or braising pan, is a commercial cooking vessel which has taken its name from the German word "brat" meaning to fry. However, it is a powerful multifunctional cooker that can be used for eight cooking functions including deep-fat and shallow frying, roasting, boiling, steaming, braising, poaching, stewing in large quantities. Thus Bratt pan replace numerous appliances and save vast

**Fig. 3.1** Boiling pan.
Courtesy of Taizy Food Machinery, https://taizyfoodmachinery.com.

**Fig. 3.2** Bratt pan.
Courtesy of Firex, www.firex.com.

amounts of space in food plants or alongside equipment such as cooking kettles. Bratt pan can be used for a variety of food products such as jams, pure´es, creams, risottos, stews, soups and sauces. The heat source in this deep rectangular pan is either electric or gas. A pressure Bratt pan has a lid that can be clamped shut for cooking at temperatures higher than 100°C. Bratt pans have a tilting feature, operating with electric motor or hand crank for easy pouring and cleaning (Fig. 3.2).

Application of Bratt pans in food plants has several advantages such as:

- Saves time
- Cleaning is easier than similar sized apparatuses because they have a central spout to quickly remove the remaining of food
- All of cooking stages is performed in one pan
- Tilting the pan is easily performed by hand-driven mechanism or electrically which facilitate the unloading of food products
- Performing a vast range of tasks, makes them ideal for little space
- Uniformity of the cooking surface
- Coupled with a swivel tap for adding liquid for cooking and washing out the instrument
- They may be programmed to operate automatically

### 3.2.3 Steaming cabinet

Steaming cabinet is a rectangular batch cooking equipment which is composed of heating system, walls, trays and a steaming unit. The steam may be provided by direct steam, steam generator or boiler. In boiler tank, water is heated using heating elements, and the generated steam enters the cabin through pipes equipped with temperature and volume control sensors. While, in a direct steam system, water is converted to steam directly by spraying on hot elements inside the cabin. The steam quality of these systems has a direct impact on the cooking quality. Foods with a softer texture with direct steam and foods with a harder texture such as proteins with a boiler have better cooking quality.

Hot air fan has the ability to blow and throw the heat in the cabin, control the temperature and cook evenly in all of the trays. The suction system of fans prevents the mixing of different food odors. If the oven opens during cooking, the fan will stop automatically to prevent temperature drop during cooking. Moreover, the fans can be used to cool the cabin if needed. The intelligent temperature and humidity sensor is responsible for the temperature and humidity control of the cabinet, coloring the food, improving the quality of cooking and preventing the burning of food. The humidity of the cabin is controlled by a proprietary humidity control system. This system continuously calculates the humidity in the cabin with the help of advanced sensors. If the humidity decreases, the steam enters the cabin with the appropriate quality, volume and temperature, and if there is excess moisture in the cabin, it is released through the outlet valve. Thanks to the humidity control system, the quality of cooked foods is always constant (Boruj Corp, 2022).

Steaming cabinets are suitable for steamed rice, bun, cakes, meat, fish, sea food, stewed food, etc. in catering companies, central kitchens, large canteens, and food processing factories. Steaming cabinets are made by stainless steel to meets hygiene requirements. They may have adjustable temperature controller, time controller, automatic washing system, auto heat-retaining and water intake function, automatic cooking, thawing and warming system and alarm indicator which can be adjusted based on product's feature. The temperature of steaming cabinets is kept constant with two double solenoid valve steam admission control. Moreover, the adjustable hinge and the integrative silica gel sealing strips of the door can ensure good sealing in steaming cabinet. Steaming cabinets are easy to operate, highly efficient and energy saving. Several foods can be prepared simultaneously in steaming cabinets provided that the cooking method is the same (hot air, steam, hot air combination and steam) (Fig. 3.3).

**Fig. 3.3** Steaming cabinet.
Courtesy of Boruj Corp, 2022. https://boruj.ir.

### 3.2.4 Sous-vide

Sous-vide is a French word which means cooking in a vacuum-sealed pouch. In this method, the food product is vacuum-sealed in an airtight food-grade plastic pouch and then submerged in a water bath to slowly cook at a controlled low temperature (Schellekens, 1996). The cooking temperature in this method ranges from 50°C to 80°C. Since the food material is not in direct contact with water, the hydrophilic compounds don't leak into water. Moreover, cooking at 100°C prevents the degradation of heat-sensitive and volatile compounds (Fig. 3.4).

Application of Sous-vide equipment for cooking of food products has several advantages:

- The products processed by Sous-vide has incomparable taste, color, flavor and texture and high nutritional quality. Because off-flavors from oxidation and loss of nutrients, moisture and flavor volatile compounds are eliminated during cooking at low temperatures.
- The aerobic bacterial growth is reduced and food is not recontaminated during storage; therefore, the shelf life of food products is increased.
- Products such as chicken and meat can be cooked in a sauce or marinade.
- The pressure is maintained during cooking operation with a pump. Times, temperatures, and pressure are precisely controlled in Sous-vide systems; therefore, the results will be easily replicable.
- It is a fully automatic operation; therefore, the food product can be left to cook unattended with guaranteed results.
- It is considered as a dual purpose machine because cooking and cooling is performed in this instrument

**Fig. 3.4** Sous-vide.
Courtesy of DC Norris Food Machinery Manufacturers, www.dcnorris.com.

Sous-vide is used in the food industry and some of the top restaurants in order to cook or prolong the shelf life of food products. There are several published papers about the influences of Sous-vide processing on different food products. Application of sous-vide for meat is so popular because cooking meat at low temperatures for long times improves its sensory attributes. Roldán et al. (2013) studied the influence of different cooking temperatures (60°C, 70°C, and 80°C) and times (6, 12, and 24h) on physicochemical, textural, histological, structural and microbiological attributes of sous-vide processed lamb loins and reported that weight loss was increased and moisture content was decreased by increasing the cooking temperature. Lightness and redness decreased and yellowness increased by cooking temperature. Whereas cooking time had limited effects on these parameters. The textural parameters showed obvious interactions between cooking time and temperature. Increasing the cooking time decreased the textural parameters at all of the temperatures. Cooking temperature up to 70°C decreased the hardness of meat samples; however, heating at 80°C increased the hardness. This behavior might be due to denaturation of actin and myosin and contraction of collagen. Also they observed granulation and gelation of connective tissue at 60°C and 70°C in the SEM images. The microbial population of Sous-vide processed lamb loins decreased significantly even at less intense treatments (60°C for 6h).

Supaphon et al. (2021) investigated the influence of Sous-vide cooking at different temperatures (60°C, 70°C, and 80°C) for 2 to 36h on microstructure of Thai beef. The muscle fibers exhibited a shrinkage during the first 6h of heating at 60°C and 70°C and the first 2h at 80°C. The Microstructure of meat samples was more variable at 60°C to 70°C which indicated a larger cross-sectional myofibrillar mass area (CSA). However, contraction of fibers was observed at all of the temperature-time treatments. Jiang et al. (2022) evaluated the influence of Sous-vide and conventional cooking on the flavor, digestion and microstructure of braised pork. The conventionally cooked pork had the highest malondialdehyde and fat content, but the lowest protein and total sulfhydryl content. The sulfhydryl content in sous-vide samples decreased with increasing the heating time, but their malondialdehyde content increased. Moreover, the α-helix content in conventionally cooked samples was higher and β-sheet content was lower than Sous-vide samples. Transmission electron microscopy (TEM) illustrated that the bright band of sarcomeres increased in conventionally cooked samples while the dark band reduced. This might be due to the excessive contraction of actin and myosin at high temperature and long heating time. Also, the differences in Sous-vide conditions had significant effects on microstructure of pork. Increasing the temperature and time in Sous-vide weakened the connection between Z-lines and increased the gaps between muscle fibers. This can be attributed to the denaturation of proteins in the bright band and the Z-line. However, the denaturation in all of the Sous-vide samples was lower than the conventionally cooked pork samples. TEM images of conventionally cooked and sous-vide braised pork is presented in Fig. 3.5.

In a study by Park et al. (2020), the effect of temperature and time of Sous-vide processing on hardness, moisture loss, and sensory quality of chicken breast was evaluated. They sous-vide the samples at 60°C and 70°C for 1, 2, and 3h and compared them with the chicken breasts cooked in a convection oven. They stated that cooking

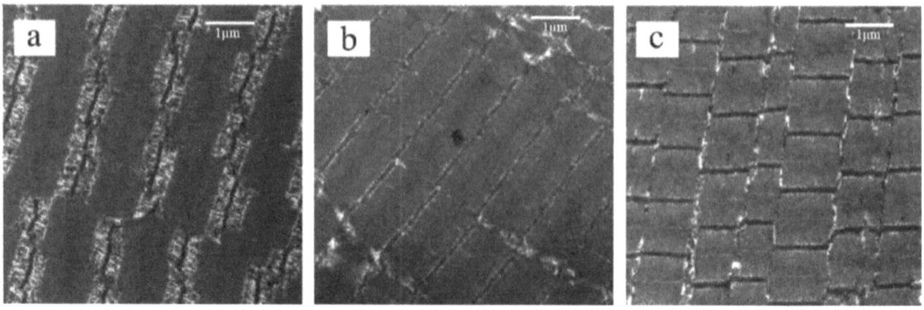


**Fig. 3.5** TEM images of conventionally cooked (A), sous-vide at 65°C for 8 h (B), and sous-vide and 75°C for 12 h (C) braised pork ( Jiang, et al., 2022).

time and temperature in sous-vide have significant effects on the physicochemical and organoleptic properties of chicken breast and the best settings for sous-vide were 60°C for 2 to 3 h. At these conditions sensory quality characteristics were improved without decreasing the water holding capacity of chicken breast. Oz and Zikirov (2015) investigated the influence of different cooking methods (Sous-vide, boiling and pan-frying) on the formation of heterocyclic aromatic amines (HCAs) in beef and reported that the Pan-fried samples had the highest HCAs content. However, sous-vide is a low temperature long time (LTLT) treatment, and Maillard reactions are not occurred at these conditions on the surface of meat. Therefore, Sous-vide processed meat does not have a brown surface and roasted flavor. To overcome this problem, meat should be processed in two stages, and meats are subjected to roasting before or after sous-vide. Ruiz-Carrascal et al. (2019) investigated the influence of Sous-vide followed by roasting or roasting before sous-vide on lamb loin quality and stated that there were not significant differences between two treatments in moisture content, cooking loss and sensory parameters. However, surface browning was more intense, and rate of oxidation to Maillard compounds was lower in lamb loins processed by Sous-vide followed by roasting.

In recent years the demand for high convenience products has increased and consumers are looking for ready-to-eat product with minimum changes of the fresh product. Therefore, finding a suitable cooking condition to preserve the fresh appearance of ready-to-eat vegetables should be considered. The influence of Sous-vide processing on different types of vegetables has been addressed in recent years. The effect on cooking methods (boiling, steaming, microwaving and Sous-vide with hot water or microwave) on asparagus spears was evaluated by Gonnella et al. (2018). They determined the chemical, physical, and sensory properties of fresh asparagus and the samples cooked by different methods. The results revealed that microwaving caused the highest total color difference and dry weight increase in cooked asparagus compared to raw samples. The Sous-vide-microwaving caused an increase in greenness and showed the lowest chlorophyll content reduction. Sensory evaluation results showed that all cooked asparagus samples were acceptable. Nevertheless, the samples prepared by Sous-vide-microwaving were more acceptable for consumers.

The antioxidant activity of different vegetables after Sous-vide and conventional cooking (boiling) was investigated by Kosewski et al. (2018). They observed that vegetables such as tomato, red onion, broccoli, cauliflower, shallot and parsley root had a higher antioxidant activity after Sous-vide processing than conventional method. This result is due to the destruction of vegetable cells during traditional cooking and the transfer of antioxidant compounds into the boiling water. While in the Sous-vide, the cellular structure is not damaged due to the vacuum packaging and low cooking temperature. Therefore, the antioxidant compounds are preserved within the vegetables (Schellekens, 1996). Araya et al. (2009) compared the quality attributes of raw, Sous-vide and boiled carrot and reported that the visual appearance, sweetness and flavor of Sous-vide carrots were more similar to raw carrots. The Cryo-SEM images revealed that the tissue structural changes were more evident in boiled carrots. The photographs and Cryo-SEM images of raw, Sous-vide and boiled carrots are presented in Fig. 3.6.

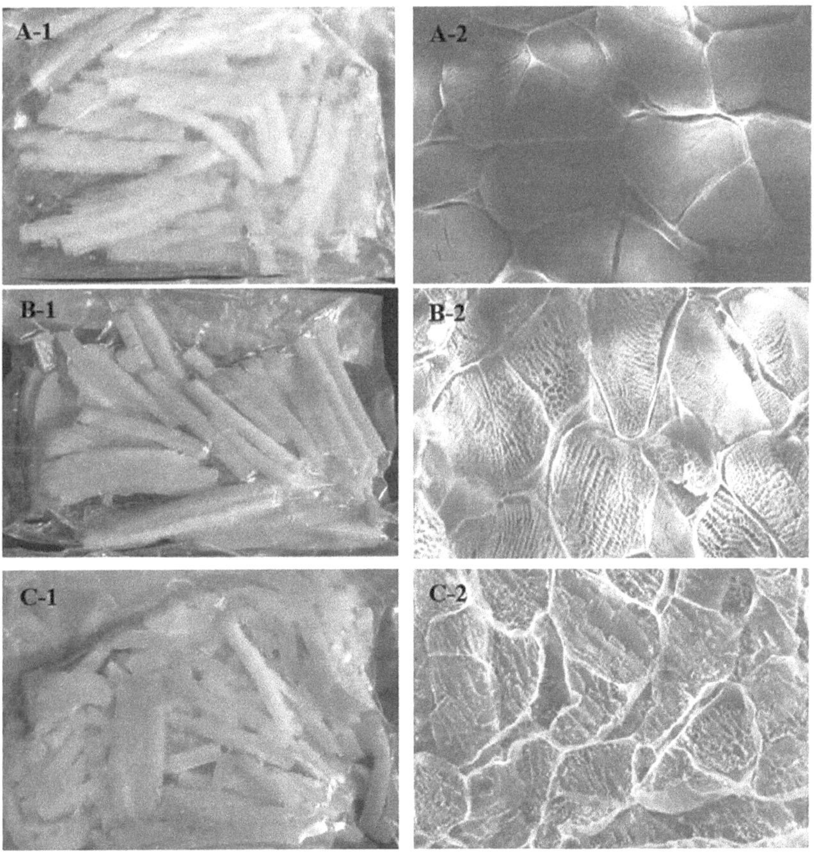


**Fig. 3.6** Photographs (1) and Cryo-SEM images (2) of raw (A), sous-vide (B) and boiled (C) carrots (Araya, et al., 2009).

### 3.2.5 Jet cookers

Cooking equipment such as kettles, heat exchangers, and similar heating systems require high extent of energy, space, and maintenance. Moreover, the control of temperature and processing conditions is difficult in these devices. Therefore, production of safe and high-quality food is challenging in these equipment. Direct steam injection is an alternative heating system for water heating, pasteurization and in-line food cooking. In this method, the heat transfer from the steam to the food product is 100%; therefore, it saves energy and the operating cost. Additionally, maintenance cost is much lower than indirect heating systems and precise control of mixing turbulence and steam flow is possible.

Jet cooker is a direct heating and cooking apparatus which is appropriate for liquid foods with different solid contents and viscosities. Jet cookers are used for cooking, pasteurization and can topping. They can be used for continuous heat processing of water miscible liquids or slurries containing bite sized particles with particulate integrity reserved. Industrial jet cookers are used for cooking a wide range of foods and beverages such as fruit Juices and purees, jams and jellies, baby foods, cans, puddings, fillings, dressings, sauces, soups, stews and gravies, oils, eggs, milk, and pet foods. Additionally, Jet cooker is key equipment for the starch cooking, modified starch production, starch liquefaction and production of glucose from starch. In this system, precisely measured amounts of steam are directly injected to the liquid through hundreds of small orifices and produce fine steam bubbles which are instantly are absorbed by the product and result in 100% heat transfer. Jet cookers use low velocity, variable orifice steam injectors, and helical flights to ensure gentle and intimate blending of steam and food product and provide a thorough cook without high shear and degradation of the product (Fig. 3.7).

Jet cookers have several advantages over indirect heating systems:

- The steam and energy consumption are reduced
- Thermal Efficiency is 100% and both sensible and latent heat of the steam are used
- Energy losses from condensate return is eliminated
- The heat transfer barrier of heat exchangers is eliminated which improves their efficiency
- Cooking is fast, thorough and uniform.
- The external steam valves are not required
- Temperature control is very accurate
- Response to temperature set-point changes is very fast
- Costs for purchase, installation and maintenance are much less than heat exchangers or kettles.
- The process is nonshearing; therefore, the liquid foods with particles can be heated without damage to texture or consistency.

### 3.2.6 Belt cooker

Belt cooker is a continuous cooking equipment, in which the product can be treated with steam under overpressure or cooked by immersion in water. Generally, horizontal belt cookers are enclosed by feeding and discharge units. The product is loaded on a horizontal conveying belt and passes through the cooker in which the steam pressure

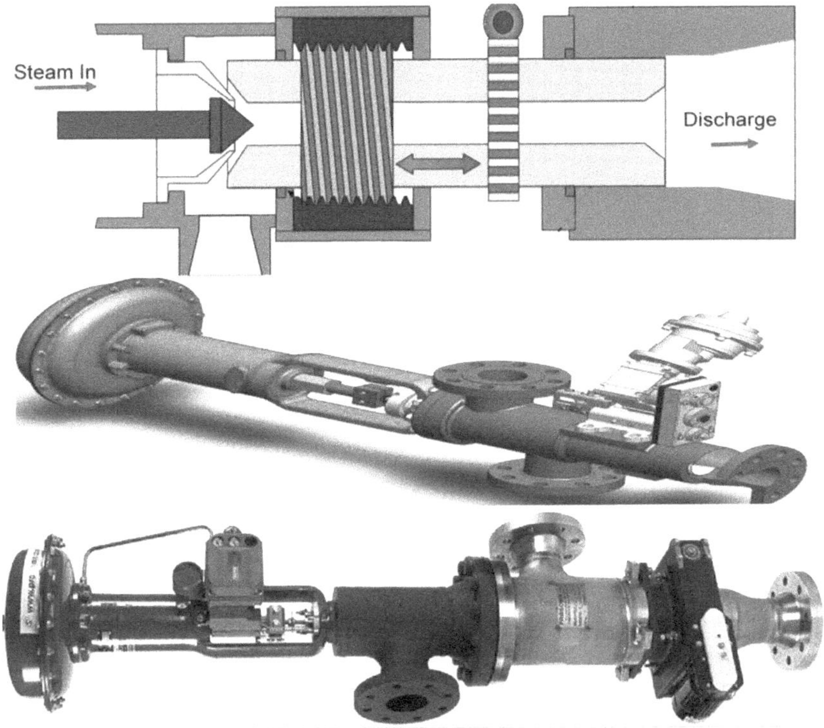


**Fig. 3.7** Jet cooker.
Courtesy of Hydro-Thermal, www.hydro-thermal.com.

and the resulting high temperature ensure the heat treatment of the product and finally reach the discharge unit that ensures uniform continuous emptying of the cooker. The retention time in the cooker can be adjusted via the speed of the feeder and the discharger. Horizontal belt cooker is used for starch gelatinization, production of quick-cook products (e.g., instant rice), enzyme inactivation, preparation of the product for flaking process, cooking and processing of different food products such as bulgur, vegetables, fruits, fish, legumes, pasta and also for steam sanitization. This equipment is composed of cooking tank, washing and cooling zones. The water in the cooking tank can be heated either by direct steam or steam-fed heat exchangers. To prevent the waste of steam and make belt cookers an energy-effective instrument, in some of the instruments the amount of steam is adjustable and can be balanced with the quantity of product (Fig. 3.8).

Belt cookers have several advantages such as:

- Hygienic food processing
- High production yield
- Low steam and water consumption
- No product damage
- Low maintenance costs and long life

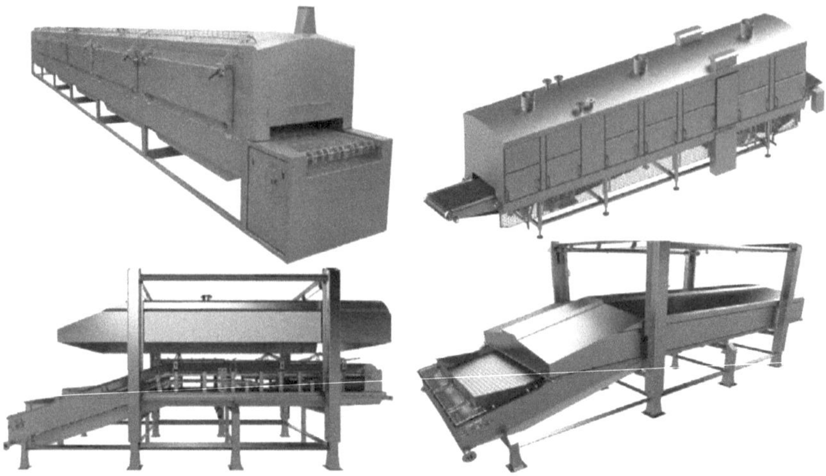

**Fig. 3.8** Belt cooker.
Courtesy of Cabinplant, www.cabinplant.com.

### 3.2.7 Autoclaves (retorts)

Sealing and cooking food products in cans, jars, pouches or other flexible containers is a very popular method of preserving flavor and extending shelf life for a wide range of products such as convenience foods, baby meals, dairy products and pet food. Pressure cooking is the most efficient method for production of these products. Retort is a pressure cooking equipment used for sterilizing or pasteurizing canned products after they are sealed. Retorts can be batch or continuous.

#### 3.2.7.1 Batch retorts

Generally, batch retorts are found at canning facilities where large, continuous systems are not justified. These cookers can be constructed in carbon steel as well as stainless steel. The advantage of batch retorts is that they can handle a range of products, have great operational flexibility for different products and all sizes and shapes of cans. Moreover, they can process smaller runs of product than continuous systems. Thus they are still quite popular in many canning factories. However, the batch systems don't have the operational efficiency of continuous retorts and have higher production costs because their loading and unloading is labor intensive and processing time in batch retorts is longer than continuous systems due to heating and cooling cycles. Batch retorts may be horizontal or vertical the former is easier to load and unload but occupies more space. In vertical retorts, cans are placed into retort chamber with baskets that are lifted by chain hoists (Teixeira, 2019; Fellows, 2017). Usually each retort holds 3 or 4 baskets. While in horizontal retorts trolley tracks are used to transfer the crates to the retorts. Generally, in food canneries a battery of individual batch retorts may be operated to increase the production rate.

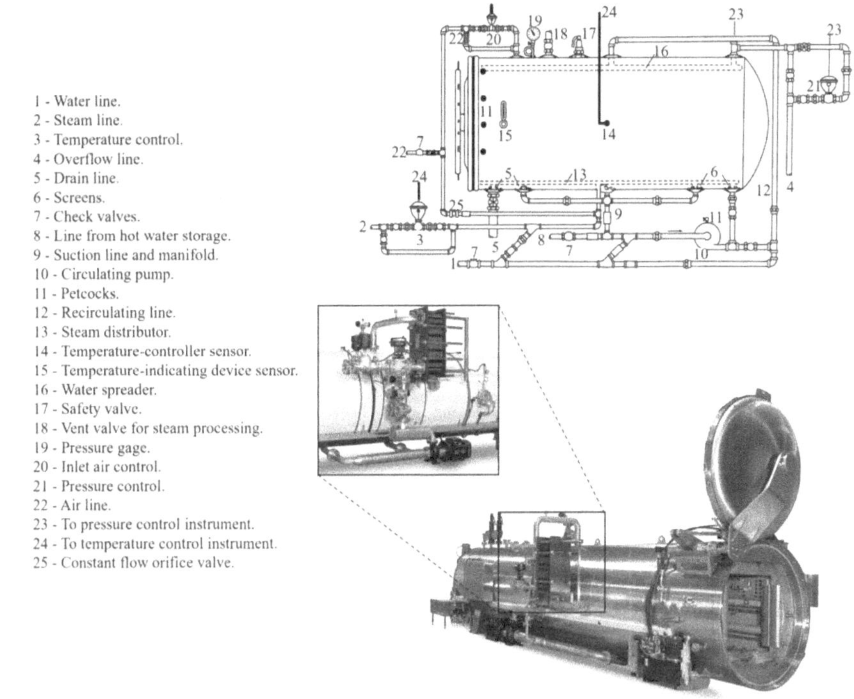


**Fig. 3.9** Horizontal batch retort.
Courtesy of CFT Group Company, www.cft-group.com.

Old batch systems are static and has disadvantages such as inefficient thermal transfer, risking cold-spots or burning. Therefore, retorts are equipped with facilities for agitating the containers. The movement of the products is achieved by axial rotation or end-over-end agitation. The rotary motion provides even heating of cans and accelerates heat transfer to the coolest part of any container particularly useful for products of high viscosity. This product agitation increases the heat penetration rate. The reduced process time has positive effects on the organoleptic and nutritional quality of the products compared to the quality of products obtained by static retorts. To facilitate the movement of hot air in the retort chamber, it is jacketed with steam tubes and may be fitted with electric fans. After heat treatment, the temperature and pressure in the retort are reduced and then the chamber is opened. The operation of each retort is controlled either manually or automatically to make sure that the process is delivered for each retort (Fig. 3.9).

### *3.2.7.2 Continuous retorts*

Continuous retorts have been developed to lower the labor costs, increase the production rate and the rate of heat transfer in food products. During continuous retorting, cans are continuously transferred from the atmospheric conditions into the pressurized steam through a conveyer to pass the processing time and then returned to the atmospheric environment. Continuous retorts may be static or agitating.

In continuous retorts the pressure inside cans is gradually changed thus can seams are exposed to less strain compared to batch retorts. The temperature and pressure inside different sections are constant as the products pass through different parts of the equipment; therefore, it has great thermal efficiency. Additionally, not requiring heating/cooling cycles reduces the processing time. However, the capital cost of continuous equipment is higher than batch systems, and in-process stock is higher which could be lost if a breakdown occurs (Teixeira, 2019). Some of the commercially available continuous retorts are cooker-coolers, crateless retorts, rotary pressure cookers, hydrostatic sterilizers, and flame sterilizers.

### Cooker-cooler continuous retort

Cooker-cooler instrument can be used for cooking, blanching or pasteurization based on retention time and operating temperature. These devices can be static or agitating (rotary). In a static cooker-cooler system, a chain conveyor or roller is used to transfer cans into a tunnel with three sections or three interconnected pressure chambers that have different pressures to preheat, sterilize and cool the product. Cooker-cooler provides consistent heating and cooling of food products which allows for even cooking and better quality yields. The heating medium in cooker-cooler systems can be water, saturated steam, or a mixture of steam and air. Usually turbo fans are used to produce a homogenous mixture of air and steam. Cooker-cooler equipment offers optimal temperature, securing increased shelf-life and quality of food products and is widely used for fruit, vegetable, meat, and seafood, processing. Cooker coolers are applicable for processing of rigid, semirigid, or flexible containers such as glass jars, metal cans, bowls, semirigid plastic, and retortable pouches. The system is very flexible and allows for heat processing of containers with different shapes and sizes with the same platform. Heat processing in can provide no agitation or axial rotation. The Continuous cooker-coolers can be equipped with loading and unloading systems which are integrated into the system to make installation faster, easier, and requiring less overall human resource for operation.

### Rotary cookers (sterilizer)

The rotary cooker is a continuous horizontal retort for cooking and sterilization of food packed in cans or retortable plastic trays and bowls. Rotary cooker system is composed of heating and cooling shells for continuous handling of containers with high production rate. The continuous axial agitation of the product through a spiral route with rotational reel mechanism during heating and cooling processes prevents from any burnt product. To achieve the required heat treatment, the residence time of cans is controlled by adjustment of rotation speed of reel. Moreover, pressure loss and steam escape should be prevented when cans enter the cooker. Rotary cooker allows high temperature short time (HTST) which improves organoleptic quality and color of product. Cans are transferred into the sterilizer from a rotary transfer valve through a synchronized feeding device. When cans enter the cooker, they pass through the area between the reel and inner circumference of the shell. There are spines on the reel and a spiral or helical track on the shell to hold the cans and impart an axial rotation.

After traveling the length of a shell, cans may either transfer to the next shell by means of a rotary transfer valve for more cooking or cooling or may exit through a discharge valve.

Generally, three shells in series including a pressure cool shell, an atmospheric cool shell and a cooker or sterilizer are required to achieve controlled cooling. While for cold-fill foodstuffs five shells are required to provide the atmospheric preheating, pressure preheating, pressure cooking, pressure cooling, and atmospheric cooling (Teixeira, 2019). Due to the design and operation, a continuous rotary sterilizer can only accommodate a specific size of can. Thus, several rotary sterilizers may be observed in a food canning plat to be devoted to a different sizes of cans. Usually, these sterilizers are fitted with, 2 to 3 vent lines and the venting schedule is performed by opening the vent valves and drain valve. Also, bleeders, drain, and purge lines can be used for air removal as an alternative to the main vent lines. However, removing air in this method is more time consuming than the traditional venting system.

## Crateless retorts

Crateless retorts are still steam retorts which are composed of several individual retorts. Each retort is operating in a batch mode and is considered as an automatic cook room. In the first stage, the equipment is filled with preheated cushion water and then the cans are conveyed into the retort and fall into the hot cushion water. Then the hatch is closed and water is replaced by steam, and the products are processed under constantly monitored temperature and pressure. When the cooking operation finished, the retort is refilled with cushion water to accomplish cooling. Subsequently, the processed cans are transferred to the canal system for final cooling and conveyed to the unscrambler to be oriented. It is worth noting that, all of the conveyor transfers take place under water.

Crateless retorts are continuous, high speed, adaptable, and extremely efficient which have overcome many of the technological bottlenecks in heat processing and has several advantages over other types of retorts such as:

- Loading, unloading, and process operations are performed automatically without the use of crates.
- The process preparation time is short because the products enter the retort without basket loading and waiting.
- Handles high temperatures (up to 132°C), and the vent technology ensures the uniform temperature distribution
- Reduces the steam consumption (up to 50%) due to the steam injection from top and lower steam waste from bleeders, into the retort vessel.
- Occupies 25% to 40% less space compared to other continues retorts.
- -The capacity of crateless retorts is higher than normal vertical retorts
- Containers with different sizes and shapes can be processed by crateless retorts with uniform heat distribution.
- Is very flexible and often is configured to fit the existing facilities. Usually, size and configuration of crateless retorts are determined according to the line speed, size of containers, heating requirements, and available space.
- Equipped with safety valve, pressure sensing control and alarm system to ensure its safety

Therefore, the crateless retorts can meet the requirements of mass production and are used in many food plants for heat processing of different food products. Typical applications of crateless retort include processing of fruits and vegetables, pet food, fish, and meat products.

### Hydrostatic cooker (sterilizer)

Hydrostatic sterilizers are continuous pressure cookers or sterilizers in which the steam pressure is maintained hydrostatically. Hydrostatic sterilizers are agitating retort with axial or end over end agitation of cans. These cooking appliances are composed of four sections: two water columns known as "bring-up" and "bring-down" legs, a sterilizing steam dome, and a cooling chamber. The water columns are in either side of a steam chamber and the height of water columns forms a hydrostatic pressure that balances the steam pressure and seals the steam chamber. The hydrostatic sterilization is performed under a column of water with sufficient depth to provide the required pressure and maintain it in the sterilizing zone. Hydrostatic retorts are used for continuous sterilization of canned foods on a large scale. The height of water column is approximately 35 ft; therefore, the hydrostatic retorts are often located outdoors adjacent to the canning plant.

Containers are loaded horizontally on a carrier chain and enter the retort from the top of the water column (water leg) and travel downward and the water temperature increases as the containers move down. Cans enter the bottom of the steam dome to pass through the sterilizer with a controlled speed to achieve the required residence time in the steam dome. The steam dome is considered as the heart of hydrostatic sterilizer, and the containers are subjected to high temperature that is maintained by the hydrostatic pressure. Afterward, the containers exit the steam dome, pass through the water seal interface at the bottom and travel upward through the leg in which the cans encounter progressively warmer water until they exit at the top. Finally, cans are cooed with water and travel down the outside of the sterilizer on their return to the discharge conveyor station.

Hydrostatic sterilizers are used for different sizes and types of cans such as metal, glass, plastic and pouches. These sterilizers are designed for different types of food products especially the products which require long process times at high throughput with little/no agitation. However, some types of hydrostatic sterilizers have agitation during cooking and cooling operations. High-volume products which need long cook times such as meat, fish, soups, low-acid drinks, dairy and soy drinks, vegetables and ready-to-eat meals, and moist pet food can be processed with these sterilizers.

The advantages of hydrostatic sterilizers are:

- High energy usage efficiency
- High product quality due to consistent and uniform process conditions
- High-volume process capabilities of cans, glass, plastics, and pouches
- High reliability and low maintenance cost
- Minimum operator involvement
- Minimal floor space usage

However, hydrostatic retort has a high capital cost and requires very high production volumes which need long cook times. The hydrostatic sterilizers are usually working for periods of up to several weeks and may be shut down and cooled only when repair or maintenance is required.

### Flame sterilizers

Flame sterilization is a high temperature, short time (HTST) process using direct flame for heating the sealed metal cans. The flame sterilizers are composed of three sections: (1) Steam preheater: the containers are heated to 100°C. (2) Gas burners: the cans roll through a series of open flames at 1100°C. (3) Holding section: the containers pass through an intermittently heated burner-holding section for about 4 to 5 min. After flame sterilization the containers are cooled down by water spray (Ramesh, 2007). In flame sterilization, the containers are positioned in close proximity to the burners. The high-temperature differences between heat source and food product increase the rate of heat penetration into the containers and provide rapid heat processing. To increase the heating rate, reduce the probability of superheating of the inside surface of cans and burn-on in flame sterilizers, cans should be kept in constant and high rate agitation. The short processing times in flame sterilizer improve the quality of food products and reduce the energy consumption. However, high temperature increases the internal pressure of containers which may damage can seams. Therefore, flame sterilization should be used for small and fairly rigid cans or a high vacuum should be applied to prevent the distortion of containers. Moreover, the brain content should be minimized to prevent the increase of internal pressure. Flame sterilizers have been used for fish, beef cubes, rice pudding, custard and canned vegetables such as sweet corn, green beans, peas, carrots, and mushrooms with minimum brine content (Holdsworth, & Simpson, 2016; Fellows, 2009).

## 3.3 Conclusion

Large volume production cooking environments require different cooking equipment. The cooking equipment should be selected by considering several items including production volume, cost and efficiency. Cooking appliances are primary factors in energy consumption and may not provide high-quality products. Therefore, the main objectives for cooking equipment manufacturers are faster cooking, better temperature control and distribution, safer operations, better sanitation, less energy consumption, lower labor costs, and flexibility. Moreover, innovative cooking equipment can improve the appearance, color, texture, and nutritional properties of foods. Emerging cooking appliances should be designed to provide high-quality food products with low energy consumption.

## References

Aguilera, J.M., 2018. Relating food engineering to cooking and gastronomy. Compr. Rev. Food Sci. Food Saf. 17 (4), 1021–1039.

Araya, X.I.T., Smale, N., Zabaras, D., Winley, E., Forde, C., Stewart, C.M., Mawson, A.J., 2009. Sensory perception and quality attributes of high pressure processed carrots in comparison to raw, sous-vide and cooked carrots. Innovative Food Sci. Emerg. Technol. 10 (4), 420–433.

Boruj Corp, 2022. https://boruj.ir.
Featherstone, S. (Ed.), 2015. A Complete Course in Canning and Related Processes: Volume 1 Fundemental Information on Canning. Woodhead Publishing.
Fellows, P.J., 2009. Food Processing Technology: Principles and Practice. Elsevier.
Fellows, P.J., 2017. Food Processing Technology Principles and Practice, fourth ed. Elsevier.
Gonnella, M., Durante, M., Caretto, S., D'Imperio, M., Renna, M., 2018. Quality assessment of ready-to-eat asparagus spears as affected by conventional and sous-vide cooking methods. LWT 92, 161–168.
Holdsworth, S.D., Simpson, R., 2016. Thermal Processing of Packaged Foods, third ed. Springer.
Imholte, T.J., 1984. Engineering for Food Safety and Sanitation: A Guide to the Sanitary Design of Food Plants and Food Plant Equipment. Technical Institute of Food Safety.
Jiang, S., Xue, D., Zhang, Z., Shan, K., Ke, W., Zhang, M., et al., 2022. Effect of sous-vide cooking on the quality and digestion characteristics of braised pork. Food Chem. 375, 131683.
Katsuyama, A.M. (Ed.), 1993. Principles of Food Processing Sanitation, second ed. The Food Processors Institute, Washington, DC.
Kosewski, G., Górna, I., Bolesławska, I., Kowalówka, M., Więckowska, B., Główka, A.K., et al., 2018. Comparison of antioxidative properties of raw vegetables and thermally processed ones using the conventional and sous-vide methods. Food Chem. 240, 1092–1096.
Oz, F., Zikirov, E., 2015. The effects of sous-vide cooking method on the formation of heterocyclic aromatic amines in beef chops. LWT-Food Sci. Technol. 64 (1), 120–125.
Park, C.H., Lee, B., Oh, E., Kim, Y.S., Choi, Y.M., 2020. Combined effects of sous-vide cooking conditions on meat and sensory quality characteristics of chicken breast meat. Poult. Sci. 99 (6), 3286–3291.
Ramesh, M.N., 2007. Canning and sterilization of foods. In: Handbook of Food Preservation. CRC Press, pp. 603–642.
Roldán, M., Antequera, T., Martín, A., Mayoral, A.I., Ruiz, J., 2013. Effect of different temperature–time combinations on physicochemical, microbiological, textural and structural features of sous-vide cooked lamb loins. Meat Sci. 93 (3), 572–578.
Ruiz-Carrascal, J., Roldan, M., Refolio, F., Perez-Palacios, T., Antequera, T., 2019. Sous-vide cooking of meat: a maillarized approach. Int. J. Gastron. Food Sci. 16, 100138.
Schellekens, M., 1996. New research issues in sous-vide cooking. Trends Food Sci. Technol. 7, 256–262.
Supaphon, P., Kerdpiboon, S., Vénien, A., Loison, O., Sicard, J., Rouel, J., Astruc, T., 2021. Structural changes in local Thai beef during sous-vide cooking. Meat Sci. 175, 108442.
Teixeira, A., 2019. Thermal processing of canned foods. In: Handbook of Food Engineering. CRC Press, pp. 951–984.

# Section B

# Baking and roasting operations

# Fundamentals of food baking processes

4

*Noemi Baldino, Francesca Romana Lupi, and Domenico Gabriele*
Department of Information, Modeling, Electronics and System Engineering (D.I.M.E.S.), University of Calabria, Rende, CS, Italy

## 4.1 Introduction

Baking is a method of cooking foods by putting all the mixed ingredients in an oven and exposing them to high temperatures in different ways for an appropriate time. The baking process is used to obtain foods like cakes, bread, biscuits, snacks, and so on, and, due to the importance that this process has in conferring the right final texture and flavor, it can be considered one of the most important process operations for food. The prominence of baking is also due to the importance that cereals play in food nutrition as a source of dietary energy and nutrients. In fact, bakery products such as bread (leavened or unleavened), cakes, biscuits, and cookies, which are generally cereals based, have been consumed since ancient times and the practice of cooking them in an oven to obtain a food with longer shelf-life, good taste, and high nutritional value is very old.

The baking method is an operation during which dough or batter—generally a starchy food—goes through changes to achieve the desired final characteristics, which are generally: volume increase, color, alveolar texture, and flavor. As it is possible to imagine, this operation is very complex because numerous physicochemical and biochemical changes are involved, taking place due to heat and mass transfer. The heat phenomena can involve radiation, convection, and conduction, while the mass transfer involves water diffusion and evaporation. Moreover, all the ingredients, generally flour, sugar, fat, baking soda, raising agent, or yeast, and salt, used in the mixing operation play a crucial role in the success of the baking process, which is used to increase the temperature inside the food, but also cause some reactions and water loss. The process has to be controlled to avoid certain reactions and to obtain only the desired ones, and furthermore, because during baking the structure of the baked food is developed together with the color. Specifically, when the temperature inside the food is increased, water diffuses from the bulk to the interface with the hot air and then evaporates, causing the desired water loss.

Another important change, which generally occurs before the diffusion and evaporation of water, is the evolution of gas bubbles (alveoli) inside the food during the first stages of the baking. The evolution of these bubbles occurs due, above all, to the decomposition of the raising agents at around 30–40°C and/or water evaporation. This last phenomenon causes the typical alveolar structure of the majority of baked

**High-Temperature Processing of Food Products. https://doi.org/10.1016/B978-0-12-818618-3.00008-2**

starchy food (Baldino et al., 2014, 2020); however, it must be said that the structure evolution, which is one of the factors appealing to the consumer, is influenced by many variables, including ingredients, rheology, starch gelatinization, proteins coagulation, and also by the heat and mass transfer phenomena controlling the entire baking process.

As can be seen, there are many factors involved in the success of a bakery product, so it is important to understand how these are interconnected in order to devise a mathematical model capable of simulating the industrial process of baking. In fact, good raw materials, correct mixing conditions, and cooking processes with the right operating conditions are the key factors for the correct success of a baked product with a suitable texture and flavor (Sablani et al., 1998; Baldino et al., 2014). Therefore, in this chapter we will discuss the role of ingredients and rheology, including the most suitable mechanical properties, in the success of the cooking process and the transport phenomena involved in cooking, referring to the main classes of baked materials such as bread, cakes, and cookies.

## 4.2 The Baking process in foods

Nowadays, big confectionery companies satisfy the ever increasing demand for bakery products, be they bread and bakery products or biscuits and cakes, by placing a lot of emphasis on the characterization of the ingredients, both in terms of quality and proportions in the recipe. In fact, a minimum change in raw materials or process conditions is sufficient to obtain a different product, which, potentially, may allow the company to gain greater market share. Depending on the final product to be obtained, both the process and the ingredients assume considerable importance in the production of an industrial food product. Industrial products, such as biscuits, are the final result of a series of operations, the optimization and control of which have substantial effects on the final food both in terms of taste and texture. The main goal is to create a product capable of satisfying the needs of consumers, which involves the search for the right ingredients, the right quantities, and the appropriate processing methods. In particular, the growing need to reduce production costs pushes companies to use low-cost ingredients and research is focused on minimizing the impact of these substitutions in recipes on the final product.

The initial phase of the production of a baked product is the choice of ingredients and quantities: each ingredient and its quantity has a significant influence both on the dough and on the quality of the final product. In the next section, we will discuss in detail the ingredients and factors that influence rheological behavior and final texture. This first phase is followed by the mixing phase. Usually, the ingredients are mixed in mechanical mixers, and the mixing time and the rotation speed of the blades have a considerable influence on the final dough, because this phase must lead to the formation of a structure that allows the development of gas bubbles, in most baked products, during the subsequent leavening phase and/or during the cooking phase. The structure obtained must be able to maintain the desired height during the final cooking phase. The leavening stage is a phase during which leavening agents are introduced inside the dough, batter, and similar products to obtain a porous structure. This stage can happen before baking, maybe introducing air or steam during the mixing phase, or introducing the so-called leavening agents, like yeast, which release gases during fermentation to cause expansion of gas cells, or during baking, if chemical leavening agents, like a combination of baking soda (sodium bicarbonate) and dry acids or acid salts, are used. The leavening phase becomes very important when it is intended to make products characterized by a certain height and porosity. This phase is greatly influenced by the temperature at which it is carried out: low temperatures require long leavening times and little development of gas bubbles in the dough; higher temperatures reduce leavening times and favor the development of bubbles. Naturally the leavening times and the temperatures at which this must take place depend on the final characteristics wanted for the product.

Finally, the cooking phase, especially in the case of bread products, must be such as to favor the formation of an external crust and a soft crumb inside: temperatures that are too high lead to the overly rapid formation of an external crust that, by excessively slowing down the transport of material and heat towards the inside, causes poor cooking conditions of the "core."

## 4.2.1 The role of the ingredients

Baking is a very complex food process that depends on different factors, of which the first one is the ingredients and their role. Generally, the baking of a food product after an ingredients change can produce an enormous impact on the final product, and this can be due to the specific role that each of them has. If we are to try grouping the main ingredients of the most important baked goods, such as bread, cakes, and biscuits, into categories, we can say that they are: flours, water or other liquids, fats, eggs, yeast or raising agents, salt, sugars, and hydrocolloids.

The ingredients listed above play different roles. One of the principal ingredients is normally the flour because, when mixed with water, this forms an elastic network, able to trap gas and then to resist during baking against the increment in volume of the bubbles. This is a key factor in texture formation because, by using the correct type of flour with the appropriate protein content, it is possible to obtain a good texture, a correct hardness when used for biscuits, a big increment in volume for bread, or a fine foam and soft texture for cakes. Flour is generally chosen on the base of protein content, and wheat flour, with high protein content (12%–13%), named as strong flour, helps to obtain good gas-retention when it is used for bread, while for biscuits and cakes the desired protein content is about 7%–10% and 7%–8%, respectively (Sai Manohar, 2014). Flours having a protein content ranking of about 8% can be defined as soft flours, while flours with a protein content of about 10% are medium. The protein is mainly gluten, and the percentage of protein determines the flour strength and the dough extensibility. When bread dough is processed, the correct gluten content and mixing conditions allow good workability and make it possible for the bubbles to expand within the system itself during the baking, obtaining a good final texture. While strong flour can be used for baking products such as bread, for biscuits it is preferable to use mixtures to obtain medium-strong flours. Furthermore, it must be emphasized that the flour or mixture of flours must be chosen according to the product to be obtained.

When used for bread, among its functions flour also has to guarantee the growth of bubbles during the leavening and baking phase and guarantee good workability of the dough. The flour that is generally used for soft biscuits has a protein content of less than 10% while the flour for cake has around 8% of protein content because it is not necessary to have a strong structure, even if has to retain gas during the first stage of baking, unlike that used for bread. In fact, a cake's structure has to be soft and delicate. The principal problem in cake making is sugar because it is a water-absorbing competitor of starch and can create some problem when it is present in high quantities in a recipe.

In the baking industry, the treatment of flour with chlorine gas is used to lower the pH and also to help with improving the functional flour components. In the last few years, chlorination has been substituted with heat-treatment of flour, as this procedure can also produce some benefits to the cake making process and it is not chemical. Specifically speaking, dry heat treatment was introduced to substitute for chlorination in the preparation of high ratio cake flours.

Another important role is played by the liquids added to the dough. In general, water is an ingredient that is frequently used in different types of recipes as it plays an important role in the reactions that occur during the early phases. It is absorbed by different ingredients, such as flour or sugar, or it can dissolve some ingredients such as sugar, if present in low quantities, or salt. The water also makes the dough the right consistency and it is responsible for the partial gelatinization of the starch. Obviously, the consistency of the dough is connected to the proportion between proteins present and starches. Water is a necessary ingredient in the formation of a dough. The addition of water causes, a sharp decrease in viscosity and a small reduction in the relaxation time of dough, or for biscuits, an elongation and a decrease in thickness. If the water content is too low, a brittle, inconsistent dough is obtained, which exhibits, during cooking, a marked "crust" effect due to the rapid dehydration of the surface (Maache-Rezzoug et al., 1998).

In Fig. 4.1 are reported the complex modulus and the phase angle for biscuit dough with three different water contents. The data are compared in terms of G*, complex modulus, and angle phase. It is possible to see that, comparing the rheological data with a standard base biscuit recipe, increasing the water content in the recipe, the dough consistency decreases (lower G*), while the consistency increases removing the 30% of water. Moreover, water contributes during baking to an increase in bubbles because of the formation of steam with temperature. Milk is also an important liquid that is used in cake recipes. This liquid ingredient helps, thanks to the presence of proteins, to obtain larger bubbles, a brown crust, and in development of gluten.

Oils or fats are used in dough recipes for some type of bread and biscuits, for which they are among the principal ingredients, and for cakes. Generally, with the term shortening it refers to oils and fats that can be added to modify the mouth-feel of baked products. Fat contributes to the plasticity of the mixture as a lubricant. When present in large quantities, the effects are so pronounced that a small amount of water is

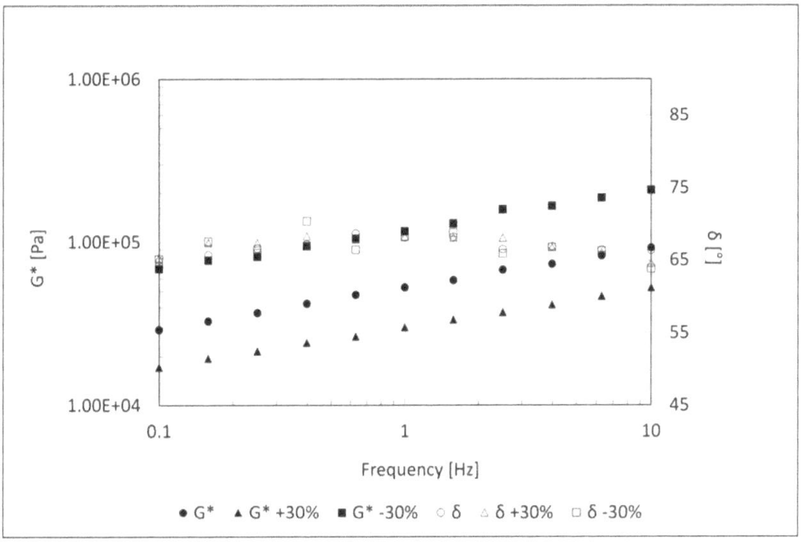


**Fig. 4.1** Complex modulus and phase angle trend for a dough biscuit at three different water content levels.
Unpublished data.

needed to achieve a soft consistency. When the fat is mixed with the flour before hydration, it prevents the formation of a gluten network and produces a less elastic dough. In general, highly elastic dough is not desired for making certain types of biscuits, because it creates difficulties in the lamination process (Maache-Rezzoug et al., 1998). The addition of fat causes a reduction in viscosity and relaxation time of dough, or for biscuits, a reduction in thickness and weight and an increase in length. Fat also makes the biscuit crumbly and favors chewing by the consumer.

Fats can have other roles. As an example, the addition of solid fats in bread stabilizes the rise of bubbles, avoiding their coalescence and giving a softer crumb. In biscuits, the usage of fat is necessary to confer the typical flavor in the mouth and friable texture. Furthermore, during the mixing phase, the added fats compete with the liquids present, be they water or milk, in the absorption of the flour on the surface and, therefore, also disturb the formation of the gluten network.

The competition that is created with water, in particular, can reduce the hydration and gelatinization of starches. The type of fat used, whether it is liquid or solid, hydrogenated or not, can lead to an increase in spreading and a change in the final height. All this is directly related to the rheological characteristics of the mixture. A typical example of the trend of the rheological moduli at room temperature is shown in Fig. 4.2. It can be seen that a hydrogenated solid fat has a higher modulus than a non-hydrogenated solid fat. This results in a change in the final texture, so for example, an increase in the final height for the sample with less consistency.

Shortenings are also used for cake doughs as they are fundamental in the creaming phase, increasing the ability to trap air and, therefore, because they make the final

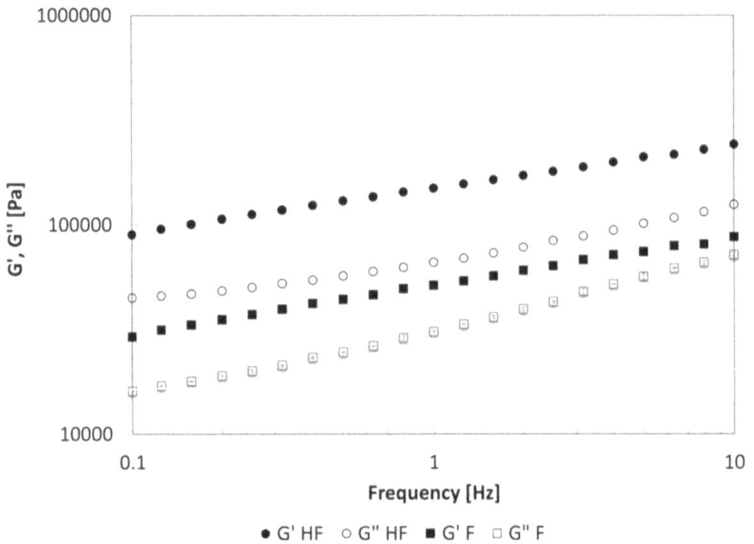


**Fig. 4.2** Elastic and viscous moduli for a shortcake dough made with solid hydrogenated fat (HF) and solid fat (F) with the same quantity of all ingredients.
Unpublished data.

texture more delicate and soft. This latter aspect is due, as mentioned before, to the fact that they interrupt the continuity of the gluten network, avoiding the formation of a final crunchy structure.

Mono- and diglycerides of fatty acids are conventionally used as emulsifiers. They are able to form complexes with the starch, slowing down the retrogradation process during storage as concern the baked food. Emulsifiers improve also the dough handling properties. These are also used to extend shelf life. At a structural and textural level, emulsifiers seem to improve the gas retention capacity, favoring an increase in volume and improving the properties of the crust and/or the dough. Among the most widely used emulsifiers, we note glycerol monostearate, which acts on the bulk mechanical properties of the dough rather than on the increase in volume, favoring its elasticity because it acts on retarding starch retrogradation. When emulsifiers are used for the production of biscuits, an important role is played by the proportion between the fat phase and the water present and the possible presence of other substances; while for cake production, they can be a key factor in obtaining a fine texture with small and well-dispersed bubbles and fats.

In a traditional sponge cake recipe, eggs are commonly used because they seem to improve the formation of small air bubbles during the mixing step. The lipoproteins are the principal reason for the stability of the air bubbles and the homogeneous distribution in the bulk. Important properties imparted by the addition of eggs to the dough are the color that they can confer to the final product and the "hardness" caused by the partial contribution by the egg white to the gelatinization of the starches and the development of the gluten network.

Sugars or sweeteners, in general, are nutritive carbohydrates used by the baking industry to add sweetness and flavor. They perform key functions in doughs and batters that improve the overall quality of the finished products. Specifically speaking, sugar affects the starch gelatinization and coagulation temperature, which move upwards, allowing a greater growth of bubbles inside the dough before the continuous phase is completely rigid. It has been observed that sugar-rich biscuits are characterized by a crunchy texture and that the sugar makes the biscuit fragile as it controls hydration and tends to disperse proteins and starch molecules, preventing the formation of a continuous mass. Furthermore, studies on the effects of the quantity of sugar and granulometry on spreading (the phenomenon according to which a dough, after being dosed, for example on a cooking surface, flows under the action of its weight, increasing its diameter at the expense of height) of the biscuit show that it is favored by small grain size and a high concentration of sugar (Maache-Rezzoug et al., 1998). The quantity of sugar used depends on the granulometric distribution and also influences the development of gluten in short doughs, due to the competition in the water absorption of the system. Different types of sweeteners are used, including maltose and lactose, glucose, and fructose, but only these last two are capable of giving a golden color to the baked product.

Other important ingredients are yeasts and raising agents, which can generate carbon dioxide by biological or chemical methods. Yeasts can be compressed or dried, and their main role is to produce, together with the sugars, carbon dioxide and alcohol, which contributes to the expansion of the dough during cooking. The natural yeasts produce biological leavening, thanks to the intervention of organisms of a microbial nature, they are real living yeasts and bacteria. In this category, we find brewer's yeast (dried, fresh, powdered), and sourdough. Brewer's yeast is not a bacterium, but a fungus called *Saccharomyces cerevisiae* and produces only an alcoholic fermentation, while we will see that the sourdough is made of many microorganisms, such as lactic ferments, absorbed from the environment. The optimum temperature at which yeasts work is around 40–43°C. Leavening agents are also significantly responsible for the increase in the volume of the baked product, giving lightness to the final product and a honeycomb structure, even if an important role is played by the aeration phase during the mixing phase.

Finally, salt is also employed in the bakery industry to confer flavor and sapidity, and to control the water activity. Salt enhances the golden color of bread crust by reducing the destruction of sugar in the dough and thus increasing caramelization. Salt strengthens the gluten in the bread dough by providing uniform grain while giving consistency and strength to the dough, allowing the dough to expand without breaking. It is used in low percentages (1%–1.5%). Salt is also used in sweet baked products as a flavor enhancer and as a reinforcing element in leavened pastries. In fact, the addition of salt in moderate percentages (from 1% to 2.5%) improves the resistance and elasticity of the dough by having a positive effect on the formation of gluten. In leavening, if used correctly, salt optimizes kneading times and helps to keep fermentation under control; however, it must be used in low quantities, as it turns out to be a real yeast inhibitor. Finally, salt is also an acidity regulator and its use in leavened desserts, especially in those with sourdough, regulates the production of some acids, preventing any abnormal formation of acidity. Moreover, when it is also used in the cake or biscuit industry, salt can adjust the color of the crust.

### 4.2.2 Mixing phase and structure development

The water and flour, with or without yeast, are the basic and typical ingredients to obtain bread dough. The mixture obtained by mixing water and flour can be considered as a particular state of aggregation, after mixing with water, and an almost solid system, consisting approximately of a three-dimensional protein network of gluten (insoluble in water), which incorporates all the other components, including the hydrated starch granules, mainly responsible for the mechanical behavior, and gas bubbles, responsible for the increase in volume during leavening, fermentation, or cooking, stabilized by a protein membrane that prevents collapse.

In addition to the air incorporated during mixing, the bubbles distributed in the dough mass contain water in thermodynamic equilibrium with the water-bound within the dough, and possibly other substances (reaction products of yeast), always in thermodynamic equilibrium. As time passes, the equilibrium is broken by the process conditions due to both the thermal and humidity profiles to which the system is subjected, and to chemical reactions. It can, therefore, be assumed that the bubbles initially grow in an infinite medium, and then begin to interact with each other, finally reaching coalescence conditions with the possibility of collapse towards the outside. From this complex growth-collapse mechanism, which in particular takes place in the first phase of cooking, an alveolar structure is determined with a particular texture (Baldino et al., 2014). For this type of dough, the mixing, rheology, and cooking steps are interrelated: each of these phases is characterized by mechanisms that have effects on the gluten network and, therefore, on the final properties. In general, it is possible to move from a first phase in which the structure is underdeveloped, in terms of resistance and toughness (under mixing), to one in which there is the optimal consistency, to arrive at that in which the structure now formed collapses, passing to an overmixing condition. Therefore, the mixing operation must be characterized by an appropriate duration time and an appropriate energy value being supplied (Letang et al., 1999).

Rheological properties of dough during mixing is affected greatly by the flour composition (i.e., low or high protein content), processing parameters (mixing time, energy, temperature, and speed rate), and ingredients (for example, water and salt). Different flours have different optimum mixing times. Mixing time and energy input are normally used to change dough properties. The widespread belief is that stronger dough requires longer mixing time and more energy input. To achieve optimum dough development, the mixing time and work input must be above the minimum critical level (Angioloni and Dalla Rosa, 2005), and also the mixer design, operation conditions, and water level influence the final quality. In fact, an optimum in work input and mixing time have been related to optimum bread-making performance, which varies according to mixer type, composition, and ingredients.

Work input and mixing intensity and speed are two critical factors for optimal dough development. High mixing speed leads to a large disorientation and destruction of dough structure, and hence, reduced viscosity. On the other hand, water is responsible for hydrating the protein fibrils and interactions between the protein cross-links with the disulfide bonds during dough mixing and an optimum level is required to develop dough with optimum gluten strength. During mixing, the energy flow and the hydration processes are accompanied by a temperature increase. Temperature is another factor that influences the elastic properties of dough. The temperature increase is determined by the speed of mixing. The literature considers 28–30°C to be the optimum temperature range for dough mixing (Sai Manohar, 2014).

Another important class of baked products is biscuits, which can be broadly divided into crackers, hard sweet, and semi-sweet biscuits; and because they have different characteristics and recipes, they need different mixing, forming, and baking processes. In general, all types of dough for biscuits require a fermentation phase or a standing time after mixing. Only after this phase is it possible to move on to the forming phase and the subsequent baking. Crackers are a special class of salted biscuits. They are characterized by a low amount of fat and sugar. Crackers can be made from both fermented and non-fermented elastic and hard dough and the flour type has to be weak, with a protein content of about 8%–9%. Doughs for hard sweet biscuits have a strong, well-developed gluten network, which is elastic and suitable to be sheeted and cut. This type of dough often shrinks in the first stage of baking. In general, this biscuit has low sugar and fat content and the water level is typically around 12%–15%. The flour used for these biscuits should not exceed 9.0% protein content.

In contrast to these types of biscuits, there are the soft or sweet biscuits. For this last type of biscuits, the development of gluten is minimal and only serves to promote cohesion to facilitate the processing of the dough and to give it shape. They are extensible but not elastic. The main ingredients of a short dough are flour, fat, sugar, and only a small amount of water (<20%). The mechanical and rheological properties of a mixture of this type strongly depend on the quantity and quality of fat, on the water and sugar content. The water has the effect of increasing the plasticity of the dough, decreasing the gelatinization temperature, and can generate biscuits that exhibit a friable behavior; fat also contributes to plasticity of the dough (Baltsavias et al., 1999). Adding sugar to a short dough causes it to soften and greatly affects the final texture of the biscuit.

One type of cracker can be obtained by a single mixing step, during which the ingredients can be mixed "all in" on a horizontal mixer at about 33°C, with the possibility of adding an enzyme. Another type of cracker, named "saltine" or "premium," can be obtained by a two-stage mixing process followed by a long fermentation and fast baking. To produce crackers with an open texture it is advisable to use a flour with a medium-high percentage of proteins (10%–11%). Another method exists based on a two-stage process, wherein a "sponge" dough is prepared, obtained by mixing part of the raw materials in a special mixer. The resulting mix is left to rest in a leavening cell at a controlled temperature and humidity. After leavening, the remaining raw materials are added to the dough until a soft dough is obtained, which is placed back in the leavening cell to rest. When the second leavening is complete, the dough is sent for processing. The first sheet of dough is obtained, which is superimposed several times and then thinned to the desired thickness, by passing through graded series of rolling cylinders. This is followed by the printing of the desired shape and the cut.

Hard sweet biscuits are obtained by an "all-in-one method" using a horizontal mix. All the ingredients are mixed at the same time and the ingredients are kneaded for an optimal time. The mixing phase can confer the right rheological characteristics and, thanks to the mixing phase, it is possible to obtain an optimal consistency, which will then guarantee the final quality and rheological properties of the dough. The mixing time must guarantee the correct development of the gluten network, which will give the dough extensibility. Generally, the mixing time is around 20 min and the dough must not reach too high a temperature, as otherwise it becomes unstable. The short dough (molded biscuits) is a biscuit category with many forms. The dough has high fat and sugar content, as said before, and low water content. This type of biscuit is obtained in two stages and, during the mixing phase, the gluten network does not develop. Generally, the entire mixing phase is about 10–15 min. In the first step, all the ingredients, except flour and soda, are placed in the mixer, with the intention to dissolve as much sugar as possible in the water first, and then adding the other ingredients. During the second stage, flour and soda are added. In this way, a homogeneous mixture is obtained and, during the whole process, it is important to keep the dough temperature low. The process takes place at low speed and the final temperature of the dough is approximately 18–22°C. The final dough is sheeted after pressing and the right shape is conferred by a rotary molder.

The main ingredients in a cake mixing process that make up a batter are flour, sugar, water, egg, oil, raising agents, and glycerol. Doughs of this type have a different structure from those seen previously. For this type of dough, the development of the gluten network is completely avoided, preventing the water from binding to the flour; this is done by paying close attention to the order in which the various ingredients are mixed (in the industrial process, flour is the last ingredient added). These mixtures exhibit a liquid-like behavior and, therefore, are structurally weaker than both water-flour and short mixtures, with really different rheological and mechanical properties.

The main purpose of the mixing process is to incorporate air into the mixture and this phase is called the aeration phase. During this phase, the bubbles are uniformly incorporated and subsequently, their growth and stabilization take place. To obtain a batter dough it is possible to proceed in four different ways: creaming the butter and sugar together, also called the creaming method, flour batter or blending method, single-stage method, and emulsion method. The first one is the classic method and the most common. The first stage starts with beating the butter or another fat and sugar together until they are lightened in color and fluffy. This phase allows the incorporation of air. Only then are dry solids and liquids added. The correct mixing time is very important to prevent a loss of air incorporated during the first stage. Generally, the second stage includes the addition of eggs. Only at the end, milk and flour are added (which therefore would represent the third stage). Particular attention must be paid to adding milk or other liquid, which must take place with the right speed and alternation with flour. Adding flour and liquids alternately ensures all the liquid will be thoroughly absorbed into the batter. If there's a high amount of butter or other fat in the batter, it's hard to get the liquid mixed in. Moreover, if added too quickly, the milk can cause an inversion of the emulsion.

In the flour batter method, an equal quantity of flour and fat are creamed together, also eggs and sugar are whipped to obtain a firm foam. The egg is beaten while adding sugar gradually until it is light and frothy. This foam is added to the creamed flour and fat cream gradually. This type of method gives rise to a less aerated product, which has a more developed gluten network. The simplest method is the single-stage one, where only raising agents are added at the end. The total time required is approximately 8–10 min. To obtain a cake with a high volume, the emulsion method can be applied. The fats and sugar are mixed and milk is gradually added until a soft mass is obtained. Finally, the flour and eggs are added. Generally, the process takes about 12–15 min.

## 4.3 Modeling of simultaneous heat and water transport in the baking process

The baking step is probably the most relevant one in determining the final properties of baked goods; therefore, it should be controlled carefully with the aim of obtaining the expected characteristics. The control of industrial ovens, based on empirical procedures, can be very complex owing to the large number of transformations occurring in the system during baking. Therefore, a deep knowledge of the physical phenomena involved in it is necessary to describe the most relevant transformations, intending to predict final properties such as texture, color, nutritional properties, etc., as a function of both raw ingredients and operating conditions.

During baking, simultaneous mass and heat transfer occurs, causing a change in moisture content and transforming a viscoelastic liquid (having a prevalent liquid-like or solid-like behavior according to the specific use) into an alveolar solid structure. In addition, different biochemical reactions occur, yielding changes in color (because of the non-enzymatic browning or Maillard reaction), gas cell expansion (because of the raising agents' decomposition), inactivation of enzymes and microorganisms, and changes in ingredients' digestibility, etc. These changes are related to temperature and humidity profiles (in addition to other variables such as pH) within the system and can often be described by introducing the proper kinetic equations to estimate the rate of the specific phenomena. Nevertheless, the use of these kinetic equations require the determination of temperature and concentration as a function of temperature and position; therefore, it is necessary to describe in the proper way the physical phenomena related to mass and energy transfer both within the bakery product and outside it (i.e., within the oven).

Owing to the relevance of this issue, in recent years a large number of models concerning baking of cereal products (both fine products such as biscuits or thick products such as bread) have been proposed in the literature (Broyart and Trystram, 2003). A preliminary rough distinction among them can be based on the focus of the analysis (Broyart and Trystram, 2003). The first category of models discusses in detail the transport phenomena between the oven elements and the bakery product, giving less relevance to phenomena occurring within the product, where uniform conditions are often assumed. Even if these models can be interesting to optimize the working conditions of the oven, they do not discuss in details physical changes such as gas cell expansion, changes in color and dimensions, etc.

The second type of model discusses mainly the transport phenomena within the product, simplifying the description of mass and heat flux between the product and the oven. In this case, attention is given to internal temperature and moisture profile, trying to identifying the controlling mechanism.

It is worth noting that this second type of model can probably give more information about the effects of operating conditions (considered as boundary conditions) on product properties; therefore, it can be more useful in optimizing and standardizing the final product quality. As a consequence, in the following discussion, after a brief preliminary description of transport phenomena in the oven, attention will be given mainly to internal heat and mass transfer to investigate the involved mechanisms and the effects on product characteristics.

### 4.3.1 External transport phenomena

Heat is transferred to bakery products according to different mechanisms: radiation from the oven walls to the product surface, convection (and, in a minimal amount, conduction) from the heating medium (humid air), and conduction from the oven belt or tray on which the product is placed to the product surface (Fellows, 2000). Different studies have shown that radiation is the most relevant mechanism for heat transfer in industrial ovens, representing 50%–80% of the total heat transfer. Radiation is the emission or transmission of energy through electromagnetic waves between two bodies having different temperatures. In baking, radiation comes from gas burners, electric elements, and from the oven walls.

The heat transfer depends on different factors, such as the surface temperatures of involved bodies, their shapes, and their surface properties (Fellows, 2000). The heat flux ($q_R$), i.e., the rate of heat flow per unit area, from a "black body" (a perfect radiator or absorber) can be estimated according to the Stefan-Boltzmann equation (Bird et al., 2007):

$$q_R = \sigma T^4 \tag{4.1}$$

where $\sigma$ is the Stefan-Boltzmann constant and $T$ the absolute temperature.

Nevertheless, oven elements are not perfect radiators and foods are not perfect absorbers. This is taken into account by introducing the "gray body" and the emissivity $\varepsilon$ (ranging between 0 and 1):

$$q_R = \varepsilon\sigma T^4 \tag{4.2}$$

The net energy adsorbed by a body is the difference between the adsorbed energy and the emitted one; therefore, if $T_1$ and $T_2$ are the temperatures of emitter and receiver, respectively, the net heat flux will be (Bird et al., 2007):

$$q_R = \sigma\left(\varepsilon_2 T_2^4 - \alpha_1 T_1^4\right) \tag{4.3}$$

where $\alpha_1$ is the absorptivity of the receiver, i.e., the amount of radiation absorbed by a gray body; it is usually assumed to be equal to the emissivity of the same body at temperature $T_1$.

In addition, a view factor, i.e., the fraction of radiation leaving the emitter and directly intercepted by the receiver, should be introduced to take into account the reciprocal position of the bodies. Emissivity values can be found in the literature (Marcotte, 2007). Further details on the radiation mechanism can be found in (Bird et al., 2007).

Convection is the transport of energy by the bulk motion of a fluid, which can be caused by differences in density (natural or free convection) or by mechanical agitation (forced convection). The amount of heat transferred by convection per unit area ($q_{cv}$) can be estimated as the product of the driving force of the process, i.e., the temperature difference between the heating medium ($T_f$) and the surface of the bakery product ($T_s$), and the transfer coefficient ($h$), which can be seen as the specific rate of heat transport (Bird et al., 2007; Marcotte, 2007):

$$q_{CV} = h(T_f - T_s) \tag{4.4}$$

The transport coefficient, $h$, depends on fluid properties and on the transport mechanism (i.e., natural or forced convection) and it can be estimated according to the typical semi-empirical equations of chemical engineering, relating the dimensionless groups involved in heat transport phenomena, i.e., Nusselt ($Nu$), Prandtl ($Pr$), Reynolds ($Re$), and Grashof numbers (Bird et al., 2007):

$$Nu = \frac{hL}{k} \tag{4.5a}$$

$$Re = \frac{\rho v L}{\mu} \tag{4.5b}$$

$$Pr = \frac{c_P \mu}{k} \tag{4.5c}$$

$$\mathrm{Gr} = \frac{L^3 \rho^2 g \beta (T_s - T_\infty)}{\mu^2} \tag{4.5d}$$

where $\rho$, $\mu$, $k$, and $c_P$ are viscosity, density, thermal conductivity, and specific heat of the heating medium (typically air), respectively, $v$ is the air velocity, $g$ is the standard gravity, $\beta$ is the coefficient of thermal expansion of air, $T_s$ is the surface temperature, $T_\infty$ is the bulk temperature, and $L$ is a characteristic length that has to be specified for each flow system (it is a characteristic length of the specific bakery product).

Different relations can be found in the literature (Bird et al., 2007; Marcotte, 2007) depending on the specific system conditions (e.g., geometry, free or forced convection, laminar or turbulent flow, etc.); for instance, for laminar flow around a flat plate the following equation can be used:

$$Nu = 0.664\,\mathrm{Re}^{\frac{1}{2}}\mathrm{Pr}^{\frac{1}{3}} \tag{4.6}$$

Whereas for free convection over a flat surface (a condition sometimes used because air velocity in the industrial oven is often very small), it holds that:

$$Nu = 0.28(Gr\mathrm{Pr})^{\frac{1}{3}}\mathrm{Pr}^{0.034} \tag{4.7}$$

In addition to the use of literature correlations to estimate radiative and convective heat flow, heat flux sensors can be used to measure heat flow directly within the specific oven. In this case, it is assumed that heat transfer between oven and bakery product is the same as that occurring between oven and sensor. Heat sensors can be used to measure the total heat flow or to estimate the radiative and the convective fractions; in the latter case, the sensor is covered either with a radiation reflector to measure convective flux only or with a black radiation absorber to estimate the total radiative and convective flow.

Conduction is the transport of energy through a molecular mechanism. The heat flux ($q_{CD}$) is determined by the difference of temperature between different spatial positions and the resistance to heat transfer; it is described by Fourier's law (Bird et al., 2007), which, for one-dimensional phenomena, can be written as:

$$q_{CD} = -k\frac{\partial T}{\partial x} \tag{4.8}$$

where $x$ is the direction of transfer and k the thermal conductivity (i.e., the reciprocal resistance to heat transfer).

If the temperature varies in all three directions, assuming that transport occurs in isotropic media (i.e., without a preferred direction), Fourier's law can be written in a three-dimensional form as:

$$\underline{q}_{CD} = -k\underline{\nabla}T \tag{4.9}$$

Simultaneously with the transport of heat, mass transport occurs between the bakery product and the heating medium, whereas it can be assumed that no transport occurs at the surface placed on the oven belt or tray. Mass transport involves mainly water, evaporating at the external surface, although carbon dioxide and ammonia, potentially developed by raising agents' reactions, can also be present.

Transport phenomena are mainly based on a convective mechanism (even if a small contribution of diffusion could be present) and the rate of transfer of water per unit area (mass flux, $N_i$) can be estimated as the product of the driving force of the process, i.e., the difference of molar fraction between the air ($y_{if}$) and the surface of the bakery product ($y_{is}$), and the mass transfer coefficient ($k_y$), which can be seen as the specific rate of mass transport (Bird et al., 2007):

$$N_i = k_{yi}\left(y_{if} - y_{is}\right) \tag{4.10}$$

The concentration at the product surface can be computed through the use of water activity, which can be related to water concentration by different semi-empirical models (Baldino et al., 2014, 2017).

The mass transport coefficient $k_{yi}$ can be determined according to the analogy between mass and heat transfer phenomena (Bird et al., 2007):

$$j_H = j_D \tag{4.11}$$

where $j_H$ and $j_D$ are the Chilton-Colburn factors for heat and mass transfer, respectively, yielding (Baldino et al., 2014):

$$k_{yi} = \frac{hc}{\rho c_P}\left(\frac{c_P \rho D_{i/air}}{k}\right)^{\frac{2}{3}} \tag{4.12}$$

where $h$ is the heat transport coefficient, $c$ the air concentration, $\rho$, $c_P$, and $k$ density, specific heat, and thermal conductivity of air, respectively, and $D_{i/air}$ the diffusivity of the $i$-th component in air.

### 4.3.2 Internal transport phenomena

Heat transport within biscuits occurs by conduction from the external (both upper and lower) surfaces to the inner parts; at the same time, water is diffused from the inner to the external (upper) surface where it evaporates and is removed by the air flow. Conductive heat flux is described by the already cited Fourier's law (Eq. 4.8) where the proper value of thermal conductivity has to be used. Diffusive mass flux ($J$) is described by Fick's law, which, for one-dimensional phenomena, can be written as (Bird et al., 2007):

$$J = -D\frac{\partial c}{\partial x} \tag{4.13}$$

where $x$ is the direction of transfer and $D$ the diffusivity.

Fick's law can be written in a three-dimensional form for diffusion within an isotropic material as (Bird et al., 2007):

$$\underline{J} = -D\underline{\nabla}c \tag{4.14}$$

Even if conduction and diffusion are the transport mechanisms involved in baking, the description of phenomena occurring within products is quite complex because they are mainly heterogeneous systems constituted by a dense cereal paste containing many gas cells that might be either closed from or open to the ambient conditions. This would require modeling transport phenomena in gas and liquid phases, combining them through the proper equations. This approach would be very complex and most of the approaches proposed in the literature are based on a pseudo-homogeneous system ( Jayapragasam et al., 2021; Lostie et al., 2002; Nicolas et al., 2017; Sablani et al., 1998) where a continuous medium characterized by equivalent transport properties is considered. Even if this approach can make easier the discussion of the problem, the relevant phenomena affecting product evolution (such as the gas cell expansion and reduction and the mass transfer between the cereal paste and the gas bubbles) cannot be properly taken into account and they appear only in the estimation of equivalent properties (such as an effective diffusivity or an effective conductivity depending on the porosity). Rheological aspects related to the expansion of gas cells towards a viscoelastic paste cannot be considered and the material expansion, often, is modeled as the result of water vaporization assuming a deformation depending on overpressure (Lostie et al., 2002).

A different approach has been proposed by de Cindio et al. (Baldino et al., 2014, 2020; de Cindio et al., 2006; de Cindio and Correra, 1995) and in a different way by (Bikard et al., 2012). These works model the evolution of aerated systems during baking, describing the changes at two different scales: a "gas cell scale," defined as "microsystem" by (de Cindio and Correra, 1995), and a "product scale," defined as "macrosystem." At the microsystem scale, phenomena related to a single gas bubble are described in detail by considering the thermodynamic and mechanical equilibrium at the interface between gas and paste (assumed as a viscoelastic liquid). This makes it possible to describe the flow of water vapor and other potential gases produced by raising agents' decomposition from the paste to the gas cell and the subsequent expansion hindered by the resistance of the viscoelastic paste surrounding the bubble. The final bubble radius will be the result of the mechanical equilibrium between these opposite effects and will determine the local void fraction.

At macrosystem level, a continuous medium is considered, which makes it possible to write the usual mass and heat transfer equations, however the material properties change with the position, according to the local void fractions obtained by the local microsystem and providing that mass exchange between the paste and the gas phase is properly taken into account in equations. In this way, the macrosystem describes the evolution of the global structure using boundary conditions of the oven operating parameters. This approach was used initially to describe the baking behavior of a thick cereal product (bread type) with closed gas cells (de Cindio and Correra, 1995). In further studies, it was used to model the evolution of sponge-type goods (de Cindio et al., 2006) and semi-sweet short dough biscuits (Baldino et al., 2014). Furthermore, different criteria, depending on the characteristics of the product, were proposed to also assess gas cell stabilization.

As an example, the main equations describing the microsystem and macrosystem for biscuits will be reported here, whereas further details can be found in (Baldino et al., 2014, 2020).

### 4.3.3 Microsystem

The bubbles introduced inside the dough during the mixing grow as a consequence of different combined actions: raising agent development, water evaporation, and temperature increase. At the beginning of the process, bubbles may be considered as submerged in an infinite medium of paste; therefore, the system can be assumed to be constituted by a single closed spherical gas bubble surrounded by an infinite mass of a viscoelastic paste. In a spherical coordinate system, the mechanical equilibrium can be written as:

$$\frac{P_G - P_\infty}{\rho} - \frac{2\sigma}{\rho R} - \frac{3}{\rho}\int_{-\infty}^{t} \frac{G(t-t', x_w T)\dot{R}'(R')^2 \ln(R'/R)}{(R')^3 - R^3} dt' = 0 \tag{4.15}$$

where $\sigma$ is the surface tension, $P_\infty$ the pressure at infinite distance from the bubble, $\rho$ the liquid density, $P_G$ the bubble pressure described by the ideal gas equation, $R$ the bubble radius, and $G$ the relaxation modulus of the viscoelastic paste assumed in linear conditions.

Internal pressure $P_G$ depends on the amount of gas within the bubble, which, in turn, depends on the thermodynamic equilibrium of water, carbon dioxide, and ammonia (developed by raising agents' decomposition) in the gas phase (within the bubble) and in the liquid phase (i.e., in the water in the paste surrounding the bubble):

$$P_{H_2O} = P_G \frac{n_{H_2O}}{n_{tot}} = a_w \cdot P^0_{H_2O} \tag{4.16a}$$

$$P_{CO_2} = P_G \frac{n_{CO_2}}{n_{tot}} = H_{CO_2} \cdot x_{CO_2} \tag{4.16b}$$

$$P_{NH_3} = P_G \frac{n_{NH_3}}{n_{tot}} = P^0_{NH_3} \cdot x_{NH_3} \tag{4.16c}$$

where $P_i^0$ (i = $H_2O$ or $NH_3$) is the vapor pressure, $H_{CO_2}$ is the Henry's constant for carbon dioxide equilibrium, and $a_w$ is the water activity of the paste.

The mechanical and thermodynamic equilibrium equations are solved to compute the bubble radius (and, therefore, a local void fraction) at any time; in addition, $CO_2$, $NH_3$, and $H_2O$ mass distribution between gas and paste phase are obtained, which makes it possible to estimate the net molar flow from the paste to the bubble (to be used in the macrosystem equations).

No limitation is imposed on the bubble growth in this model, whereas experimental evidence suggests that, depending on the specific product, two different situations can occur: a maximum bubble size is reached and cannot be overcome (cell stabilization) or the layers between adjacent bubbles break and cells become interconnected or opened to the external surroundings (cell opening and coalescence). In the former case, a criterion for cell stabilization is necessary. For instance, (van Vliet, 2008) proposed, for bread production, a criterion based on dough strain hardening, to be investigated with rheological biaxial extension tests. For semi-sweet short dough biscuits, (Baldino et al., 2014) suggested a criterion based on the onset of a critical temperature, corresponding to the protein cross-linking condition, estimated from rheological tests, whereas for sponge-type products, (de Cindio et al., 2006) proposed a criterion based on a critical thickness of the layer surrounding the gas cell estimated by a mechanical test. In a similar way, different criteria for cell opening are proposed in the literature. For instance (Gan et al., 1995) suggested that coalescence occurs as a consequence of the breakage of a thin lamellar film that surrounds gas cells, which becomes thinner and thinner during the cell expansion; (Baldino et al., 2020), analyzing the mechanical phenomena in the paste layer around the bubble, presented a criterion based on the elastic energy stored in the material during the deformation. When a critical value is reached, the paste layer, no longer able to bear it, breaks and coalescence occurs. More recently, (Nicolas et al., 2017) proposed a very simple criterion based on a critical temperature, assumed to be equal to 60°C, corresponding to the condition of cells opening.

### 4.3.4 Macrosystem

At the macroscopic level, heat and mass transport equations are written on a pseudo-homogenous system where local effective properties are estimated according to the microsystem results. Assuming a cylindrical coordinate system and angular symmetry conditions, only variations along axial ($z$) and radial ($r$) coordinates are considered; therefore, the following mass balance equations for water or $i$-th component ($i = CO_2$ or $NH_3$) can be obtained (Baldino et al., 2020; Bird et al., 2007; de Cindio et al., 2006):

$$\frac{\partial[c_w(1-\phi)]}{\partial t} = -\frac{1}{r}\frac{\partial}{\partial r}(rN_{wr}) - \frac{\partial}{\partial z}(N_{wz}) + R_w(1-\phi)\rho_P - S_w \tag{4.17a}$$

$$\frac{\partial[c_i(1-\phi)x_i]}{\partial t} = -\frac{1}{r}\frac{\partial}{\partial r}(rN_{ir}) - \frac{\partial}{\partial z}(N_{iz}) + R_i(1-\phi)\rho_P - S_i \tag{4.17b}$$

where $\phi$ is the void fraction, $N_i$ the molar flow of $i$-th component, $R_i$ the rate of change in the number of moles involved in chemical reactions (given by proper kinetic equations), and $S_i$, the net molar flow between gas cells and paste, given by the microsystem.

The equations differ because water diffuses in the paste whereas carbon dioxide and ammonia diffuse within the water. The molar flow of single components can be computed, assuming Fick's law, as:

$$N_{wz(r)} = -D_{w,D}\frac{\partial c_w}{\partial z(r)} \tag{4.18a}$$

$$N_{iz(r)} = -D_{i,w}\frac{\partial}{\partial z(r)}[c_w(1-\varphi)x_i] + x_i \cdot N_{wz(r)} \tag{4.18b}$$

Boundary conditions are obtained by imposing the continuity of mass fluxes and chemical potential at the interfaces between biscuit and air. Mass fluxes in the air are computed according to Eq. (4.10) previously described. The energy balance equation can be written as:

$$\rho c_p\frac{\partial T}{\partial t} = \frac{1}{r}\frac{\partial}{\partial}(q_r) + \frac{\partial}{\partial z}(q_z) - \sum_i \lambda_i S_i \tag{4.19}$$

where $\rho$ and $cp$ are density and specific heat of dough, respectively.

Heat fluxes $q_r$ and $q_z$ can be estimated assuming Fourier's law (Eq. 4.9). The term $\lambda_i S_i$ takes into account the latent heat related to the phase transition of some component at the gas cell/paste interface. A similar equation can also be written for the oven tray (Baldino et al., 2014). Eq. (4.9) can be solved using the proper boundary conditions where the continuity of temperature and heat fluxes between biscuit and air and biscuit and tray are assumed. Heat fluxes in the air are estimated according to Eq. (4.3) and Eq. (4.4) to take into account both radiation and convection, whereas tray side conduction (Eq. 4.8) and radiation (Eq. 4.3) can be assumed.

Finally, equations able to describe the volume reduction caused by moisture loss (i.e., the shrinkage caused by drying) can be added and, often, empirical linear relations are used (Baldino et al., 2014):

$$\frac{V(t)}{V(t_0)} = 1 + k(U(t) - U(t_0)) \tag{4.20}$$

where $V(t)$ and $U(t)$ are volume and moisture content as a time function, respectively, whereas the subscript "0" refers to a specific time used as starting point (often it is the time corresponding to peak height), and the parameter $k$ describes how the volume changes with time—usually, it is determined by fitting experimental data obtained on industrial or lab scale.

Considering that, usually, biscuit thickness is very small when compared to biscuit diameter, it is often possible to neglect changes along the radial coordinate and a 1D model along the z coordinate can be sufficient to describe the main phenomena. On the other hand, when thicker systems, such as small sponge-type products, are considered it is better to solve the 2D model to have a more accurate description of the material evolution.

### 4.3.5 Simulation and optimization of baking processes

Numerical procedures proposed to solve physical models can be based on finite difference methods (Baldino et al., 2014) or finite elements methods (see for instance, Feyissa et al., 2011; Nicolas et al., 2017). The solution of Eqs. (4.17a), (4.17b), (4.18a), and (4.18b) makes it possible to compute the moisture and temperature profile within the biscuit, whereas the solution of Eq. (4.15), yielding the bubble radius, allows estimation of the final biscuit void fraction and height. Therefore, the model can be used to investigate the effects of operating conditions or formulation aiming at determining a range of optimal values of variables able to yield the expected final properties.

For instance (Baldino et al., 2014) simulated the effects of different operating conditions (in terms of increase or decrease of heat fluxes concerning a value chosen as "standard") and different formulations (changing water content or raising agent level). They observed that a change in heat flux in the first baking zone (see Fig. 4.3 for the definition of baking zones and standard fluxes) affects both peak and final heights.

An increase in heat flux yields faster reactions, therefore, a rapid bubble growth is observed (as evidenced by the onset of height increase occurring at shorter times); nevertheless, an earlier protein cross-linking occurs accelerating the stabilization

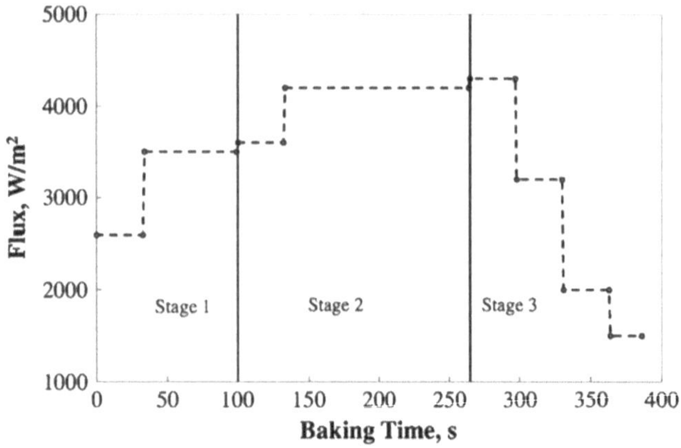


**Fig. 4.3** Top and bottom heat flux oven profile.
Reprinted with permission from Baldino, N., Gabriele, D., Lupi, F.R., de Cindio, B., Cicerelli, L., 2014. Modeling of baking behavior of semi-sweet short dough biscuits. Innov. Food Sci. Emerg. Technol. 25, 40–52, Copyright (2014), Elsevier.

phenomena and yielding, as a result, a lower peak height (Fig. 4.4). In addition, an extensive drying also occurs. Therefore, very high fluxes in the first zone seem inappropriate to obtain high-quality products.

Changes in formulations, such as the water amount, modify the rheological properties of the paste and, as a consequence, the mechanical equilibrium of bubble growth. An increase in water amount (+20% with respect to the standard recipe)

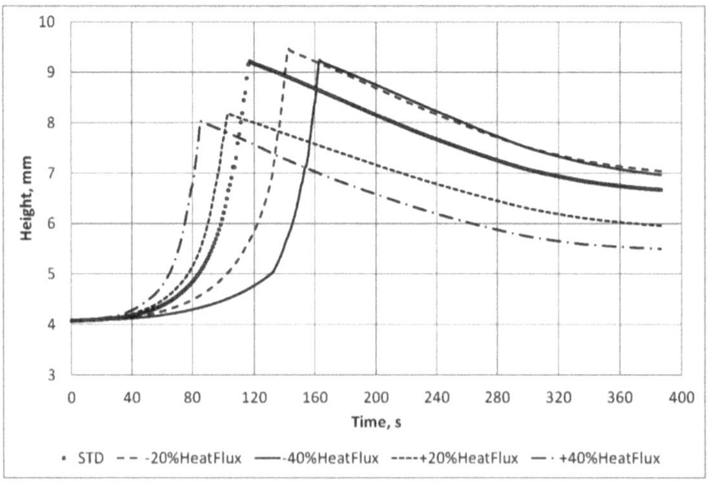


**Fig. 4.4** Simulated height biscuit profiles for different oven conditions in the first step.
Reprinted with permission from Baldino, N., Gabriele, D., Lupi, F.R., de Cindio, B., Cicerelli, L., 2014. Modeling of baking behavior of semi-sweet short dough biscuits. Innov. Food Sci. Emerg. Technol. 25, 40–52, Copyright (2014), Elsevier.

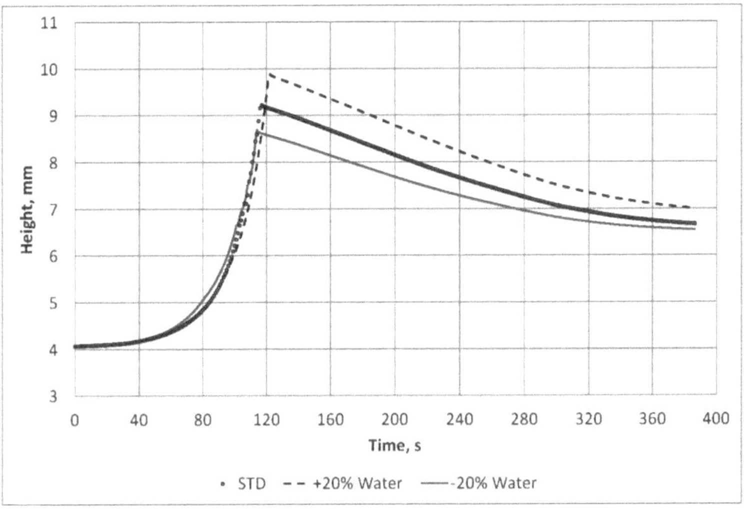


**Fig. 4.5** Height biscuit profile changing initial water level.
Reprinted from with permission from Baldino, N., Gabriele, D., Lupi, F.R., de Cindio, B., Cicerelli, L., 2014. Modeling of baking behavior of semi-sweet short dough biscuits. Innov. Food Sci. Emerg. Technol. 25, 40–52, Copyright (2014), Elsevier.

decreases the relaxation modulus and this makes the bubble expansion easier, giving a higher void fraction and larger biscuit heights (Fig. 4.5).

Similar effects are obtained when changing raising agents content, keeping constant the other conditions, which modifies the amount of developed gases (i.e., carbon dioxide and ammonia) affecting the driving force for bubble growth. An increase in raising agents yields greater height of biscuit both at peak and at the end of the process (Fig. 4.6).

Therefore, different simulations make it possible to estimate the "optimal" conditions (oven settings and formulation) to obtain products with the expected macroscopic characteristics.

## 4.4 Summary and perspectives

Baking is a relevant unit operation in the transformation of cereal-based goods, such as biscuits, breads, sponge cakes, etc. The properties of final products depend significantly on both formulations and operating conditions. The main ingredients, present in the large majority of bakery products, are flour, fat, and water; in addition, eggs, salt, sugars, and hydrocolloids can be used to modify the texture and/or the organoleptic properties, yielding a wide range of products with different characteristics. These products are characterized by a specific porosity related to the presence of gas cells introduced within the dough during the mixing step and growing either by the effects of yeasts or by the decomposition of chemical raising agents. Gas bubble expansion is hindered by the resistance of the viscoelastic medium surrounding the

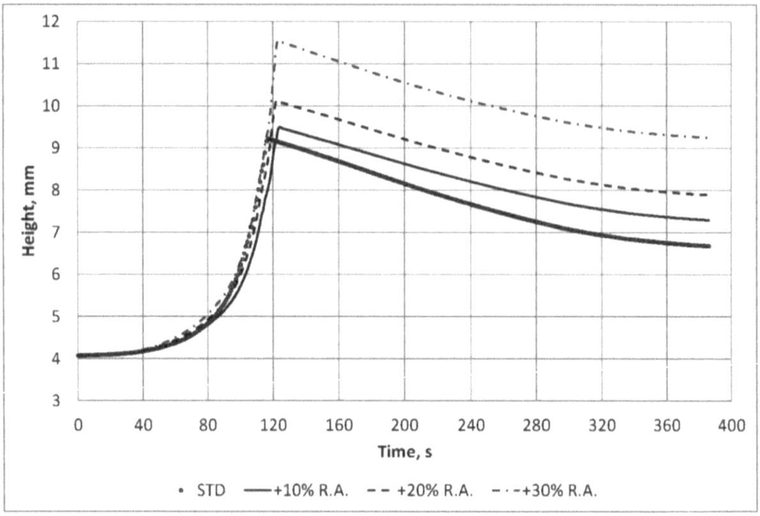


**Fig. 4.6** Height biscuit profile changing raising agents level.
Reprinted with permission from Baldino, N., Gabriele, D., Lupi, F.R., de Cindio, B., Cicerelli, L., 2014. Modeling of baking behavior of semi-sweet short dough biscuits. Innov. Food Sci. Emerg. Technol. 25, 40–52, Copyright (2014), Elsevier.

cells; therefore, the final void fractions is the result of a mechanical equilibrium between the internal pressure and the viscoelastic stresses of the paste.

During the baking step, in the oven, the simultaneous transfer of mass and heat occurs and, because of it, the temperature within the product increases and water diffuses towards the surface where it evaporates. In addition, many reactions occur, triggered by the temperature increase, modifying the consistency, the color, the nutritional content, and the already discussed void fractions. The final organoleptic properties depend on these transformations and, therefore, on temperature and moisture profiles; therefore, operating conditions play an important role and they should be chosen carefully to obtain the desired properties. To avoid empirical procedures, the modeling of transport phenomena and kinetic aspects is very important to be able to predict the effects of changes in operating conditions. Therefore, this issue has been widely investigated in the literature and different models, based on different initial assumptions, have been proposed. The final common aim is the description of key characteristics (such as internal profiles or final height) as a function of both oven conditions and formulations (described in terms of material and transport properties such as diffusivity, conductivity, rheological properties, etc.).

Even if several models have been proposed so far, probably a more accurate description of phenomena occurring at the local level (i.e., at the interface between gas cells and liquid paste) is necessary to take account properly of rheological and interfacial contributions. This would allow a better estimate of bubble evolution and interactions among adjacent bubbles.

## References

Angioloni, A., Dalla Rosa, M., 2005. Dough thermo-mechanical properties: influence of sodium chloride, mixing time and equipment. J. Cereal Sci. 41, 327–331. https://doi.org/10.1016/j.jcs.2004.10.004.

Baldino, N., Gabriele, D., Gentile, L., Carnevale, I., Lupi, F.R., de Cindio, B., De Luca, M.C., 2017. Drying of sausages made from the meat of black and white pigs: numerical modeling and structural investigation. Dry. Technol. 35, 724–735. https://doi.org/10.1080/07373937.2016.1209681.

Baldino, N., Gabriele, D., Lupi, F.R., de Cindio, B., Cicerelli, L., 2014. Modeling of baking behavior of semi-sweet short dough biscuits. Innov. Food Sci. Emerg. Technol. 25, 40–52. https://doi.org/10.1016/j.ifset.2013.12.022.

Baldino, N., Lupi, F.R., Gabriele, D., de Cindio, B., 2020. Use of mathematical modelling of dough biscuits baking behaviour. In: Sevda, S., Singh, A. (Eds.), Mathematical and Statistical Applications in Food Engineering. CRC Press, pp. 294–306.

Baltsavias, A., Jurgens, A., Van Vliet, T., 1999. Properties of short-dough biscuits in relation to structure. J. Cereal Sci. 29, 245–255. https://doi.org/10.1006/jcrs.1999.0250.

Bikard, J., Coupez, T., Della Valle, G., Vergnes, B., 2012. Simulation of bread making process using a direct 3D numerical method at microscale: analysis of baking step. Int. J. Mater. Form. 5, 11–24. https://doi.org/10.1007/s12289-010-1018-3.

Bird, R.B., Stewart, W.E., Lightfoot, E.N., 2007. Transport Phenomena, second ed. John Wiley and sons, Inc., USA.

Broyart, B., Trystram, G., 2003. Modelling of heat and mass transfer phenomena and quality changes during continuous biscuit baking using both deductive and inductive (neural network) modelling principles. Food Bioprod. Process. 81, 316–326. https://doi.org/10.1205/096030803322756402.

de Cindio, B., Correra, S., 1995. Mathematical modelling of leavened cereal goods. J. Food Eng. 24, 379–403. https://doi.org/10.1016/0260-8774(95)90052-D.

de Cindio, B., Gabriele, D., Migliori, M., Baldino, N., 2006. Rheological modelling of sponge-type goods baking. In: Fischer, P., Erni, P., Windhab, E.J. (Eds.), Proceedings of 4th International Symposium on Food Rheology and Structure. Laboratory of Food Process Engineering Institute of Food Science and Nutrition ETH Zürich, Zurich, pp. 543–547.

Fellows, P.J., 2000. Food Processing Technology - Principles and Practice, second ed. Woodhead Publishing Limited and CRC Press LLC.

Feyissa, A.H., Gernaey, K.V., Ashokkumar, S., Adler-Nissen, J., 2011. Modelling of coupled heat and mass transfer during a contact baking process. J. Food Eng. 106, 228–235. https://doi.org/10.1016/j.jfoodeng.2011.05.014.

Gan, Z., Ellis, P.R., Schofield, J.D., 1995. Gas cell stabilisation and gas retention in wheat bread dough. J. Cereal Sci. 21, 215–230. https://doi.org/10.1006/jcrs.1995.0025.

Jayapragasam, P., Le Bideau, P., Loulou, T., 2021. Approximation of heat and mass transport properties for one sided cake baking. J. Food Eng. 290. https://doi.org/10.1016/j.jfoodeng.2020.110211, 110211.

Letang, C., Piau, M., Verdier, C., 1999. Characterization of wheat flour-water doughs. Part I: Rheometry and microstructure. J. Food Eng. 41, 121–132.

Lostie, M., Peczalski, R., Andrieu, J., Laurent, M., 2002. Study of sponge cake batter baking process. II. Modeling and parameter estimation. J. Food Eng. 55, 349–357. https://doi.org/10.1016/S0260-8774(02)00132-2.

Maache-Rezzoug, Z., Bouvier, J.M., Allaf, K., Patras, C., 1998. Effect of principal ingredients on rheological behaviour of biscuit dough and on quality of biscuits. J. Food Eng. 35, 23–42. https://doi.org/10.1016/S0260-8774(98)00017-X.

Marcotte, M., 2007. Heat and mass transfer during baking. In: Yannotios, S., Sunden, B. (Eds.), Heat Transfer in Food Processing - Recent Developments and Applications. WIT Press, pp. 239–265, https://doi.org/10.2495/978-1-85312-932-2/08.

Nicolas, V., Glouannec, P., Ploteau, J.-P., Salagnac, P., Jury, V., 2017. Experiment and multiphysic simulation of dough baking by convection, infrared radiation and direct conduction. Int. J. Therm. Sci. 115, 65–78. https://doi.org/10.1016/j.ijthermalsci.2017.01.018.

Sablani, S.S., Marcotte, M., Baik, O.D., Castaigne, F., 1998. Modeling of simultaneous heat and water transport in the baking process. LWT Food Sci. Technol. 31, 201–209. https://doi.org/10.1006/fstl.1997.0360.

Sai Manohar, R., 2014. Baking. In: Bhattacharya, S. (Ed.), Conventional and Advanced Food Processing Technologies. John Wiley & Sons, Ltd, pp. 159–195.

van Vliet, T., 2008. Strain hardening as an indicator of bread-making performance: a review with discussion. J. Cereal Sci. 48, 1–9. https://doi.org/10.1016/J.JCS.2007.08.010.

# Fundamentals of food roasting process

# 5

*Pranjal P. Das*[a], *Prangan Duarah*[b], *and Mihir K. Purkait*[a]
[a]Department of Chemical Engineering, Indian Institute of Technology Guwahati, Guwahati, Assam, India, [b]Centre for the Environment, Indian Institute of Technology Guwahati, Guwahati, Assam, India

## 5.1 Introduction

Food offers the nutrients required for the body's optimal functioning, development, and repair. Processing is defined as the change in food constituents from its natural state to a more refined condition for the convenience of customers or to extend its shelf life. Dehydrated, canned, frozen, roasted, baked, and fortified foods are some of the examples of processed foods. The processing methods that cause raw foods to convert into an edible form include heating and cooking. Roasting is termed as a common food processing method that involves dry heating, in which hot air is blown over the food, thereby evenly cooking it from all directions. In an oven, the food products are heated simultaneously through convection, conduction, and radiation. At different phases of the roasting process, the influence of individual heat transmission mechanism may change. Through structural and physicochemical alterations of the food product, heat treatment methods such as roasting can increase the palatability and digestibility of food items, while also improving the bioavailability of the components (Bhattacharya, 2014). Moreover, roasting promotes the formation of color, fragrance, flavor active chemicals, and antioxidants. Roasting can also gelatinize or cook the food components, puff/expand them, and transform them into a more compliant, delicious, and appealing shape. Roasting is also used to decrease or eradicate the natural toxins, enzyme inhibitors and microbial load. Nevertheless, the types of food and its processing circumstances have a significant impact on the nutritional value of the roasted products.

Roasting is a high-temperature treatment method where the food is cooked in an oven at a temperature ranging from 150°C to 300°C or even more. The treatment time might vary depending upon the desired properties of the final product. For example, the coffee beans are typically roasted at a temperature range of 160°C to 240°C for a period of 8 to 25 min (Król et al., 2020). The roasting process is divided into three stages: drying, roasting, and flavor/color formation. Water evaporation from the food cells occurs in the first phase, in which the temperature remains above 150°C, thereby reducing the high initial moisture content of the food to a much lower one. Chemical alterations viz. oxidation, reduction, hydrolysis and other pyrolytic processes take place during the successive roasting phase, that begins at around 185°C to produce flavor, fragrance, and color components (Sruthi et al., 2021).

**High-Temperature Processing of Food Products. https://doi.org/10.1016/B978-0-12-818618-3.00005-7**

To achieve the desired characteristics, roasting causes heat-induced changes on the surface of the food while retaining the inner moisture content at the same time. The roasted foods are reported to have modified properties as a result of being subjected to extreme temperatures. To increase the quality of end products without affecting the overall flavor, various novel roasting methods have been extensively tested. Also, the influence of roasting on the stability and quality of several food products have been studied. Moreover, previous research has suggested that the roasting of food products causes a series of metabolic events that produce a variety of compounds having both positive and negative effects. Maillard reaction products (MRPs), viz., hydroxymethylfurfural and acrylamide have been shown to be harmful and are contemplated as one of the accurate markers for identifying the final attribute of heat-processed foods (Bertuzzi et al., 2020). Moreover, heat-induced processes are accountable for the development of flavor and color characteristics associated with the roasting of foods, thereby making them very crucial.

Several investigations have also found that thermal processing of temperature-time combinations can have a significant impact on the final properties of treated foods. The characteristics of the treated food products were also reported to differ with regards to the roasting process used. Conventional roasting methods are frequently linked to longer processing times and uneven heating, resulting in either under-heated or over-heated foods within the same batch. However, novel roasting processes result in more precise heating and easier control (Gunel et al., 2020). As such, development of appropriate roasting processes for food products requires a thorough understanding of the roasting mechanism, related by-products, and their effects on various food bioactive signatures. To date, a complete evaluation of the roasting processes and their associated changes in food products have not been reported, thereby necessitating an organized and systematic effort to outline several investigations and develop rational conclusions. As such, this chapter compiles the data on various features of roasting and its impact on the structural, biochemical and physical attributes of food. Furthermore, this chapter discusses the recent scientific literature associated with several roasting methods, including the effect of advanced and conventional techniques on different food products along with its benefits and drawbacks.

## 5.2 Heat and mass transfer during roasting

Roasting is a complicated technique that involves energy and mass transfer operations, and is followed by volatile and moisture loss. Oxidation, reduction, polymerization, hydrolysis and decarboxylation are some of the chemical processes involved during roasting. Parameters such as product composition, flow rate, air temperature, treatment duration, air velocity, and chamber pressure significantly affect the quality of roasted foods and the efficiency of the overall process. The level and choice of the individual parameters affect the rate of heat and mass transfer as well as the dissipation of volatile compounds that are essential for achieving the required product qualities. Interpretation of the fundamental operations of heat and mass transfer in roasting process considerably assists in the modification of existing methods along with the development of novel and more efficient techniques (Fabbri et al., 2011).

Since the operating temperature, reaction time and other roasting parameters have a significant impact on the characteristics of the final treated food products; therefore, mathematical models for simulation of heat and mass transfer conditions, prediction of moisture content distribution and treatment time, as well as obtaining an appropriate control over the complete roasting process must be developed. The modeling of heat-assisted processing operations are often difficult to perform due to parameters such as the presence of both vapor and liquid inside the solid matrix, reliance of the transport properties on operating variables, change in the product shape over time and the complicated fluid-structure interconnections caused by airflow around the food product. Several studies have been conducted that provide ideas for modeling the heat and mass transfer operations throughout the roasting process. Most of the work concludes that diffusion is the most crucial water transport mechanism. Although the terms roasting and baking are sometimes used correspondently, nonetheless, the former is used for vegetables and nuts, while the latter is used for fruits and flour-based products. As a result, the heat and mass transfer equations used throughout baking are identical to those used in the roasting process. In conventional ovens, heat is transferred via (i) conduction from the plate surface to the product, (ii) convection from the heating medium and (iii) radiation from the oven walls to the surface of the food product (Sruthi et al., 2021). The overall heat transfer via radiation is determined by the following equation:

$$Q_R = \varepsilon\, \sigma\, A\, \left(T_1^4 - T_2^4\right) \tag{5.1}$$

Here, $Q_R$ denotes heat transmission through radiation (W/m$^2$), $\varepsilon$ denotes the gray body emissivity (as radiant heaters are not ideal), $\sigma$ denotes Stefan-Boltzmann constant ($5.67 \times 10^{-8}$ W/m$^2$K$^4$), $A$ denotes surface area (m$^2$), $T_1$ (K) and $T_2$ (K) denotes the temperature of emitting body surface and the food product respectively (Chhanwal et al., 2010). The net heat transfer via natural convection is governed by the equation below:

$$Q_{CV} = h\, A\, (T_s - T_a) \tag{5.2}$$

Here, $Q_{CV}$ denotes heat transmission through convection (W/m$^2$), $h$ denotes convective heat transfer coefficient (W/m$^2$K), and $T_a$ denotes the air temperature (°C). To ensure proper roasting, the air is circulated forcefully via the use of fans to induce effective turbulence, which allows for rapid heat exchange. Phenomenon such as drop-wise or film condensation takes place when the saturated vapor comes in contact with colder surfaces due to the migration of evaporated water into the food product.

The transient heat conduction during the phase change can be described as:

$$\rho_p\, C_p\, \frac{\partial Ts}{\partial t} = \frac{\partial}{\partial z}\left(k_p\, \frac{\partial T}{\partial z}\right) - \rho_s\, R_{\text{evp}}\, H_{\text{evp}} \tag{5.3}$$

Here, $k_p$ denotes thermal conductivity (W/mK), $t$ denotes operating time (s), $C_p$ denotes specific heat capacity of the food product (J/kgK), $R_{\mathrm{evp}}$ denotes evaporation rate (kg of water/kg of solids), $H_{\mathrm{evp}}$ denotes latent heat of evaporation (J/kg) and $\rho_p$ and $\rho_s$ denotes the product density and dry solid density (kg/m$^3$), respectively (Feyissa et al., 2011). Likewise, heat transfer via the plat contact surface is given as:

$$\rho\, C_{p,a}\, \frac{\partial T}{\partial t} = k\, \frac{\partial^2 T}{\partial^2 z} \tag{5.4}$$

Here, $k$, $C_{p,a}$, and $\rho$ denote the thermal conductivity, specific heat capacity, and density of the plate.

Furthermore, the conservation of mass is used to indicate the water transport phenomenon within the food product. The mass transfer occurring in a typical roasting oven is governed by the following equation:

$$\frac{\partial C}{\partial t} + \nabla(Cu_w) = \nabla(D\nabla C) \tag{5.5}$$

Here, $C$ denotes moisture content (kg of water/kg of the sample), $u_w$ denotes velocity of water (m/s), and $D$ denotes the moisture diffusion coefficient (m$^2$/s) (Feyissa et al., 2013). Nevertheless, the model that best outlines the roasting technique changes with varying food matrices, process conditions and equipments. It rather highlights the limitations of a single model in elucidating both the energy and moisture transport during the roasting process.

## 5.3 Overview of roasting processes

### 5.3.1 Conventional processes

Dry heat is used for a short period of time in conventional roasting methods, with continuous mixing or stirring to guarantee consistent heating throughout the process. Fig. 5.1 depicts the schematic diagrams of some of the most frequently used conventional roasting processes for a wide range of food products. This section discusses about few such conventional roasting processes in brief.

#### 5.3.1.1 Pan roasting

The pans used in the conventional pan roasting process include coated enamelware, clay pots, stainless steel and cast iron. Surface contact phenomenon is used in pan roasting process to direct the transfer of heat from either a hot plate or a flame. Mixing or stirring is done to ensure that heat energy is distributed evenly and quickly, which in turn amplifies the roasting method. During pan roasting process, the surface temperature of the food is higher than the core temperature, resulting in an over-roasted outer surface with no significant influence on the center (Ozel et al., 2014).

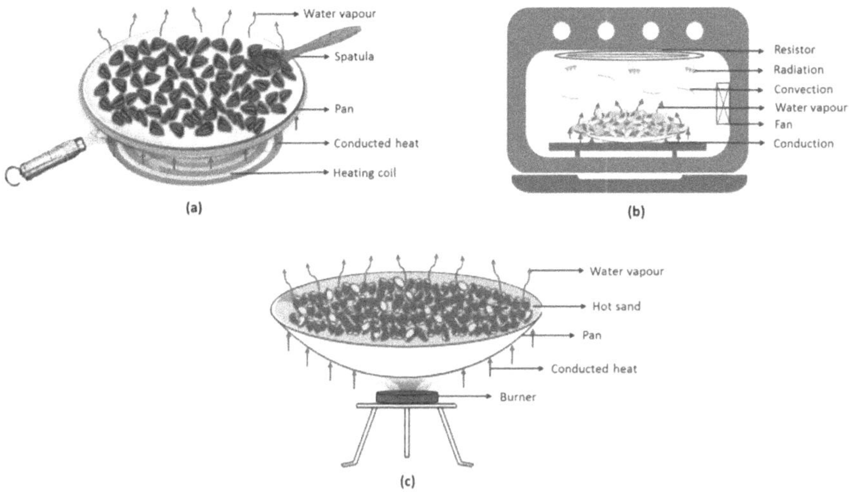


**Fig. 5.1** Conventional roasting processes (A) Pan roasting, (B) Oven roasting, (C) Sand roasting.
Obtained with permission from Sruthi, N.U., Premjit, Y., Pandiselvam, R., Kothakota, A., Ramesh, S.V., 2021. An overview of conventional and emerging techniques of roasting: effect on food bioactive signatures. Food Chem. 348, 129088. https://doi.org/10.1016/j.foodchem.2021.129088.

As compared to dry heat treatment, when barnyard millet and rice samples were dry roasted in a pan, the process resulted in significant moisture loss and color alteration, with higher redness and lower lightness values. The important aroma compounds created in coffee include furanone, pyrazines, furans, and benzene derivatives, which account for 4.65% of volatile components present in the thoroughly roasted terebinth fruits (Gogus et al., 2011). The pan-roasted sample contains the highest quantity of pyrazine and is the best approach for maintaining the aroma components. Nevertheless, the conventional roasting process is comparatively more severe, resulting in considerable changes in the nutritional composition of foods. The reason can be attributed to the challenges in maintaining the process parameters. Kanagaraj et al. observed a charring phenomenon during the pan roasting process of barnyard millet and rice samples, carried out at temperatures exceeding 190°C for an extended period. It was concluded that the open pan roasting process resulted in chance admixture of impurities in the food products and was considered unhygienic (Kanagaraj et al., 2019).

### *5.3.1.2 Oven roasting*

The conduction and convection roasting ovens can be heated with either electricity or gas, having different heat distributions. The phenomenon of conduction involves the use of a uniform source of heat, whereas convection assists in the circulation of heat within the oven via use of fans. The heat transfer operation in oven roasting consists of three processes: conduction from surfaces in direct contact with the product, convection from hot air, and radiation from the heat source. Even so, faster air currents deform the shapes of dry out food samples, thus, affecting their life span, texture and overall characteristics. Alternately, slow air currents reduce the rate of heat transmission and consequently extend the roasting period. Roasting ovens are categorized as either direct or indirect heating ovens based on the heating conditions. In direct heating ovens, natural convection or fans are used to recirculate the air and other combustion products. The benefits of direct heating ovens include high thermal efficiency, rapid start-up, short roasting periods and appropriate control over roasting conditions (Fellows, 2017). However, the major concerns involved in direct heating are the contamination of food products from combustion and the need for periodic servicing of gas burners in order to endure the combustion efficacy.

In small roasting plants and residential places, batch ovens are often utilized, where the food is placed in roasting ovens either in trays or as a whole. Nonetheless, the batch ovens face few challenges including a lack of homogeneity throughout the roasting process, which is further aggravated by loading and unloading delays. In addition, tunnel ovens, tray ovens, revolving hearth ovens and reel ovens are considered either continuous or semicontinuous, in which the food product is carried throughout the oven with or without the use of a fan, thereby confirming steady roasting. The lack of heating zones and complication in automating the loading and unloading delays are two major drawbacks of continuous heating ovens. Impingement ovens, in which hot gases flow perpendicular to the product surface at a higher velocity, have recently been developed. An integration of air impingement and steam injection can considerably improve the heat transmission rate and the product yield by abruptly enhancing the heat transfer coefficients (Berk, 2018).

#### *5.3.1.3 Sand roasting*

Sand roasting is a conventional and fast technique of grain processing, often used by street food vendors to produce value-added food products. Food grains such as flaxseed, chickpeas, rice, barley and groundnuts are heated in hot sand at temperatures ranging from 150°C to 350°C to produce ready-to-eat products (Swarnakar et al., 2019). Roasting is carried out in small, moderate, and industrial pan roasters utilizing sand as a medium. During roasting, the far-infrared rays permeate the food, thereby triggering protein, fat and carbohydrate degradation. The conversion of micro- and macronutrients occurs as a result of rapid starch gelatinization and concurrent drying, which significantly improves the digestibility of food products. The high temperature transforms the grain moisture into superheated vapor, thereby resulting in puffing due to the induction of pressure within the grain ( Jogihalli et al., 2017). The internal pressure formed largely depends upon the moisture content of the grain, temperature profile of the sand and the duration of roasting.

The roasted products show improved crispiness and puffiness along with enhanced flavor, color, texture, and aroma. However, factors such as nonuniform temperature distribution, difficulty in temperature control and operation, energy inefficiency, low productivity, and sand contamination in food products are some of the major limitations of sand roasting process. The food samples are also exposed to direct flame and smoke and are governed by the high silica concentration in the sand (Chikelu et al.,

2015). Thus, it necessitates the creation of high-performance and energy-efficient roasting methods with minimum contamination.

### 5.3.2 Advanced roasting processes

The increasing requirement for roasted fortified food products, along with the necessary food safety concerns has resulted in the development of novel roasting equipment and processes. Fig. 5.2 represents the schematic diagrams of some of the most frequently used advanced roasting processes. In the following section, a few of these advanced roasting processes have been discussed in details.

#### 5.3.2.1 Microwave roasting

Microwaves are termed as nonionizing radiations with frequencies ranging from 300 MHz to 300 GHz in the electromagnetic spectrum. The utilization of microwave ovens has created new dimensions in the area of food preparation and management, and they have become increasingly popular in both domestic and industrial applications. Unlike the traditional hot-air roasting process, which necessitates an extended roasting period, thereby resulting in the degradation of product quality, microwave roasting, on the other hand, requires less time (Goszkiewicz et al., 2020). As the microwaves permeate the food surface, interaction of the electromagnetic field with the chemical compounds of food promotes the water within the product to heat up, owing to the excitation and molecular friction. Due to pressure gradient, the vapor generated in the interior of the food product via volumetric heating is forced to migrate outside, thereby allowing rapid moisture removal from the food.

One of the most important features in microwave roasting is the inside-out heating of the food products, caused by the friction forces between water molecules. This feature provides many advantages over the conventional heating process such as low energy consumption, selective heating, easy process control, lower start-up and process time and higher nutritious end product. Microwave roasting has been widely studied for its effects on functional, sensory and physicochemical aspects of a number of food grains, including peanuts, chickpeas, millets, flaxseeds and poppy seeds (Ghafoor et al., 2019; Kumar et al., 2020). Also, microwave heating has been found to enhance the oil output in apricot kernels and increase the oil pressing efficacy of peanut oil and rape seeds (Rękas et al., 2017). During the comparison of convection and microwave roasting effects on the quality parameters of sunflower seeds, Goszkiewicz et al. observed that the utilization of microwave heating considerably enhanced the quality of fat in seeds and led to a substantial reduction in the peroxide value (Goszkiewicz et al., 2020). Nevertheless, a combination of hot air/microwave roasting of carob powder produced superior product quality while using less energy than separate treatment processes. In microwave roasting, the core temperature is higher than the surface temperature; however, in conventional roasting, the outer surface of the food products may be over-heated, while heat may not reach the core properly. Even so, microwave roasting has been linked to the generation of fatty acid

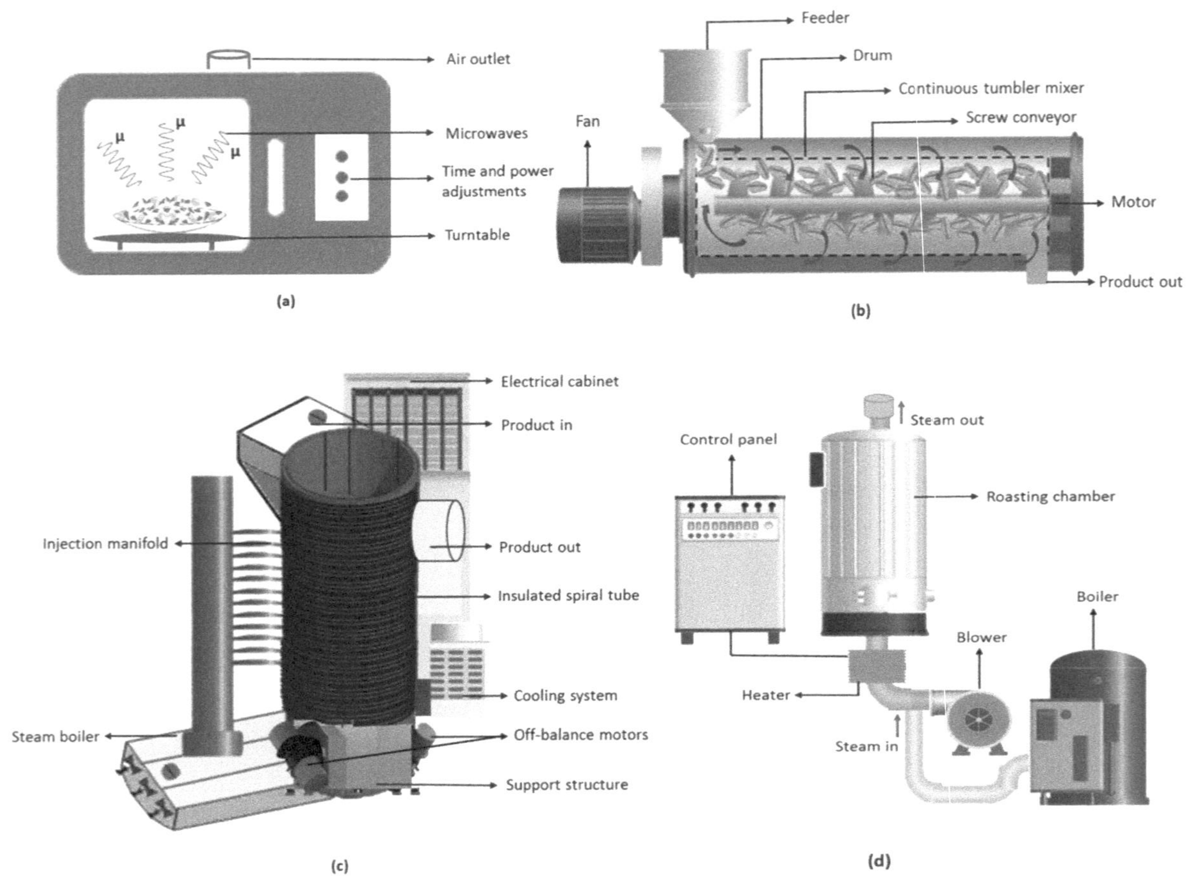


**Fig. 5.2** Advanced roasting processes (A) Microwave roasting, (B) FCCT roasting, (C) Revtech roasting, (D) Superheated steam roasting. Modified with permission from Sruthi, N.U., Premjit, Y., Pandiselvam, R., Kothakota, A., Ramesh, S.V., 2021. An overview of conventional and emerging techniques of roasting: effect on food bioactive signatures. Food Chem. 348, 129088. https://doi.org/10.1016/j.foodchem.2021.129088.

conjugates and oxygenated molecules along with a reduction in oil stability and an increase in oil rancidity (Gunel et al., 2020).

### 5.3.2.2 Superheated steam roasting

The use of superheated steam during roasting is one of the modern heating techniques. It involves the addition of sensible heat to water which increases the temperature beyond boiling point at a certain pressure, thereby resulting in the generation of superheated steam. An electric heater, a cylindrical roasting chamber, a rear curved fan and an electric steam boiler capable of producing saturated steam constitutes the superheated steam roasting system (Chindapan et al., 2019). Electric heaters assist in the heating of saturated steam from the boiler to a predetermined superheated steam temperature. A blower circulates the steam throughout the roasting chamber. Direct contact between the product and superheated steam inside the heating chamber leads to roasting. At the same operating temperature, superheated steam shows higher thermal characteristics than air, thus, allowing for better heat penetration into the food products without compromising the quality parameters.

Since the use of hot air as a roasting medium has its limitations, therefore, superheated steam-based roasting has been preferred as an alternate process. It can be employed to a wide range of food products such as peanuts, coffee beans, rice and cumin seeds (Idrus et al., 2017; Zzaman et al., 2017a). Superheated steam roasting provides high rate of heating than the standard hot-air roasting, leading to shorter roasting periods, which in turn reduces the loss of heat sensitive compounds. Superheated steam roasting was reported to reduce the level of unsaturated aldehydes generated via oxidation of lipid in coffee beans due to the unavailability of oxygen during the process (Yodkaew et al., 2017). Zzaman et al. observed smaller crystals during the melting of superheated-steam-roasted cocoa butter, which are beneficial in commercial applications to manufacture high-quality cocoa-based products. Furthermore, the other benefits of superheated steam roasting process include improvement in thermal degradation, heat transfer rate, energy efficiency and drying rate (Zzaman et al., 2017b).

### 5.3.2.3 Revtech roasting

Revtech processing system was founded in 1997 for continuous and automated thermal treatments. The applications of Revtech processing system include roasting, pasteurization, steam sterilization, drying, and heat treatment of pharmaceutical and chemical products. During the roasting process, low-moisture content food products are conveyed from a hopper into a closed heated spiral tube via the use of a screw feeder. The food products are conveyed and mixed inside the tube by vibration, owing to the spinning of two off-balance motors attached on the opposite side of each other at a predefined frequency. Change in the motor angle regulates the vibration, while the product flow is controlled by adjusting the screw speed. The product remains in close contact with the heated tube, thereby ensuring a consistent and efficient heat transmission with higher thermal property (Young et al., 2020).

The temperature distribution of the pipe is affected by parameters such as the thermal characteristics of the raw material, residence time and the effect of convection at the entrance and exit of the pipe (Keppler et al., 2018). Once the product temperature reaches the desired level, around 3% to 10% of steam is injected within the process for effective treatment of organic compounds. Condensation is prevented by the incorporation of sensible heat to the product and steam, which assist in maintaining the steam in a dry saturated state. As such, the process is also known as dry steam process or controlled condensation steam process (Anderson, 2019). Cold air is pushed via a separate spiral tower attached to the top of the heating spiral to cool the sample as it is transported through the system, thereby allowing the treated product to be discharged at room temperature. Some of the major benefits of Revtech roasting process include homogenous product treatment, high energy savings, low maintenance and cleaning costs and reduced air treatment. The process can also be used for pasteurization of barley grains, treatment of whole yellow peas and beans to improve the baking and functional properties as well as heat treatment of soft wheat flour to produce high ratio cakes (Bourré et al., 2019; Keppler et al., 2016a). In a study conducted by Keppler et al., it was reported that a thin layer deposition of particle on the interior of the pipe integrated with electrostatic forces, greatly enhanced the residence time over the processing time, thereby necessitating the regular cleaning of pipes to ensure consistent behavior and control of residence time (Keppler et al., 2016b).

### 5.3.2.4 Forced convection continuous tumble roasting

Forced convection continuous tumble (FCCT) roasting is an energy efficient process that can be utilized to alter the microstructure of cereal grains. It comprises of a semi-closed roasting chamber with continuous recirculation of heated air throughout the roasting process. The moisture content of the product is used to produce superheated steam, resulting in rapid and uniform heat transfer. A screw conveyor coupled to a perforated rotor assist in the transportation of products within the chamber. The product gets suspended in heated air inside the rotating cylinder, resulting in uniform exposure of the entire food surface to heat. As a result, the complete product surface area is used for heat transmission, which provides uniform roasting to the treated sample. Continuous recirculation of heated air and thermal insulation significantly assist in minimizing the heat loss and energy consumption of the process (Sruthi et al., 2021).

During the investigation of 3D microstructure of roasted whole wheat, Schoeman et al. found that traditional oven roasting process provides highly porous and enlarged structures compared to FCCT roasting, thereby implying minimal degradation effect of the latter process (Schoeman et al., 2016). FCCT roasting process resulted in negligible structural changes, steady material density and highly acceptable internal structure, which assist in minimizing the heating effect on milling yield. In addition, oven roasting process displayed more harmful effects on functional, structural and physicochemical characteristics of milled flour compared to FCCT roasting process under similar roasting conditions (Schoeman and Manley, 2019; Schoeman et al., 2017). As such, it can be inferred that FCCT roasting can be better alternative to traditional grain roasting in the production of improved functional and structural quality foods.

Nonetheless, FCCT roaster is generally used for large scale commercial applications with considerably higher cost of investment. Moreover, the servicing of such roasters is unfavorable due to the complexity of equipment design.

## 5.4 Effect of roasting process on food quality

### 5.4.1 Functional properties of food

#### 5.4.1.1 Flavor and volatile compounds

The roasting process consists of various complex phenomena such as Maillard reaction, Strecker degradation, caramelization, oxidation, and thermal deterioration of lipid, which produces discrete volatile flavor compositions. The roasted aroma of treated food samples is attributed to the presence of N-heterocyclic compounds such as pyrroles and pyrazines, and O-heterocyclic compounds such as pyranone and furans (Schlörmann et al., 2019). Formation of pyrazines via Maillard reaction under thermal condition involves the reaction of carbonyl group of reducing sugars and amine group of amino acids. Moreover, the substitution of one or more methyl groups in pyrazines with the ethyl groups significantly decreases its odor threshold. As such, acetylpyrazines provide highly nutty, toasty, and woody aroma. The pH of the product has been reported to impart a considerable influence on the formation of volatile compounds.

The formation of pyrazines is favored at lower pH ($\leq 5$), while Strecker aldehydes and methyl pyrazines are generated at higher pH ranges. Functional groups such as ester, alkane, aldehydes, alkene, ketones and alcohols are produced via the oxidation of unsaturated fatty acids in roasted products, thereby imparting a fatty or oily flavor to the treated food (Zou et al., 2018). In a study carried out by Majcher et al., it was found that roasting of barley at 190–200°C resulted in the generation of 2-furfurylthiol, 3-(methylthio)-propanal and 2,3-butanedione, which provide a distinctive nutty and coffee flavor to the roasted barley brew (Majcher et al., 2013). Furthermore, compounds such as maltol, 2-furaldehyde, and methylpyrazine were identified in barley malt products after being roasted at 220°C for 2h. However, Young et al. reported that the incorporation of roasted peas in bread delivered a much lesser aroma and unique pulse flavor compared to the untreated peas, regardless of steam addition or roasting conditions. Due to the development of alternate flavor compounds, the unpleasant off-flavor constraint of peas can be easily controlled by the roasting technique (Young et al., 2020).

#### 5.4.1.2 Color changes

Color is considered as an indicator of roasting degree owing to the browning reactions and caramelization process. The water activity of the product decreases throughout the roasting process, which stimulates the generation of pigments such as melanoidins. These pigments provide a distinctive brown color. According to Bhinder et al., the roasting temperature showed a distinctly positive correlation with browning index

(Bhinder et al., 2019). During the roasting process, Zhang et al. reported that the operating time and the roasting temperature showed nonlinear effects on the $L^*$, $a^*$, and $b^*$ color characteristics of pepper seeds (Zhang et al., 2019). The reason was attributed to the deterioration of capsorubin and capsanthin pigments throughout roasting.

Moreover, Berk et al. reported a decrease in the $L^*$ (lightness) value of sesame seeds with an increase in the thermal load, thereby suggesting the development of a darker color (Berk et al., 2019). The progress in roasting process is indicated by the formation of brown pigments via Maillard reaction which led to an increase in $a^*$ (redness) values. Besides, a decrease in the $b^*$ (yellowness) value owing to the roasting of products at 180°C resulted in the formation of brown color. Roasting has been shown to disintegrate pigment and protein complexes in some cases due to protein denaturation. This increases the concentration of pigments viz. carotenoids and chlorophyll in pistachios. Also, an increase in the negative a* value which indicates a bright green color was related to the increased concentration of chlorophyll after roasting (Suri et al., 2019).

#### 5.4.1.3 Water and oil absorption capacities

The water absorption index (WAI) is based on the availability and presence of hydrophilic groups along with the ability to create macromolecular gels. The water solubility index (WSI) is a parameter that accounts for the quantity of released polysaccharides from starch granules and is used to measure the transformation rate of starch during processing. The quantity of absorbed oil in a sample is determined by the oil absorption capacity (OAC). OAC has a significant effect on the organoleptic characteristics of food such as mouthfeel and flavor, which plays a major role in food formulations. In a study conducted by Hatamian et al., it was found that roasting of chia seed flours improved the OAC, WAI and WSI considerably (Hatamian et al., 2020). The increase in OAC is likely due to the degradation and denaturation of protein, which exposes the hydrophobic groups present in oil hydrocarbon chains, while an increment in WAC might be due to the exposure of more polar amino acids to water. WAC can also be improved by the formation of porous structure and capillaries in the endosperm, which is assisted via the breakdown of starch by gelatinization. Moreover, Jogihalli et al. reported that, an improvement in WAI of sand-roasted chickpea was due to the higher macromolecular gel-forming ability during the roasting process ( Jogihalli et al., 2017). On the other hand, a reduction in WSI was associated with the development of insoluble components during the process.

### 5.4.2 Food macro-nutrients

#### 5.4.2.1 Carbohydrates

During the roasting process, fructose and glucose are degraded. Raffinose and stachyose, on the other hand, are unaffected. This finding was ascribed by Redgwell et al. that reducing sugars, such as glucose and fructose, combine with amino acids to form specific compounds in thermal processes, whereas nonreducing sugars remain unreactive (Redgwell et al., 2003). The incorporation of chemical as well as microstructural alterations owing to the formation of MRP is another vital component of the roasting process on carbs. There are three stages of the Maillard reaction: the first, middle, and final stages, each of which involves the creation of distinct chemicals. Starting with the condensing of amino acids and carbonyl groups, Schiff bases and the release of water molecules are formed as the result of this condensation. This is accompanied by Schiff base cyclization, which results in the formation of a condensation product, N-substituted glycosylamine, which is then reorganized to form an initial MRP with greater stability, such as Heyns rearrangement products (HRP, 2-amino-2-deoxyaldose from ketose sugars) or Amadori rearrangement products (ARP, 1-amino-1-deoxy-2-ketose from aldose sugars) (Nooshkam et al., 2019). After the Amadori rearrangement, reducing sugars combine with amino acids to make 5-Hydroxymethylfurfural (HMF), which is a roasting intermediate molecule that is then condensed with nitrogenous chemicals and polymerized to produce brown pigments (Hamzalıoğlu and Gökmen, 2020).

While researching the effects of roasting hazelnuts at various temperatures and times, the relevance of the 1,2-enolization phase in the interconversion of glucose and fructose was discovered by Göncüoğlu Taş and Gökmen (2017). They observed that HMF is produced by dehydration of fructofuranosyl cation rather than 3-deoxglucosone, with glucose participating more to the initial phase of the Maillard reaction than fructose and fructofuranosyl cation. Aldol and aldehyde-amine condensations of reductones, fission products, including Strecker degradation products yield brown, nitrogen-containing polymeric compounds with macromolecules known as melanoidins in the last stage of the Maillard process. Browning from the formation of melanoidins is dependent on the Maillard process. Color changes with varied roasting settings are significant because browning development also involves thermal process pollutants. The available lysine, decrease of ophthaldialdehyde reactivity, the fluorescence of intermediate MRP and tryptophan (FAST index), and browning index are all used to quantify Maillard reaction changes. Roasting buckwheat groats enhanced the FAST index considerably, according to Wronkowska et al. (Wronkowska et al., 2016). The reason can be attributed to an increase in MRP (FIC) fluorescence and a decrease in tryptophan fluorescence.

### 5.4.2.2 Proteins and lipids

Thermal reactions at high temperatures over long periods of time alter proteins' secondary and tertiary structures, produce protein aggregates, and cause refolding and unfolding. Major and minor protein fractions are denaturized during roasting. Protein content was lowered by infrared roasting on Tartary buckwheat cultivars, with the greatest degradation happening at 170°C, according to Bhinder et al. (2019). When the intensity of roasting treatment is increased, the intensity of fluorescence associated with tryptophan decreases significantly. The roasting temperature also exhibited a substantial negative relationship with proteins. Ji et al. reported that, when the roasting temperature and duration of time increased, the total amino acid concentration in sesame seeds considerably decreased ( Ji et al., 2019). At 200°C for 60min, the amino acids, arginine, lysine, cysteine, and serine, all decreased significantly. These four amino acids were thought to be important in the production of the sesame oil scent. To prove that heat treatment including roasting had no effect on nutritional value, Simsek et al. showed that unroasted chickpeas had a lower total amino acid content than roasted chickpeas (Simsek et al., 2016). Histidine, glutamic acid, aspartic acid, lysine, arginine, and phenylalanine have been demonstrated to improve the amino acid profile of chickpea because of their ability to scavenge free radicals.

Heat-induced changes in cell wall structure resulted in increased oil content when seeds were roasted. It increased the porosity of cell walls, allowing oil to escape. Several studies have also highlighted the influence of roasting on the functional characteristics of proteins, resulting in oil ejection. Using a roll crusher, Li et al. found that roasting peanuts at 150°C for 20min with roll crusher grinding gave a maximum free oil production of 92.2% (Li et al., 2016). Changes in the peanut protein fractions were thought to be responsible for this. In other protein-rich roast samples, moisture content differential was connected to increased fat content in roasted chickpeas compared to raw chickpeas, as reported by Simsek et al. (Simsek et al., 2016). These two samples exhibited moisture levels of 93.79% and 31.76%, respectively, when they were uncooked and roasted. Microstructural alterations in the seed cell wall are likely to have been caused by the roasting process, resulting in an increased release of fat from the chickpea matrix. At 180°C for a time period of 10min, infrared and dry air roasting of black cumin seeds increased the oil production by 34.3% and 35.1%, respectively, according to Suri et al. (2019). It has been shown by Ahmed et al. that roasting Samh seeds alters the fatty acid profiles of the seeds, increasing the amount of behenic acid, stearic, linoleic, and arachidic acids and decreasing the amount of arachidic and linolenic acids in the seeds (Ahmed et al., 2020). The breakdown of diacylglycerols, monoacylglycerols, and triacylglycerols, as well as the accumulation of fatty acids in roasted seeds, were hypothesized by Yodkaew et al. to be the cause of an increment in fatty acid concentration after roasting (Yodkaew et al., 2017).

## 5.4.3 Food bioactive compounds

### 5.4.3.1 Antioxidant activity

During the first stages of the autoxidation process, antioxidants inactivate reactive radicals, which in turn prevent the radical chain reaction from spreading. Chen et al. investigated the phytochemical profiles in addition to the antioxidant properties of barley malts under various roasting circumstances and found that roasting had a favorable effect on barley malts' biological activity (Chen et al., 2019). Schouten et al. investigated the antioxidant capabilities of Robusta and Arabica coffee varieties after the roasting process (Schouten et al., 2021). In dark roasted coffee samples, they calculated a loss of antioxidant components (caffeoylquinic acids) of 90% 5-O-caffeoylquinic acid, 70% 3-O-caffeoylquinic acid, and 70% 3,5-O-dicaffeoylquinic acid in contrast to the maximum value. In addition, the rise in 3-O-caffeoylquinic acid in moderately roasted green beans relative to raw green beans has been linked to the isomerization of chlorogenic acids that occurs during roasting.

Roasting of nongerminated peanut kernels, as demonstrated by Aljuhaimi and Özcan, increased the antioxidative activity through MRP, whereas a recent study by Ghafoor et al. identified that roasting of poppies decreased the antioxidant property, with roasted samples providing low antioxidant properties than the control samples (Ghafoor et al., 2019; Aljuhaimi and Özcan, 2018). Antioxidant potential was shown to be reduced in foods cooked in an oven. Similar findings were observed by Kinge et al., who discovered that standard roasting reduced the antioxidant activity of Djansang seeds considerably when compared to boiling (Kinge et al., 2019). Time and power of microwave roasting had a detrimental influence on the total flavonoid content of chickpea flour, according to Jogihalli et al. (2017). Flavonoids in chickpea have antioxidant effects due to their interaction with metal ions. Wani et al. found that roasting arrowheads increased reducing power, an indicator of antioxidant properties, when compared to pan roasting (Wani et al., 2016). Oxidative reducing power is connected to reductase, an enzyme that donates hydrogen atoms to break free-radical chains and reduces peroxide formation. According to the investigation, microwave roasting had a higher reduction power than pan roasting. After roasting, it's probable that MRPs improve antioxidant properties and increase reducing potency.

### 5.4.3.2 *Total phenolic content*

Stability of polyphenol is determined by its interaction mode and structure in the food matrix during heat processes. A trend for the growth of total phenolic content (TPC) after roasting has been identified by several investigations. TPC was found to be higher in heavily (microwave) roasted chickpea samples at 900 W and 15 min than in unroasted chickpea samples containing TPC, according to Jogihalli et al. (2017). Similarly, increasing the roasting air temperature, infrared power, and duration resulted in an increase in the TPC of peanut kernels. An increase in heat-inducible as well as extractable phenolic compounds after roasting was responsible for this phenomenon. Antioxidant-rich protein-phenolic complexes and MRPs that react with the Folin-Ciocalteu reagent might lead to a significant increase in TPC. Nonetheless, investigations have linked TPC decrease to the thermal and oxidative breakdown of these chemicals (Sruthi et al., 2021). Low molecular weight phenolic compounds are easily volatilized at high temperatures. TPC is also harmed by the polymerization and the formation of insoluble chemicals such as melanoidins and polycyclic aromatic hydrocarbons. Due to polyphenols' higher redox activity under such conditions, more intensive roasting operations result in a greater TPC loss (Fernández-Romero et al., 2020). The effect of different roasting methods (conventional and advanced) on the bioactive compounds of food has been enlisted in Table 5.1.

**Table 5.1** Effect of different roasting processes on food bioactives.

| Products | Type of roasting | Optimized conditions | Biochemical changes and key findings |
|---|---|---|---|
| Almond kernel | Oven roasting | 200°C; 25 min | • Increase in MRP and nonenzymatic browning index.<br>• Breakdown of unsaturated fatty acids. |
| Apricot kernel | Microwave roasting | 540 W; 5 min | • Increase in total phenolic content and antioxidant activity.<br>• Increase in anthocyanin and flavonoid content. |
| Chickpea | Microwave roasting | 900 W; 15 min | • Increase in water absorption index, water absorption capacity.<br>• Decrease in foaming capacity, water solubility index, and oil absorption capacity. |
| Chickpea | Sand roasting | 220°C; 15 min | • Increased surface area and decreased bulk density.<br>• Decrease in L (lightness) and increase in a, b color values.<br>• Increase in water absorption index, oil absorption capacity and water solubility index. |
| Coffee | Oven roasting | 200°C; 45–60 min. | • Oven roasting preserved potassium and magnesium content.<br>• Higher roast degree led to higher α-glucosidase inhibition.<br>• Compositional changes induced by roasting. |
| Peas | Oven roasting and Revtech roasting | 120°C; 30 min | • Revtech roasting without steam improved flour color.<br>• Oven roasting led to flour with higher water absorption capacity and improved functional and compositional properties. |
| Poppy seeds | Oven roasting; Microwave roasting | 130°C; 25 min 720 W; 25 min | • Decrease in phenolic content and antioxidant activity (Oven roasting > microwave roasting).<br>• Roasting negatively impacted tocopherol and flavonoids content. |
| Samh seeds | Oven roasting | 150°C; 10 min | • Enhanced antioxidant content, linoleic acid, potassium and calcium content. |

**Table 5.1** Continued

| Products | Type of roasting | Optimized conditions | Biochemical changes and key findings |
|---|---|---|---|
| | | | • Improved oil content and total phenolic and flavonoid compounds content. |
| Wheat grain | Oven roasting; FCCT roasting. | 180°C; 140 s | • Wheat kernels puffed, with decrease in bulk density, protein content and hardness in both methods. • Lower weight loss in FCCT than oven roasting. |
| Wild almond | Microwave roasting | 480 W; 4 min | • Highest total volatile compounds content achieved. • Increase in the oil content. |

Reproduced with permission from Sruthi, N.U., Premjit, Y., Pandiselvam, R., Kothakota, A., Ramesh, S.V., 2021. An overview of conventional and emerging techniques of roasting: effect on food bioactive signatures. Food Chem. 348, 129088. https://doi.org/10.1016/j.foodchem.2021.129088.

## 5.5 Roasting of food products

### *5.5.1 Nuts and coproducts*

Roasting is primarily used to enhance the flavor, color, crispiness, and crispy texture of nuts and coproducts. Roasting causes microstructural and chemical changes in the nuts, including a decrease in moisture content, lipid modification, and color change, as well as the creation of molecules responsible for the characteristic roasted nut flavor, mostly as a result of Maillard reaction products. However, when roasted nuts are compared to raw nuts, microstructural and lipid changes may result in an increased vulnerability to lipid oxidation. Additionally, roasting may alter the content of metabolites. As a result, roasting may have an effect on the development of both health-promoting and possibly harmful components. Lower heating temperatures and shorter heating times result in an increase in the phenolic compounds in nuts, whereas higher temperatures and longer heating times result in a decrease in these compounds (Chang et al., 2016).

There are several variables that affect the antioxidant property of nuts and their coproducts, such as the duration and temperature at which they are processed (stir-frying and roasting), the kind of nuts used, and how they are processed. The quantity of extractable chlorogenic acid, gallic acid, and antioxidant activity can be increased by stir-frying pecans at a low temperature for a short length of time. These values, however, declined as heating time and temperature were increased. As a result, stir-frying pecans during the preparation stage is only advantageous under particular conditions (Yang et al., 2015). Roasting significantly reduced individual phenolics but had no effect on the TPC or antioxidant property of kernels. When hazelnuts were roasted, the flavan-3-ols (epicatechin and catechin) concentration reduced dramatically, with significant variations detected between raw and roasted nuts without skin. Thermal treatment reduced the procyanidin dimers and trimers levels in hazelnut. Again, substantial changes in the stability of these chemicals were seen between natural and roasted hazelnuts, indicating that temperature does indeed impact their stability. Additionally, no polyphenols were found in roasted hazelnuts with the skin removed. This demonstrates that the majority of the polyphenols in hazelnut are found in the skin. Another study on pistachio nuts found that once raw nuts were roasted, numerous phenolic components were significantly reduced. Daidzein and chlorogenic acid were found in roasted pistachios, but not in raw pistachios (Seeram et al., 2006). The impact of roasting on the protocatechuic acid level of roasted hazelnuts was equivocal.

Similarly, roasting significantly increased the gallic acid concentration of roasted hazelnuts when compared to wild hazelnuts. This might be related to the destruction of polymerized polyphenols, particularly hydrolyzable tannins, and other glycosylated flavonoids. Other variables, including the amount of phenolic acids in the skin, their thermal breakdown, and their release from esters, glucosides, and bound forms, may also alter the phenolic acid concentration after roasting (Chang et al., 2016). Catechin and ellagic acid are similarly heat sensitive, presumably due to their highly hydroxylated structure, which is prone to redox processes, resulting in loss during heat treatment. Thus, roasting appears to lower the quantities of phenolic compounds in the majority of nuts, not only because the skin is removed but also because many phenolic compounds undergo chemical breakdown during the roasting process. The majority of them are precarious and may be destroyed during processing, especially when heat treatment is used. The thermostability of phenolic compounds is determined by their structure as well as by the food matrix ( Jinap et al., 2004).

Vinson and Cai quantified the antioxidant activity of nut and peanut butter polyphenols by determining the quantity required to prevent LDL + VLDL oxidation by 50% (Vinson and Cai, 2012). For natural (raw) nuts, the sequence of decreasing effectiveness (increasing IC50) is walnut > cashew > hazelnut > pecan nut > almond nut > macadamia nut > pistachio nut > Brazil nut > peanut. Almond macadamia pecan peanut butter crunchy > peanut butter creamy > peanut > pistachio > hazelnut walnut > Brazil nut > cashew is the order for roasted nuts and peanut butter. Roasting, on the other hand, resulted in a statistically negligible decrease in antioxidant effectiveness. The average IC50 was 4.3 2.1 mM for raw nuts and 6.9 3.9 mM for roasted nuts (excluding peanut butter). Additionally, the ability of nut polyphenols to bind to LDL and protect it from oxidation was assessed for almond (2.1 mM), peanut (5.6 mM), and walnut (1.0 mM)]. This suggests that polyphenol antioxidants tied to walnuts are twice as efficient as those bound to almonds and five times as effective as those bound to peanuts in terms of lipoprotein binding.

The lignocellulosic structure is altered and proteins are denaturized by heating or roasting plant-based foods, according to Xu and Chang (2008). This may lead to enhanced availability of plant phenolic compounds in the matrix. The release of specific bound antioxidant phenolic compounds, which operate as free radical scavengers in the cell matrix, and an increase in phenolic extractability after roasting may both contribute to an increase in antioxidant activity. Table 5.2 depicts the composition of different phenolic compounds (flavonoids and phenolic acids) in raw and roasted nuts.

**Table 5.2** Composition and phenolic compounds in natural and roasted nuts.

| Tree nut | Types of phenolic compounds | | Natural | Roasted |
|---|---|---|---|---|
| Almond | Phenolic acids | Chlorogenic acid | 1.61 | 0.4–3.22 |
| | | Gallic acid | 0.04 | 0.11–1.02 |
| | | 5-Hydroxybenzoic acid | 1.26 | 0.37–1.92 |
| | | Protocatechuic acid | 0.88 | 0.23–1.59 |
| | | Vanillic acid | 0.94 | 0.2–2.47 |
| | | Total phenolic acids | 4.82 | 1.42–12.7 |
| | Flavan-3-ols | Catechin | 0.2 | nd–0.69 |
| | | Epicatechin | 0.16 | nd–0.92 |
| | Flavanones | Eriodictyol | 0.01 | nd-0.19 |
| | | Naringenin | 0.28 | nd-0.64 |
| | Flavanols | Isorhamnetin | 0.04 | nd–1.41 |
| | | Kaempferol | 0.02 | nd–0.94 |
| | | Quercetin | 0.01 | nd–0.38 |
| Hazelnut | Phenolic acids | Gallic acid | 0.13–1.29 | 0.00–0.73 |
| | | Protocatechuic acid | 0.15–0.73 | nd–0.76 |
| | | p-Coumaric acid | 0.15–0.47 | 0.08–0.18 |
| | | Total free phenolic acids | 0.28–1.69 | 0.04–1.27 |
| | | Total bound phenolic acids | 5.93–12.62 | 3.03–9.95 |
| | | Total phenolic acids | 6.21–14.31 | 3.07–11.22 |
| | Flavan-3-ols | Catechin | 3.44–8.56 | 2.66–5.04 |
| | | Epicatechin | 0.17–1.10 | 0.09–0.30 |
| | | Procyanidin dimer 1 | 0.97–3.34 | 0.61–1.27 |
| | | Procyanidin dimer 2 | 1.25–8.70 | 0.95–1.84 |
| | | Procyanidin trimer 1 | 0.24–0.87 | 0.13–0.28 |
| | | Procyanidin trimer 2 | 0.56–1.81 | 0.25–0.86 |
| | | Procyanidin trimer 3 | 0.66–5.39 | 0.41–11.8 |
| Pecan | Phenolic acids | Caffeic acid | 96 | 0–96 |
| | | Chlorogenic acid | 350 | 17–350 |
| | | Gallic acid | 45 | 20–95 |
| | | p-Hydroxybenzoic acid | 275 | 21–250 |
| | | Protocatechuic acid | 17.5 | 0–16 |
| | | Ellagic acid | 3.4 | 0.09–3.4 |
| | | Total phenolic acids | 3.6 | 0.01–3.9 |
| Pistachio | Phenolic acids | Protocatechuic acid | 3.14 | 3.96–8.60 |
| | | Chlorogenic acid | 4.96 | 4.0–5.8 |
| | Flavanones | Eriodictyol-7-O-glucoside | 4.1 | 3.64–3.90 |
| | | Eriodictyol | 2.2 | 2.20–2.67 |
| | Flavones | Luteolin | 0.8 | 0.70–1.22 |
| | Flavanols | Rutin | 3.01 | 3.5–5.4 |
| | Flavan-3-ols | Catechin | 4.8 | 8.64–8.72 |
| Cashew | Phenolic acids | Gallic acid | 0.11 | 0.1–0.25 |
| | | p-Coumaric acid | 0.1 | 0.07–0.11 |
| | | Syringic acid | 0.61 | 0.48–0.87 |
| | Flavan-3-ols | Catechin | 11.7 | 9.6–15.6 |
| | | Epicatechin | 7.43 | 6.08–8.37 |
| | | Epigallocatechin | 4.46 | 4.2–6.54 |

Reproduced with permission from Chang, S.K., Alasalvar, C., Bolling, B.W., Shahidi, F., 2016. Nuts and their co-products: the impact of processing (roasting) on phenolics, bioavailability, and health benefits – a comprehensive review. J. Funct. Foods 26, 88–122. https://doi.org/10.1016/j.jff.2016.06.029.

### 5.5.2 Coffee beans

The roasting process affects the coffee's appearance, flavor, and scent. It takes anywhere from 3 to 20min to complete this process, which includes the following three stages: (i) raising the temperature around 180°C to remove water and reducing the moisture content from 10% to 12.5% to 2.5%; (ii) a second stage of cooking at 200–300°C to undergo physical-chemical transformations that lead to flavor development; and (iii) a cooling stage using either cold air or water jetting. Volatile substances such as alcohols, aldehydes, and amino acids are found in roasted coffee beans. These include carboxylic acids, amines, dicarbonyls, furanones, esters, enoles, furans, imidazoles, lactones, hydrocarbons, and ketones (Poisson et al., 2017). However, only pyrazines, ketones, furans, esters, phenols, and sulfur compounds have a significant influence on human perception.

The heat profile used during the roasting process can have a profound effect on these chemicals. When these molecules are formed, there are a number of chemical reactions that take place in the coffee seeds. These include cleavage and cyclization, dehydration and enolation, fragmentation, hydrolysis, isomerization, and lactonization. The Strecker degradation and the Maillard and pyrolysis reactions are the main chemical processes. Reducing sugars and amino acids are combined in the Maillard reaction, which results in the synthesis of volatile chemicals including pyridines, pyrazines, dicarbonyls, diacetyl, oxazoles, pyrroles, and imidazolones (furaneol, maltol, and cyclotene), along with formic and acetic acids, among others (Lee et al., 2015). The α-amino acids are transformed into sulfur compounds and aldehydes (e.g., 3-mercapto-3-methylbutyl formate, 3-methyl-2-butene-1-thiol, 3-methyl-2-butene-1 thiol, phenylacetaldehyde, 2-furfurylthiol, and 3-methyl-2-butene-1-thiol) in the Strecker degradation process, which contributes to the complex fragrance of coffee.

During the pyrolysis process, the coffee bean matrix's intercellular carbohydrates are destroyed, resulting in an increase in porosity and $CO_2$ emissions. Additionally, the breakdown of certain amino acids (such as proline and hydroxy amino acids), aliphatic acid (such as quinic acid), and lipids have a significant influence on the generation of volatiles during roasting. Volatile chemicals are highly dependent on the genetic variation in the coffee plant. Green coffee bean composition, geographic origin, postharvest processing, environmental conditions in preharvest processing, the existence of faulty beans, and ripening stage are other elements that determine roasted volatile constitution (de Melo Pereira et al., 2019). Fig. 5.3 shows the schematic diagram of coffee fruit structure and associated postharvest processing methods.

### 5.5.3 Tea

Tea is unique among beverages in that the grower is also involved in the production process. The tea leaf must be treated in situ as soon as it is plucked because of its fragile nature. Vertical integration between industrial production and land ownership is very cost-effective in economic terms. When compared to the higher rates of urban labor, the employment of rural labor results in further cost savings. White

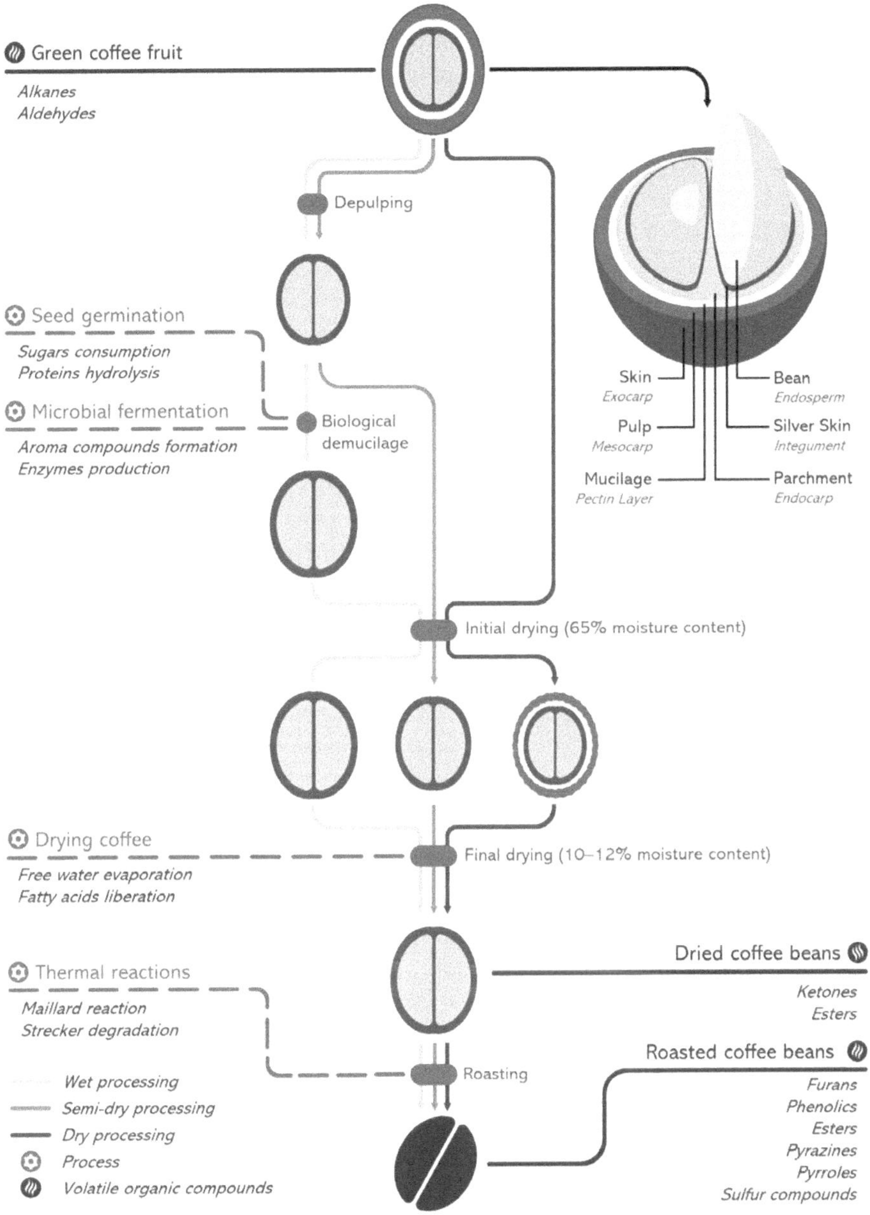


**Fig. 5.3** Schematic outline of coffee fruit structure and various postharvest techniques. Obtained with permission from de Melo Pereira, G.V., de Carvalho Neto, D.P., Magalhães Júnior, A.I., Vásquez, Z.S., Medeiros, A.B.P., Vandenberghe, L.P.S., et al. 2019. Exploring the impacts of postharvest processing on the aroma formation of coffee beans – a review. Food Chem. 272, 441–52. https://doi.org/10.1016/j.foodchem.2018.08.061.

tea, green tea, yellow tea, oolong tea, black tea, and dark tea are the most common forms of tea. During processing and after processing are the two primary types of roasted tea. Tea that is roasted during processing has a longer shelf life because the roasting reduces the amount of water in the tea. Heat can also alter the amino acids and sugars in the tea leaves to eliminate the tea's raw flavor. Each type of tea may be subdivided into subcategories based on the degree of roasting. The chemical composition and sensory quality of tea are greatly influenced by the roasting process (Zhu et al., 2021).

A study by Guo et al. examined the impact of roasting degree on the flavor quality of large-leaf yellow tea (LLYT) under various roasting degree treatments, including small fire (120–130°C), medium fire (130–140°C), and old-fire (145–155°C) (Guo et al., 2021). When it came to evaluating the flavor quality of LLYT, it was found that the roasting degree had a significant impact on the amount of nonvolatile compounds like caffeic acid and catechin and the amount of volatile compounds like total sugars and free amino acids. Tea's health benefits and scent are influenced by its makeup. While it had higher levels of volatile compounds, total volatiles, and the highest concentration of (−)-gallocatechin gallate in its LLYT roasted using an old fire roasting treatment, it had lower levels of certain polyphenols, caffeine, free amino acids, and total sugar in its LLYT roasted with small and medium fires, compared to the two other methods. Traditional fire roasting was shown to be the best method for producing LLYT, which is favored by tea drinkers, according to the study.

Qi et al. used a statistical model to identify the best conditions for roasting green tea (Qi et al., 2021). In the investigation, it was shown that the ideal temperature and time for tea roasting are approximately 200°C and 1000 s, respectively. The roasting technique for Wuyi Rock tea, a distinct subgroup of Oolong tea, was shown to be crucial in another study by Yang et al. (2021). Most fragrance molecules, particularly N-heterocyclic compounds, which are strongly linked to the Maillard process, were shown to rise in concentration with increasing roasting temperature. Some Strecker aldehydes, such as 2-methylpropanal, 2-methylbutanal, and 3-methylbutanal, showed an increasing trend in the Maillard process. As a result, the roasted scent was more noticeable at higher temperatures, while the flowery and fruity aroma was more noticeable at lower temperatures.

### 5.5.4 Oats products

An abundance of oats' high protein content has made them a popular food source for humans and cattle alike. Following corn, wheat, rice, barley, and sorghum in terms of worldwide output, oats come in at number six on the list of grain crops. A large portion is fed to cattle, although oats may be eaten by humans and have a wide range of uses, notably oatmeal and oat flour. Oats have a carbohydrate content of 60%, protein content of 14%, lipid content of 7%, and α-glucan content of 4%. Oats have a higher protein and fat content than other grain crops. Oats are an excellent source of potassium, one of the micronutrients. Oats also contain phenolics and other minor components (Schlörmann et al., 2020). Avenanthramides, a class of compounds found only in oats, have piqued researchers' attention in particular. Oat products have been processed in a number of ways to fulfill the needs of consumers, lengthen shelf life, and boost product diversity.

The glycemic response in humans can be considerably affected by the nutritional composition and physical structure of a food product that has been processed at different degrees. Preventing rancidity and extending the shelf life of oats by inactivating natural enzymes like lipase is one of the goals of heat treating them. It is also done to make items that are more user-friendly, including those that are quick and simple to prepare. Starch is partially gelatinized as a result of the heat treatment. Amylose molecules are released from the starch granules as water is absorbed, and the granules dissolve. Because of this, heat treatment boosts starch availability for enzyme hydrolysis, which results in an increase in GI. Nutritional components may also be damaged by extreme heat treatment conditions, which limit their effectiveness. Temperature, moisture, and time period all have an impact on this outcome (Zhang et al., 2021).

## 5.6 Conclusion and future prospective

Roasting is a cooking process for converting the micro and macronutrients into a more digestible state. Due to complex surface changes in the food components, the high-temperature-short time process produces a finished product with improved flavor, aroma and color, regardless of the product being roasted. The process also lowers the water activity, extends shelf life, and alters the antioxidant and functional characteristic of food products, thereby promoting better customer acceptance. This chapter discusses the roasting mechanisms and summarizes different types of emerging roasting techniques such as microwave, Revtech, superheated steam and forced convection continuous tumble roasters. It was observed that the advanced roasting techniques outperform the conventional processes in terms of easy process control, higher heat transfer rate, lower energy consumption and reduced thermal degradation of the treated products. The roasting effects on biochemical compounds of food products such as functional, compositional and nutritional attributes were also focused. Roasting was found to have both beneficial and adverse effects on different food components, depending upon the types of product treated and the processing conditions used.

While there are several studies available on the roasting of different type of food products, however, the economic aspects of the process, along with an evaluation of the transformation chemistry occurring in food matrices at various phases of the roasting process, require a detailed investigation. This also necessitates further research into a wide range of roasting conditions, in order to achieve a comprehensive understanding of the roasting effects on different quality attributes of the treated product. Furthermore, findings on novel roasting processes such as Revtech and FCCT roasting are very scant in the present literature, thereby arising the need for further experimentation.

# References


Ahmed, I.A.M., Al Juhaimi, F.Y., Osman, M.A., Al Maiman, S.A., Hassan, A.B., HAS, A., et al., 2020. Effect of oven roasting treatment on the antioxidant activity, phenolic compounds, fatty acids, minerals, and protein profile of Samh (*Mesembryanthemum forsskalei* Hochst) seeds. LWT 131, 109825. https://doi.org/10.1016/j.lwt.2020.109825.

Aljuhaimi, F., Özcan, M.M., 2018. Influence of oven and microwave roasting on bioproperties, phenolic compounds, fatty acid composition, and mineral contents of nongerminated peanut and germinated peanut kernel and oils. J. Food Process. Preserv. 42. https://doi.org/10.1111/jfpp.13462, e13462.

Anderson, N.M., 2019. Recent advances in low moisture food pasteurization. Curr. Opin. Food Sci. 29, 109–115. https://doi.org/10.1016/j.cofs.2018.11.001.

Berk, E., Hamzalloglu, A., Gökmen, V., 2019. Investigations on the Maillard reaction in sesame (*Sesamum indicum* L.) seeds induced by roasting. J. Agric. Food Chem. 67, 4923–4930. https://doi.org/10.1021/acs.jafc.9b01413.

Berk, Z., 2018. Food Process Engineering and Technology, third ed. Academic Press.

Bertuzzi, T., Martinelli, E., Mulazzi, A., Rastelli, S., 2020. Acrylamide determination during an industrial roasting process of coffee and the influence of asparagine and low molecular weight sugars. Food Chem. 303. https://doi.org/10.1016/j.foodchem.2019.125372, 125372.

Bhattacharya, S., 2014. Roasting and toasting operations in food: process engineering and applications. In: Conventional and Advanced Food Processing Technologies. Wiley, pp. 221–248, https://doi.org/10.1002/9781118406281.ch10.

Bhinder, S., Singh, B., Kaur, A., Singh, N., Kaur, M., Kumari, S., et al., 2019. Effect of infrared roasting on antioxidant activity, phenolic composition and Maillard reaction products of Tartary buckwheat varieties. Food Chem. 285, 240–251. https://doi.org/10.1016/j.foodchem.2019.01.141.

Bourré, L., Young, G., Hill, H., Fraser, S., Borsuk, Y., Sarkar, A., et al., 2019. Effect of dry and steam heating on the functional and bread baking properties of yellow pea and navy bean flours. Cereal Chem. 96, 1079–1092. https://doi.org/10.1002/cche.10218.

Chang, S.K., Alasalvar, C., Bolling, B.W., Shahidi, F., 2016. Nuts and their co-products: the impact of processing (roasting) on phenolics, bioavailability, and health benefits – a comprehensive review. J. Funct. Foods 26, 88–122. https://doi.org/10.1016/j.jff.2016.06.029.

Chen, Y., Huang, J., Hu, J., Yan, R., Ma, X., 2019. Comparative study on the phytochemical profiles and cellular antioxidant activity of phenolics extracted from barley malts processed under different roasting temperatures. Food Funct. 10, 2176–2185. https://doi.org/10.1039/c9fo00168a.

Chhanwal, N., Anishaparvin, A., Indrani, D., Raghavarao, K.S.M.S., Anandharamakrishnan, C., 2010. Computational fluid dynamics (CFD) modeling of an electrical heating oven for bread-baking process. J. Food Eng. 100, 452–460. https://doi.org/10.1016/j.jfoodeng.2010.04.030.

Chikelu, C.C., Ude, M.U., Onyekwere, D.C., Ekwereike, G.C., Okpanachi, A.I., 2015. Design of a groundnut kernel roastmaster. Int. Res. J. Eng. Technol. 2, 17–24.

Chindapan, N., Soydok, S., Devahastin, S., 2019. Roasting kinetics and chemical composition changes of Robusta coffee beans during hot air and superheated steam roasting. J. Food Sci. 84, 292–302. https://doi.org/10.1111/1750-3841.14422.

de Melo Pereira, G.V., de Carvalho Neto, D.P., Magalhães Júnior, A.I., Vásquez, Z.S., Medeiros, A.B.P., Vandenberghe, L.P.S., et al., 2019. Exploring the impacts of postharvest processing on the aroma formation of coffee beans – a review. Food Chem. 272, 441–452. https://doi.org/10.1016/j.foodchem.2018.08.061.

Fabbri, A., Cevoli, C., Alessandrini, L., Romani, S., 2011. Numerical modeling of heat and mass transfer during coffee roasting process. J. Food Eng. 105, 264–269. https://doi.org/10.1016/j.jfoodeng.2011.02.030.

Fellows, P.J., 2017. Food Processing Technology: Principles and Practices, fourth ed. Woodhead Publishing.

Fernández-Romero, E., Chavez-Quintana, S.G., Siche, R., Castro-Alayo, E.M., Cardenas-Toro, F.P., 2020. The kinetics of total phenolic content and monomeric Flavan-3-ols during the roasting process of Criollo cocoa. Antioxidants 9, 7–10. https://doi.org/10.3390/antiox9020146.

Feyissa, A.H., Gernaey, K.V., Adler-Nissen, J., 2013. 3D modelling of coupled mass and heat transfer of a convection-oven roasting process. Meat Sci. 93, 810–820. https://doi.org/10.1016/j.meatsci.2012.12.003.

Feyissa, A.H., Gernaey, K.V., Ashokkumar, S., Adler-Nissen, J., 2011. Modelling of coupled heat and mass transfer during a contact baking process. J. Food Eng. 106, 228–235. https://doi.org/10.1016/j.jfoodeng.2011.05.014.

Ghafoor, K., Özcan, M.M., Al-Juhaimi, F., Babiker, E.E., Fadimu, G.J., 2019. Changes in quality, bioactive compounds, fatty acids, tocopherols, and phenolic composition in oven- and microwave-roasted poppy seeds and oil. LWT 99, 490–496. https://doi.org/10.1016/j.lwt.2018.10.017.

Gogus, F., Ozel, M.Z., Kocak, D., Hamilton, J.F., Lewis, A.C., 2011. Analysis of roasted and unroasted *Pistacia terebinthus* volatiles using direct thermal desorption-GCxGC-TOF/MS. Food Chem. 129, 1258–1264. https://doi.org/10.1016/j.foodchem.2011.05.003.

Göncüoğlu Taş, N., Gökmen, V., 2017. Maillard reaction and caramelization during hazelnut roasting: a multiresponse kinetic study. Food Chem. 221, 1911–1922. https://doi.org/10.1016/j.foodchem.2016.11.159.

Goszkiewicz, A., Kołodziejczyk, E., Ratajczyk, F., 2020. Comparison of microwave and convection method of roasting sunflower seeds and its effect on sensory quality, texture and physicochemical characteristics. Food Struct. 25. https://doi.org/10.1016/j.foostr.2020.100144.

Gunel, Z., Torun, M., Sahin-Nadeem, H., 2020. Sugar, d-pinitol, volatile composition, and antioxidant activity of carob powder roasted by microwave, hot air, and combined microwave/hot air. J. Food Process. Preserv. 44. https://doi.org/10.1111/jfpp.14371, e14371.

Guo, X., Ho, C.T., Schwab, W., Wan, X., 2021. Effect of the roasting degree on flavor quality of large-leaf yellow tea. Food Chem. 347. https://doi.org/10.1016/j.foodchem.2021.129016, 129016.

Hamzalıoğlu, A., Gökmen, V., 2020. 5-Hydroxymethylfurfural accumulation plays a critical role on acrylamide formation in coffee during roasting as confirmed by multiresponse kinetic modelling. Food Chem. 318. https://doi.org/10.1016/j.foodchem.2020.126467.

Hatamian, M., Noshad, M., Abdanan-Mehdizadeh, S., Barzegar, H., 2020. Effect of roasting treatment on functional and antioxidant properties of chia seed flours. NFS J. 21, 1–8. https://doi.org/10.1016/j.nfs.2020.07.004.

Idrus, N.F.M., Zzaman, W., Yang, T.A., Easa, A.M., Sharifudin, M.S., Noorakmar, B.W., et al., 2017. Effect of superheated-steam roasting on physicochemical properties of peanut (*Arachis hypogea*) oil. Food Sci. Biotechnol. 26, 911–920. https://doi.org/10.1007/s10068-017-0132-0.

Ji, J., Liu, Y., Shi, L., Wang, N., Wang, X., 2019. Effect of roasting treatment on the chemical composition of sesame oil. LWT 101, 191–200. https://doi.org/10.1016/j.lwt.2018.11.008.

Jinap, M.S., Jamilah, B., Nazamid, S., 2004. Effect of polyphenol concentration on pyrazine formation during cocoa liquor roasting. Food Chem. 85, 73–80. https://doi.org/10.1016/j.foodchem.2003.06.005.

Jogihalli, P., Singh, L., Kumar, K., Sharanagat, V.S., 2017. Physico-functional and antioxidant properties of sand-roasted chickpea (*Cicer arietinum*). Food Chem. 237, 1124–1132. https://doi.org/10.1016/j.foodchem.2017.06.069.

Kanagaraj, S.P., Ponnambalam, D., Antony, U., 2019. Effect of dry heat treatment on the development of resistant starch in rice (*Oryza sativa*) and barnyard millet (*Echinochloa furmantacea*). J. Food Process. Preserv. 43. https://doi.org/10.1111/jfpp.13965, e13965.

Keppler, S., Bakalis, S., Leadley, C.E., Fryer, P.J., 2016a. Processing of barley grains in a continuous vibrating conveyor. J. Food Eng. 187, 114–123. https://doi.org/10.1016/j.jfoodeng.2016.04.010.

Keppler, S., Bakalis, S., Leadley, C.E., Fryer, P.J., 2016b. A systematic study of the residence time of flour in a vibrating apparatus used for thermal processing. Innov. Food Sci. Emerg. Technol. 33, 462–471. https://doi.org/10.1016/j.ifset.2015.12.003.

Keppler, S., Bakalis, S., Leadley, C.E., Sahi, S.S., Fryer, P.J., 2018. Evaluation of dry heat treatment of soft wheat flour for the production of high ratio cakes. Food Res. Int. 107, 360–370. https://doi.org/10.1016/j.foodres.2018.02.041.

Kinge, E.E., Tonfack Djikeng, F., Karuna, M.S.L., Zambou Ngoufack, F., Womeni, H.M., 2019. Effect of boiling and roasting on the physicochemical properties of Djansang seeds (*Ricinodendron heudelotii*). Food Sci. Nutr. 7, 3425–3434. https://doi.org/10.1002/fsn3.1163.

Król, K., Gantner, M., Tatarak, A., Hallmann, E., 2020. The content of polyphenols in coffee beans as roasting, origin and storage effect. Eur. Food Res. Technol. 246, 33–39. https://doi.org/10.1007/s00217-019-03388-9.

Kumar, Y., Sharanagat, V.S., Singh, L., Mani, S., 2020. Effect of germination and roasting on the proximate composition, total phenolics, and functional properties of black chickpea (*Cicer arietinum*). Legume Sci. 2. https://doi.org/10.1002/leg3.20, e20.

Lee, L.W., Cheong, M.W., Curran, P., Yu, B., Liu, S.Q., 2015. Coffee fermentation and flavor – an intricate and delicate relationship. Food Chem. 185, 182–191. https://doi.org/10.1016/j.foodchem.2015.03.124.

Li, P., Gasmalla, M.A.A., Zhang, W., Liu, J., Bing, R., Yang, R., 2016. Effects of roasting temperatures and grinding type on the yields of oil and protein obtained by aqueous extraction processing. J. Food Eng. 173, 15–24. https://doi.org/10.1016/j.jfoodeng.2015.10.031.

Majcher, M.A., Klensporf-Pawlik, D., Dziadas, M., Jeleń, H.H., 2013. Identification of aroma active compounds of cereal coffee brew and its roasted ingredients. J. Agric. Food Chem. 61, 2648–2654. https://doi.org/10.1021/jf304651b.

Nooshkam, M., Varidi, M., Bashash, M., 2019. The Maillard reaction products as food-born antioxidant and antibrowning agents in model and real food systems. Food Chem. 275, 644–660. https://doi.org/10.1016/j.foodchem.2018.09.083.

Ozel, M.Z., Yanik, D.K., Gogus, F., Hamilton, J.F., Lewis, A.C., 2014. Effect of roasting method and oil reduction on volatiles of roasted *Pistacia terebinthus* using direct thermal desorption-GCxGC-TOF/MS. LWT Food Sci. Technol. 59, 283–288. https://doi.org/10.1016/j.lwt.2014.05.004.

Poisson, L., Auzanneau, N., Mestdagh, F., Blank, I., Davidek, T., 2017. New insight into the role of sucrose in the generation of α-diketones upon coffee roasting. J. Agric. Food Chem. 66, 2422–2431. https://doi.org/10.1021/acs.jafc.6b04849.

Qi, K., Xu, M., Yin, H., Wu, L., Hu, Y., Yang, J., et al., 2021. Online monitoring the key intermediates and volatile compounds evolved from green tea roasting by synchrotron radiation photoionization mass spectrometry. J. Am. Soc. Mass Spectrom. 32, 1402–1411. https://doi.org/10.1021/jasms.1c00012.

Redgwell, R.J., Trovato, V., Curti, D., 2003. Cocoa bean carbohydrates: roasting-induced changes and polymer interactions. Food Chem. 80, 511–516. https://doi.org/10.1016/S0308-8146(02)00320-5.

Rękas, A., Wroniak, M., Ścibisz, I., 2017. Microwave radiation and conventional roasting in conjunction with hulling on the oxidative state and physicochemical properties of rapeseed oil. Eur. J. Lipid Sci. Technol. 119, 1600501. https://doi.org/10.1002/ejlt.201600501.

Schlörmann, W., Zetzmann, S., Wiege, B., Haase, N.U., Greiling, A., Lorkowski, S., et al., 2019. Impact of different roasting conditions on chemical composition, sensory quality and physicochemical properties of waxy-barley products. Food Funct. 10, 5436–5445. https://doi.org/10.1039/c9fo01429b.

Schlörmann, W., Zetzmann, S., Wiege, B., Haase, N.U., Greiling, A., Lorkowski, S., et al., 2020. Impact of different roasting conditions on sensory properties and health-related compounds of oat products. Food Chem. 307. https://doi.org/10.1016/j.foodchem.2019.125548, 125548.

Schoeman, L., Du Plessis, A., Manley, M., 2016. Non-destructive characterisation and quantification of the effect of conventional oven and forced convection continuous tumble (FCCT) roasting on the three-dimensional microstructure of whole wheat kernels using X-ray micro-computed tomography (μCT). J. Food Eng. 187, 1–13. https://doi.org/10.1016/j.jfoodeng.2016.04.015.

Schoeman, L., du Plessis, A., Verboven, P., Nicolaï, B.M., Cantre, D., Manley, M., 2017. Effect of oven and forced convection continuous tumble (FCCT) roasting on the microstructure and dry milling properties of white maize. Innov. Food Sci. Emerg. Technol. 44, 54–66. https://doi.org/10.1016/j.ifset.2017.07.021.

Schoeman, L., Manley, M., 2019. Oven and forced convection continuous tumble (FCCT) roasting: effect on physicochemical, structural and functional properties of wheat grain. Food Bioprocess Technol. 12, 166–182. https://doi.org/10.1007/s11947-018-2200-8.

Schouten, M.A., Tappi, S., Angeloni, S., Cortese, M., Caprioli, G., Vittori, S., et al., 2021. Acrylamide formation and antioxidant activity in coffee during roasting – a systematic study. Food Chem. 343. https://doi.org/10.1016/j.foodchem.2020.128514, 128514.

Seeram, N.P., Zhang, Y., Henning, S.M., Lee, R., Niu, Y., Lin, G., et al., 2006. Pistachio skin phenolics are destroyed by bleaching resulting in reduced antioxidative capacities. J. Agric. Food Chem. 54, 7036–7040. https://doi.org/10.1021/jf0614948.

Simsek, S., Herken, E.N., Ovando-Martinez, M., 2016. Chemical composition, nutritional value and in vitro starch digestibility of roasted chickpeas. J. Sci. Food Agric. 96, 2896–2905. https://doi.org/10.1002/jsfa.7461.

Sruthi, N.U., Premjit, Y., Pandiselvam, R., Kothakota, A., Ramesh, S.V., 2021. An overview of conventional and emerging techniques of roasting: effect on food bioactive signatures. Food Chem. 348. https://doi.org/10.1016/j.foodchem.2021.129088, 129088.

Suri, K., Singh, B., Kaur, A., Yadav, M.P., Singh, N., 2019. Impact of infrared and dry air roasting on the oxidative stability, fatty acid composition, Maillard reaction products and other chemical properties of black cumin (*Nigella sativa* L.) seed oil. Food Chem. 295, 537–547. https://doi.org/10.1016/j.foodchem.2019.05.140.

Swarnakar, A.K., Srivastav, P.P., Das, S.K., 2019. Optimization of preconditioning process of pressure parboiled brown rice (unpolished) for microwave puffing and its comparison with hot sand bed puffing. J. Food Process Eng. 42. https://doi.org/10.1111/jfpe.13007, e13007.

Vinson, J.A., Cai, Y., 2012. Nuts, especially walnuts, have both antioxidant quantity and efficacy and exhibit significant potential health benefits. Food Funct. 3, 134–140. https://doi.org/10.1039/c2fo10152a.

Wani, I.A., Gani, A., Tariq, A., Sharma, P., Masoodi, F.A., Wani, H.M., 2016. Effect of roasting on physicochemical, functional and antioxidant properties of arrowhead (*Sagittaria sagittifolia* L.) flour. Food Chem. 197, 345–352. https://doi.org/10.1016/j.foodchem.2015.10.125.

Wronkowska, M., Piskuła, M.K., Zieliński, H., 2016. Effect of roasting time of buckwheat groats on the formation of Maillard reaction products and antioxidant capacity. Food Chem. 196, 355–358. https://doi.org/10.1016/j.foodchem.2015.09.064.

Xu, Chang, 2008. Total phenolics, phenolic acids, isoflavones, and anthocyanins and antioxidant properties of yellow and black soybeans as affected by thermal processing. J. Agric. Food Chem. 56, 7165–7175. https://doi.org/10.1021/jf8012234.

Yang, J., Zhou, F., Xiong, L., Mao, S., Hu, Y., Lu, B., 2015. Comparison of phenolic compounds, tocopherols, phytosterols and antioxidant potential in Zhejiang pecan [*Carya cathayensis*] at different stir-frying steps. LWT Food Sci. Technol. 62, 541–548. https://doi.org/10.1016/j.lwt.2014.09.049.

Yang, P., Song, H., Lin, Y., Guo, T., Wang, L., Granvogl, M., et al., 2021. Differences of characteristic aroma compounds in: Rougui tea leaves with different roasting temperatures analyzed by switchable GC-O-MS and GC $\times$ GC-O-MS and sensory evaluation. Food Funct. 12, 4797–4807. https://doi.org/10.1039/d1fo00165e.

Yodkaew, P., Chindapan, N., Devahastin, S., 2017. Influences of superheated steam roasting and water activity control as oxidation mitigation methods on physicochemical properties, lipid oxidation, and free fatty acids compositions of roasted rice. J. Food Sci. 82, 69–79. https://doi.org/10.1111/1750-3841.13557.

Young, G., Bourré, L., Frohlich, P., Borsuk, Y., Sarkar, A., Sopiwnyk, E., et al., 2020. Effect of roasting as a premilling treatment on the functional and bread baking properties of whole yellow pea flour. Cereal Chem. 97, 183–195. https://doi.org/10.1002/cche.10233.

Zhang, K., Dong, R., Hu, X., Ren, C., Li, Y., 2021. Oat-based foods: chemical constituents, glycemic index, and the effect of processing. Foods 10, 1–21. https://doi.org/10.3390/foods10061304.

Zhang, R.Y., Liu, H.M., Ma, Y.X., De Wang, X., 2019. Effects of roasting on composition of chili seed and storage stability of chili seed oil. Food Sci. Biotechnol. 28, 1475–1486. https://doi.org/10.1007/s10068-019-00578-9.

Zhu, Y.M., Dong, J.J., Jin, J., Liu, J.H., Zheng, X.Q., Lu, J.L., et al., 2021. Roasting process shaping the chemical profile of roasted green tea and the association with aroma features. Food Chem. 353. https://doi.org/10.1016/j.foodchem.2021.129428, 129428.

Zou, Y., Gao, Y., He, H., Yang, T., 2018. Effect of roasting on physico-chemical properties, antioxidant capacity, and oxidative stability of wheat germ oil. LWT 90, 246–253. https://doi.org/10.1016/j.lwt.2017.12.038.

Zzaman, W., Bhat, R., Yang, T.A., 2017b. Impact of convectional and superheated-steam roasting on the physicochemical and microstructural properties of cocoa butter extracted from cocoa beans. J. Food Process. Preserv. 41. https://doi.org/10.1111/jfpp.13005, e13005.

Zzaman, W., Bhat, R., Yang, T.A., Easa, A.M., 2017a. Influences of superheated steam roasting on changes in sugar, amino acid and flavour active components of cocoa bean (*Theobroma cacao*). J. Sci. Food Agric. 97, 4429–4437. https://doi.org/10.1002/jsfa.8302.

# Baking and roasting ovens in the food industry

# 6

*Nantawan Therdthai*
Department of Product Development, Faculty of Agro-Industry, Kasetsart University, Bangkok, Thailand

## 6.1 Introduction

Baking and roasting are well-established cooking processes. In ancient time, people learnt to generate heat from a naked flame. With growing knowledge, the generated heat was used to cook foods by several processes such as baking and roasting that cook food using dry heat based on the same principle. Baking is generally used for dough/batter, while roasting is used for other foods such as meat, nuts, and snacks (Berk, 2018).

The processes have been continually developed while the oven has been in use since the Neolithic period (around 9000 years ago) to improve the process efficiency and the quality of the baked and roasted products. An ancient baking oven that may have been invented around 6500 years ago was discovered in Croatia in 2014. In Europe, baking ovens have been found in many countries such as Switzerland (dating from 5500 years ago), Egyptian (4000 years ago) and Belgium (2000–4000 years ago) (Anonymous, 2020).

Around 2000 years ago, the Romans baked bread in an oven with its own chimney. In the Middle Ages, baked goods were considered as a luxurious food in the UK as a baking oven was not standard household equipment. Only rich families might have a wood-burning stove to bake bread at home. Thus, the baking oven in this period was mainly for commercial production. Later in the 15th to 17th centuries, fashionable bread, cakes and pastries became popular. In the 18th century, a semiclosed baking oven was launched and a small bakery business could afford a small scale baking oven. In the 19th century, convenient foods became more popular with bakery products focusing on pastries and pies. The baking oven was developed in accordance with the popularity of products (Bernal, 2016).

Roasting equipment developed from simple roasting pits to a kiln-like oven structure since before the 17th century. At that time, the oven was simply built using bricks; however, later developments used iron, and a popular model was called a "five-plate stove." Then the Castrol stove was designed with the smoke contained inside, becoming the first enclosed oven. To improve roasting efficiency, the iron oven with temperature control was invented toward the end of the 17th century, but the heat source was still from burning wood. It was only in the 18th century that the heat source first changed to gas. Toward the end of the 18th century, the first electric oven was

**High-Temperature Processing of Food Products. https://doi.org/10.1016/B978-0-12-818618-3.00007-0**

manufactured in Canada and was used in commercial service in a hotel. Since the 19th century, the roasting oven has been redesigned to improve efficiency and convenience. A further development has been the use of microwaves as an alternative method to generate heat in the oven, due to the rapid heating it provides. Although a microwave oven is convenient and fast, initially, it was not popular, because a lot of users were concerned about how the radiative heating may have affected the human body (Hurwood, 2015).

Baking and roasting ovens are key equipment used transform raw materials into cooked foods using dry heat. The supplied heat can be in the form of radiant heat, convective heat, and conductive heat. The early generation of ovens used the radiant heat from a fire which involved direct heating. Later, indirect heating was proposed by the transfer of the heat from the fire through ducts that heated the surface of the ducts and provided radiant heat to products in the oven chamber. Both direct and indirect heating systems are still used for heating supply in today's baking and roasting ovens. To increase the uniformity of heat in the oven chamber, a convective fan was proposed to circulate airflow uniformly within the chamber. an oven with a forced air flow is called a convective oven. To reduce baking and roasting times, the airflow velocity was increased to enhance the heat transfer coefficient. This type of oven was the so-called impingement oven. In addition, other heating regimes have been developed to generate heat food quickly, such as infrared heating and microwave heating. Due to their rapid heating, infrared and microwave radiation can cause burning on the surface and inner layers of products, respectively. To overcome this problem, a combination of heating regimes has been introduced in the baking and roasting industry to produce high quality product within a short processing time.

This chapter discusses conventional and novel baking and roasting ovens in both batch and continuous operations. The design of each oven type is described to assist readers to understand and choose the proper oven design for the specific product.

## 6.2 Heat mechanisms in baking and roasting ovens

Heat transfer mechanisms in the oven include convection, radiation, and conduction. The proper combination of heating modes can generate good baking quality (Fig. 6.1).

### 6.2.1 Convection

The convection baking oven was developed to improve the uniformity of heat distribution in the oven. Air is heated and forced to circulate throughout the oven to reduce the difference between maximum and minimum temperatures. Air can be heated by hot surfaces such as a heating coil, duct or burner. Heating uniformity in the convection oven is dependent on the air flow velocity, forced direction, air temperature, location of the flow source, and the movement of objects in the oven. Oven designs that use the convection heating mode include a deck oven, an electrical oven, and a traveling-tray oven.

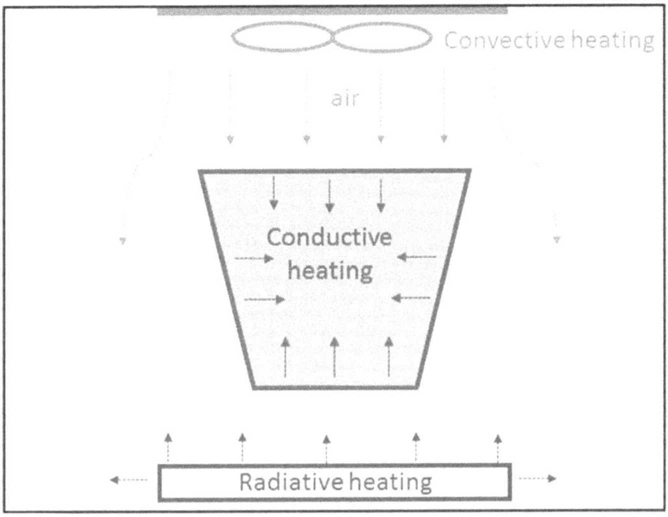


**Fig. 6.1** Heat transfer mechanisms during baking and roasting.

### 6.2.2 Radiation

Radiant heat can be generated from hot surfaces inside the baking oven such as on ducts, burners and the oven wall. The difference in temperature between the heat source and the cooler product causes heat transfer to the product. However, the amount of energy absorption is also dependent on the emissivity of the product. Furthermore, the distance between the heat source and product should be minimized to increase oven efficiency. Oven designs that use the radiation heating mode are a deck oven, an electrical oven, a tunnel oven, and a microwave oven.

### 6.2.3 Conduction

Conductive heat is transferred from bands (such as hot mesh bands and steel bands) to the dough or batter. The surfaces of the band are in direct contact with the dough or batter either with or without a container. In addition, conductive heat is transferred from the surface of hot containers (such as the mold or plate) to the dough and batter inside. With the proper amount of heat, the dough or batter is transformed into a solid. Conductive heat becomes important for proper crumb development. The heating rate is dependent on the thermal conductivity of the containers and the dough or batter, the contact area, and the temperature difference between the containers and the dough or batter.

# 6.3 Baking ovens

## 6.3.1 Batch baking ovens

### 6.3.1.1 Wood-burning oven

A wood-burning oven involves a traditional baking style that burns natural timber to generate radiant heat throughout the oven chamber. This process has been in use since the early 1800s. Initially, a wood-burning oven was an expensive household item that was affordable only to rich people. Later, as its price decreased, low-middle income households could afford it (Charrington-Hollins, 2020). Today, a modern controller system is applied to automatically control the baking temperature at the desired point (Fig. 6.2). The wood burning creates a unique aroma and flavor for baked products. However, some drawbacks of concern include the use of firewood, the presence of flames and toxic smoke, and the difficulty of controlling the temperature.

### 6.3.1.2 Rack baking oven

A rack oven consists of a number of trays on a vertical rack that facilitates loading dough or batter for cooking and unloading the baked products. The number of trays per rack is dependent on the product height. Normally, the distance between racks should equal the product height plus an additional 25 mm. This design saves space and is economic. A large amount of dough or batter products can be baked in a batch (Fig. 6.3). This type of oven is also flexible in terms of energy supply that can be gas, oil, or electricity. For gas and oil, combustion occurs in the burner to generate hot air and gas into the radiant tubes in the oven chamber. This is an indirect form of heating

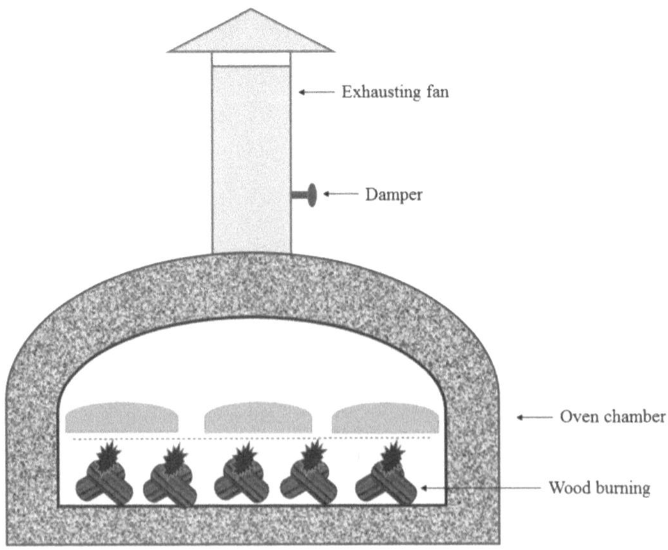


**Fig. 6.2** Wood-burning oven.

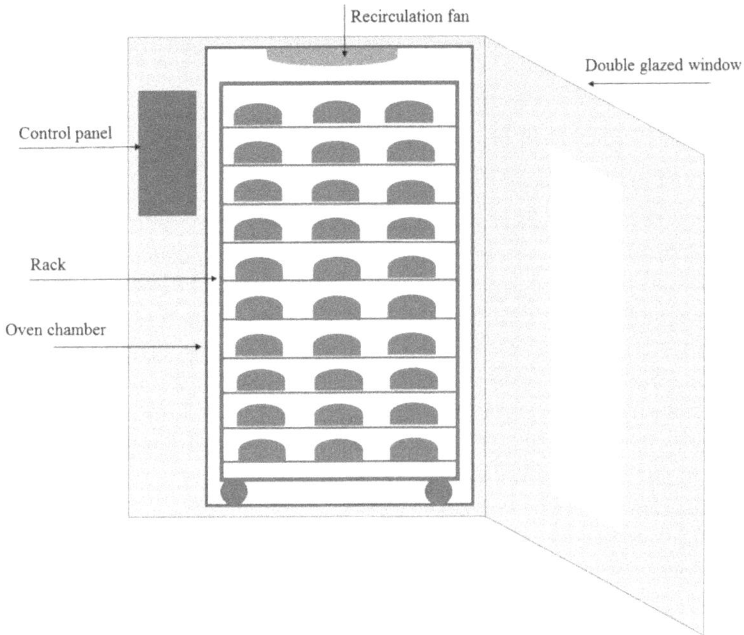


**Fig. 6.3** Rack oven.

from the burner to the products. With electricity, the heating element is connected directly to the electric power supply. Generally, the baking temperature in a rack oven is not more than 300°C. The insulation should be well designed to keep heat within the oven chamber. Normally, a rack oven is designed with a glass window to allow visual monitoring of the progression of baking without having to open the door and lost heat. Heat-reflecting double glass should be used to prevent heat loss through the window.

An axial fan can be installed on the ceiling of the oven to facilitate hot air circulation, eliminate hot and cold spots, and increase the heat transfer coefficient. A frequency converter may be installed on the fan to efficiently adjust the airflow in the oven, particularly when cooking of sensitive products such as macaroons. In addition to using a convection fan, uniformity of the temperature profile can be obtained from rotation of the rack, in the so-called "rotating rack oven." An automatic system rotates the rack of food as soon as the oven door is closed. When the door is opened, rotation of the rack must stop and can only start again after closing the door.

In some designs of a small-scale oven, a prover can be installed beneath the rack oven to save space. Dough can be proved on the bottom and then easily moved to the top for baking. In a large-scale oven, the insulation on the floor should be well designed to prevent heat loss to the cooler bakery floor. An adjustable ramp can be fitted to keep the platform level. The rack rotation speed should be adjustable. The rotation should be gradually accelerated from the beginning and gently retarded at the end of baking, especially for sensitive products. The rack oven is very flexible as it can be used for baking many kinds of products such as bread, sourdough, pastries and muffins. Therefore, the main advantages of the rack oven are space saving, flexibility for cooking almost all baked products, and uniform heat distribution in the oven chamber.

### 6.3.1.3 Deck baking oven

A deck oven is composed of several stacked baking chambers, with each chamber having its own controlling system. Thus, each chamber can be operated separately to suit a variety of baked products such as bread, pizza, and pastry that can be baked at the same time. The stacking baking chamber is designed to have a narrow space between the ceiling and bottom floor countered by a large depth. The baking surface can be stone or steel plate. The radiant tube is installed on either on the top or bottom or on both to effectively transfer the radiant heat to the dough or batter. Heat transfer in the deck oven is mainly from radiation (Fig. 6.4). Each deck is individually controlled within a range set by upper and lower temperatures using individual thermostats. Generally, the set temperature can be in the range of 50°C to 360°C. However, it can be designed to provide high-temperature baking of up to 500°C for baking a pizza within 2 to 3 min. If a glass window is installed to monitor the progress of cooking, double glazing should be used to prevent heat loss. Similar to a rack oven, a small-scale deck oven may have a prover installed beneath the deck to save space and to allow convenient loading from proving to baking.

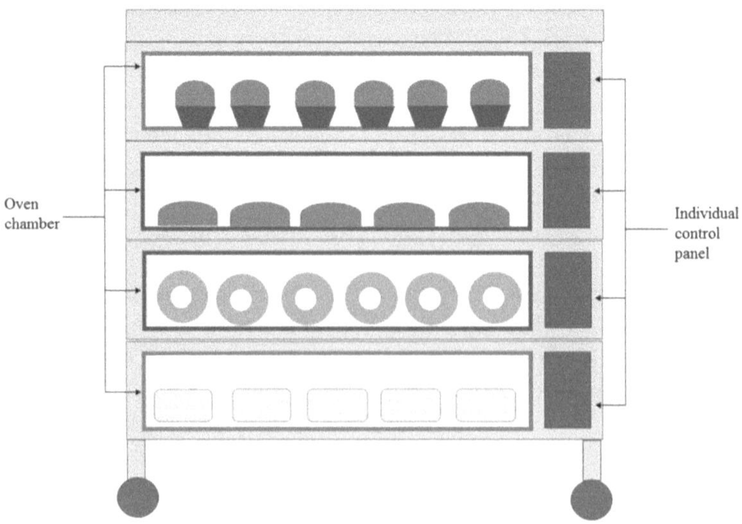


**Fig. 6.4** Deck oven.

## 6.3.2 Continuous baking ovens

### 6.3.2.1 Tunnel baking oven

A tunnel baking oven is a classical design that provides high energy efficiency and reasonable baking quality. It can be designed for both small and large scales of production. A tunnel oven with at a small production scale is designed to have a narrow space in the baking chamber to save energy and a moving mesh to transfer cooking products continuously. The heat source is most likely electricity that can be installed both on the top and bottom of the oven chamber (Fig. 6.5). This oven design is good for pizza baking. The setting temperature is approximately 400°C for baking a pizza within 2 to 15 min.

For large-scale production, a tunnel oven functions better using gas or fuel oil as the energy source. As it has a continuous configuration, it can facilitate mass production on an industrial scale. The design of the oven imitates a tunnel with the dough or batter entering and exiting from opposite ends. The distance between the top ceiling and bottom floor of the oven is reasonably narrow to save energy. However, the oven is very long and can be divided into zones. Each zone is operated under different baking conditions to obtain the optimum temperature profile for each product. While traveling in the oven, the dough or batter is heated by the hot air, by radiant heat and by combined hot air and radiant heat. The speed of the conveyor that carries the dough or batter inside the oven is suitably adjustable to manage protein denaturation, starch gelatinization, evaporation, and the browning reaction to ensure proper crumb and crust formation.

Energy can be supplied to the tunnel baking oven chamber by either direct or indirect heating. With direct heating, gas (such as natural gas or liquid petroleum gas) is burnt to directly transfer heat into the oven chamber. There are ribbon gas burners below or above the bands or both. The radiant heat from the gas flames is the main heat source supplied in the oven, while radiant heat from the oven walls and ceiling is a minor heat source.

With indirect heating, either gas or other fuel can be used for combustion in the burners to generate burnt gas and hot air that is transferred in radiant tubes installed on the top or bottom or both of the oven chamber. Once the burnt gas flows through the ducts, the surfaces of the ducts are heated and become the heat source supplying

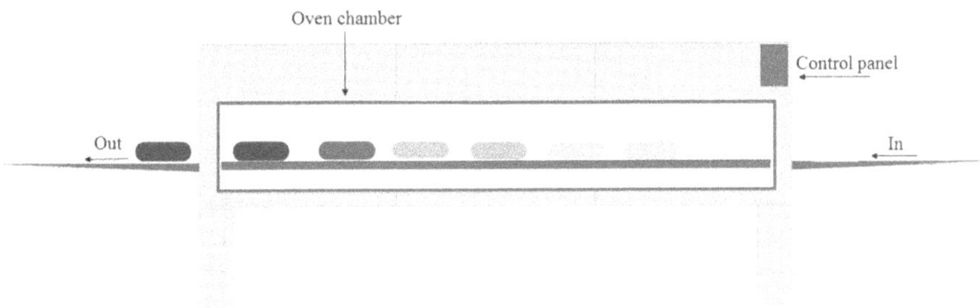


**Fig. 6.5** Tunnel oven.

radiant heat to the baking chamber. To save energy, the burnt gas is recirculated and passed back to the burner where it can be reused and thus save some energy.

For a tunnel oven without a convective fan, the heat supply is from the radiant heat from the hot surface of the duct that may be placed on bottom or top or both. After operation for a while, all the oven walls are heated and can also provide radiant heat from the walls for cooking. Heat accumulation within the oven during baking should be monitored and controlled to set points (Fig. 6.6). Although there is no fan to generate forced convection, natural convection can occur from the bottom to the top of the oven because the surrounding air is heated up while its density is decreased, resulting in the heated air naturally moving toward the ceiling.

In a tunnel oven operated with a convective heat transferring system, the burnt gas from the burner travels through the ducts to directly bake the product in the oven chamber. This type of oven is called a "direct convective oven." Another design is the "indirect convective oven" in which the hot air from the burner generates heat in a heat exchanger that then passes through the ducts. The air in the oven chamber is recirculated through the heat exchanger and then is transferred back into the oven chamber to bake the dough. In addition, a convective fan can be installed to force air in the designed direction to ensure effective heat distribution within the oven. After the hot air has transferred heat to the products in the chamber, the used air becomes cooler and is moved to the recirculating system. It can be reheated in the burner before blowing through the ducts and directly heat the product in the oven chamber (in a direct convective oven) or provide thermal energy to the heat exchanger (in an indirect convective oven). This design can save energy in a similar way to a tunnel oven operated using radiant heat.

In the tunnel oven, a band is used to transfer the dough into the oven. To provide conduction heat to the dough, the band should be made of metal and can be solid or a mesh. First, the band is preheated before the dough sample is placed on top of it. Then, conductive heat is transferred from the preheated band to the sample as it passes through the tunnel oven. A tunnel oven is good for baking biscuits, bread, and pizza. For biscuits, the operating temperature should be in the range of 300°C to 400°C for 3 to 5 min. For bread, the operating temperature varies, depending on the size of the loaf but generally, the baking temperature should be in the range of 200°C to 250°C for bread. A pizza has a broad contact area with the hot plate and large circumference

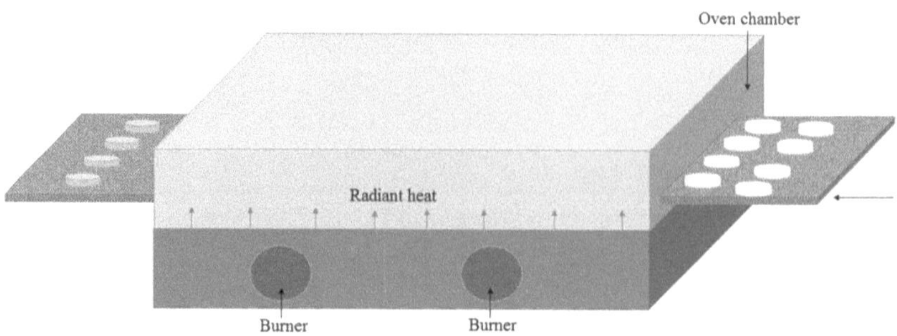


**Fig. 6.6** Tunnel baking oven with radiation heat transfer.

exposed to the surrounding heat. Thus, the operating temperature can be in a high range (250–300°C) for a short baking time. Based on the processes outlined above, the advantage of a tunnel baking oven includes continuous production (first-in, first-out) without any contamination between unbaked and baked products, high energy efficiency, zone optimization, and mass production. Consequently, the tunnel oven is very popular for industrial baking.

### 6.3.2.2 Traveling-tray baking oven

A traveling-tray baking oven is an oven with continuous operation that can serve mass production on an industrial scale. All its walls and ceilings must be well insulated to keep heat within the chamber. To save on working space, the dough or batter travels on trays that enter in the top zone of the oven followed by a U-turn to the bottom zone where the trays exit at the completion of the baking time. The U-turn design can reduce the length of oven while still allowing different baking conditions in each zone to obtain the desired baking temperature profiles (Fig. 6.7).

Direct heating is achieved using a burner to generate hot gases that are then blown through the ducts in the oven chamber to directly transfer heat to the samples. The energy source should be gas rather than diesel oil fuels. A traveling-tray baking oven may often use indirect heating to supply energy to the oven chamber, with the burners generating hot air and burnt gas that are subsequently used to supply ducts (radiant tubes) in the oven chamber. The hot air transfers heat to the duct surfaces and then these hot surfaces generate radiant heat into the oven chamber. Natural convection occurs from the hot air circulating in the ducts and mainly affects temperatures in the bottom zone. With fans installed in the oven, force convection is generated and affects temperatures in the force direction. As the burner is often installed underneath the oven chamber, the bottom zone may be the hottest. Thus, fans are often installed in

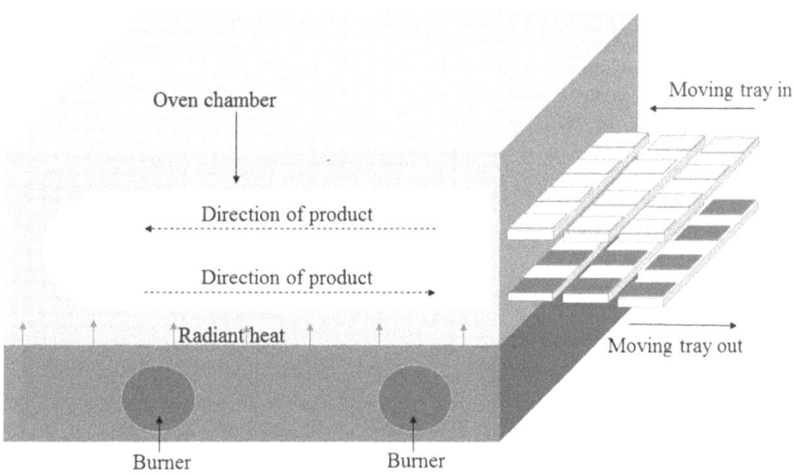


**Fig. 6.7** Traveling-tray baking oven.

the bottom zone to manipulate the hot air flow to the top zone and promote uniform air circulation within the oven.

There are several zones in a traveling-tray baking oven: top-front, top-back, bottom-front, and bottom-back. However, as each zone is connected to the others, adjusting the heating conditions in one zone possibly interferes with other zones. A traveling-baking oven is good for industrial baking of bread, biscuits, cakes, and pizza. To bake bread, the operating temperature should be in the range of 200°C to 240°C. The first zone temperature is set slightly lower than the others to allow the dough to expand during initial baking with gradually evaporation to form a crust. Then, the temperature is gradually increased in the second and third zone temperatures might be gradually increased to speed up the heating rate within the dough. With the U-turn design in the traveling-tray oven, the increased temperature in the third zone which is underneath the second zone can interfere with the second zone by increasing the temperature. The final zone temperature is similar to that in the third zone to ensure proper crust color and flavor development. The advantages of a traveling-tray baking oven include continuous production (first-in, first-out), space saving, zone operation, and flexibility for many products. A cheaper energy supplies such as gas can be used in the burner to generate enough hot air to transfer to the ducts in the oven.

## 6.4 Roasting ovens

### 6.4.1 Batch roasting ovens

A batch roasting oven can be a part of small or household equipment or be large for commercial-scale production. In household equipment, the roasting oven likely uses electricity to supply energy for oven operation. According to an International Energy Agency (IEA) report, electrical energy for household equipment was more than 3000 TWh in 2016 and was predicted to increase to over 4000 TWh by 2030 (International Energy Agency (IEA), 2020). For domestic roasting, oven operation generally requires 800 to 950 W for 2 to 10 min. In addition, a domestic oven has to meet standards such as the EU small domestic appliances standard (SDA) mainly to protect users from overheating and electrical accidents. The A13 standard specifies the maximum temperature rise limits for contact surfaces. For example, the temperature rise limits are 45°C for bare metal, 55°C for coated metal, and 60°C for glass and ceramic. However, the limits can be extended by using labeling warning of high temperature (Güneş et al., 2020).

#### 6.4.1.1 Charcoal roasting oven

A charcoal roasting oven involves the traditional roasting style where charcoal or natural timber is used to fire a burner and generate radiant heat through the oven chamber. An airflow regulator should be also installed to control temperature distribution. A built-in thermometer is installed to provide automatic control of the roasting temperature (Fig. 6.8). The oven temperature can be around 180°C and 280°C for grilling and roasting, respectively (Szterk and Jesionkowska, 2015). The charcoal roasting

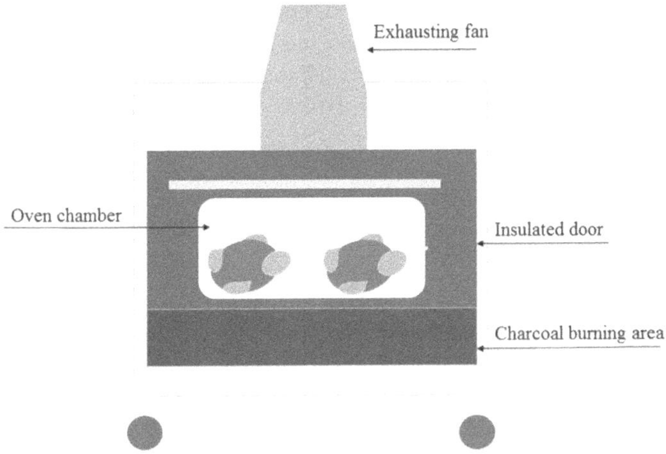


**Fig. 6.8** Charcoal roasting oven.

oven design includes stainless steel spit forks, two leg brackets, and a spine fork to hold meat in the oven. The meat can be rotated smoothly in the oven at a speed of around 4 rpm without vibrations. Due to the slow roasting process, the motor must provide rotation for a long period and this can create heat accumulation. Thus, a cooling fan may be required to prevent overheating of the motor. The charcoal roasting oven can be also simply designed by placing meat on a static rack. The side facing the fire is heated more rapidly than the other side so that flipping of the meat may be required to obtain uniform heating on both sides.

Charcoal has some drawbacks with high levels of carcinogens such as benzo(*a*) pyrene (BAP) and polycyclic aromatic hydrocarbons (PAH) being transferred to the roasted products. Charcoal roasted pork contained 10 μg/ kg PAH and 3 μg/ kg BAP on average while charcoal roasted beef contained less than 0.80 μg/ kg PAH and 0.15 μg/kg BAP on average (Chung et al., 2011). The PAH content was reported to be related to the processing time and the amount of skin (such as the skin on a chicken drumstick). A long roasting time tended to increase the risk of high carcinogens. Removing the skin before and after roasting could help to reduce the PAH intake (Chiang et al., 2020).

A design of charcoal oven for reducing those carcinogens has been developed. The design included a filter with fractionating material (York Twist packing number 779 Otto H. York Company, Inc., Fairfield, NJ, USA) placed between the fire and the product so that the product is not in direct contact with the fire. Furthermore, the filter could separate low boiling point flavoring compounds (desirable) from high boiling point carcinogenic compounds (undesirable). Thus, the produced charcoal roasted product would have a reduced content of carcinogens and is therefore healthier. From a sensory quality perspective, the roasted sample from an oven with a filter had high moistness and oily mouth coating, resulting in more preferable quality than the roasted sample from the regular charcoal oven (Sundararajan et al., 1999). In addition, avoiding meat juice directly dropping on the charcoal during roasting could reduce

PAH levels by 48%, 89%, 76%, and 85% for pork neck lean, pork belly, beef ribs, and beef loin, respectively. Installing a ventilation duct between the flame and the meat to reduce the smoke from charcoal impacting the meat could reduce PAH levels by 41%, 74%, 71%, and 73% for pork neck lean, pork belly, beef ribs, and beef loin, respectively. The reduction in PAH levels was greater for meat with a higher fat content (Lee et al., 2016).

#### 6.4.1.2 Plate roasting oven

A plate roasting oven is operated by the conduction of heat from a hot contact surface to the food. When the food is exposed to high heat flux, its physicochemical properties can be transformed, inducing a crust formation on the lower surface and deformation of the top or contact surface that causes the release of cooking exudates for the enhancement of roasted food quality. However, the crust could act as a barrier to the heat and mass transfers between the food and its surroundings (Rocca-Poliméni et al., 2019).

A plate roasting oven is composed of four major parts: cover, plate, thermostat, and heating element. The cover is on both the top and bottom of the roasting oven and the plate is under the cover. The thermostat is on the plate to regulate the roasting temperature for the optimum level. The heating element can be nickel, iron, chromium, or aluminum alloy and converts the electrical energy into heat that is then applied to cook the food (Fig. 6.9). The temperature of the heating element can vary according to its location in the oven. To increase uniformity, an insulator should be installed between cover and plate to ensure the temperature rise is uniform and within the limits of any standard (Güneş et al., 2020).

The operating temperature of the heat source should be in the range of 128°C to 255°C. After heating for 10 s, the temperature on the lower surface should be higher than 100°C to achieve the desired vaporization the energy required is 55% to 65% of the total energy (Güneş et al., 2020). Variation in the contact surface temperature causes variation in pyrazine compound generation and thereby affects the flavor and color. However, there was no clear effect of surface temperature on the tenderness of meat (Wall et al., 2019).

Similarly, a countertop electric griddle oven is designed for domestic roasting. The oven is composed of a roasting pan, a heating element, and a lid. The element is heated

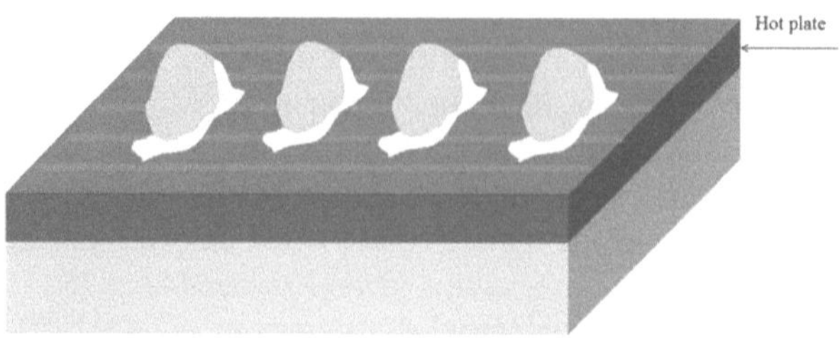


**Fig. 6.9** Plate roasting oven (countertop electric griddle).

using electrical energy that is then radiated to the food placed on the roasting pan. The lid is used to accumulate heat within the roaster. A domed lid is a newly design concept to increase the volume of food that can be covered in the roasting pan. This oven design is popular for roasting a turkey.

#### *6.4.1.3 Convection-type roasting oven*

A convection-type roasting oven can use both direct and indirect heating. For direct heating, gas is fired to generate heat into an oven chamber. Suitable convective heating media include air, gas ($N_2$), and steam (super-heated steam). For air and $N_2$, the media is injected through pipes and sprayed onto the food at 200°C to 220°C with a velocity of approximately 15 m/s for 14 min (Matsuda et al., 2013).

For convective heating using air as a heating media, the hot air can be directly circulated around the food which is normally placed at the center of the oven chamber. The direction of air circulation is controlled by a centrifugal fan to ensure uniform heat distribution in the roasting oven. An increase in airflow velocity could increase moisture and fat losses. However, variation of the air flow velocity in the range of 3.8 to 7.5 m/s did not affect the final fat and moisture losses of food when the roasting time produced a center temperature in the food of 72°C. Likewise, variation in the roasting temperature in the range of 100°C to 175°C did not have any significant effect on the final fat and moisture losses of the cooked food. Compared with a radiation type roasting oven, a convection type roasting oven has a much lower heat transfer coefficient and requires a longer cooking time. Therefore, infrared roasting prior to convective roasting could be a potential option to increase roasting efficiency. In addition, humidity control and the use of steam are recommended in a convective roasting oven to prevent too much moisture loss from the hot airflow (Braeckman et al., 2009).

To generate super-heated steam as the heating medium in the roasting oven, initially a boiler is used at 100°C to produce steam that is then gas heated to 240°C before being sprayed on the food at a velocity of approximately 27 m/s for 14 min. The heating rate of the superheated steam is not rapid and the carbonization phase begins after 12 min indicating that the roasting time should be prolonged to develop a proper Maillard browning reaction. This could be related to the absence of oxygen ($O_2$) during roasting. The oxygen is required for oxidation to induce browning coloration via a Maillard reaction. Thus, the kinetic rate constant of lightness ($L^*$) reduction during the superheated steam roasting is very low compared to a radiant heat roasting operation (Matsuda et al., 2013).

In addition, the oven can be designed to keep the food rotating in the oven chamber during roasting to improve uniformity (Fig. 6.10). Normally, roasting is operated at a not-so-high temperature, so it is a time-consuming process. The motor regulating the rotation of the product must be capable of working for long periods. With a vertical rotary design, an increase in vertical length can also increase the oven capacity per batch. Some ovens have rotating hangers to hang food for cooking. This oven design is popular for duck and chicken roasting.

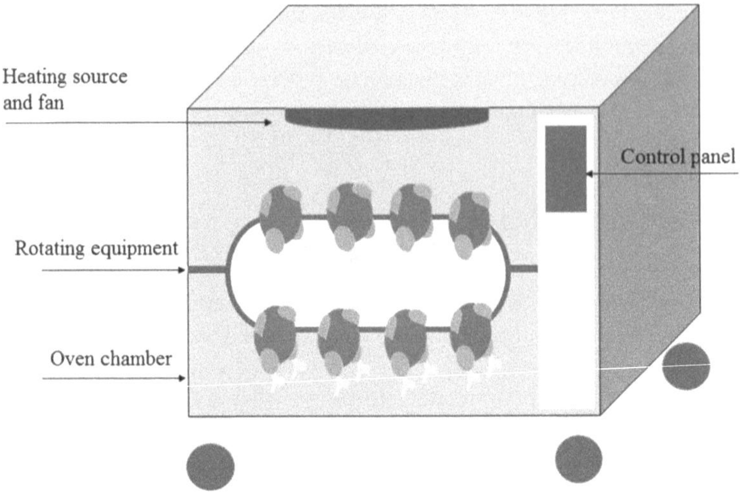


**Fig. 6.10** Convection-type roasting oven.

### 6.4.2 Continuous roasting oven

A continuous convection roasting oven or conveyor roaster or belt roasting oven can be operated by hot air which may be generated from electricity, oil-fired, LNG/LPG gas, propane or a mixture of these. The roasting can be divided into 3 zones: high-temperature roasting zone, transition zone and cooling zone. Food is placed as a layer on a perforated stainless steel conveyor (such as AISI 304 Cr Ni) that allows the hot air to pass through to the food layer. The hot air is provided by the burner. In addition, a hot air blower is installed above the conveyor to distribute the hot air evenly throughout the oven. Another hot air blower can be installed on the bottom of the conveyor to control the hot airflow to the underside of the food (Fig. 6.11). To reduce energy

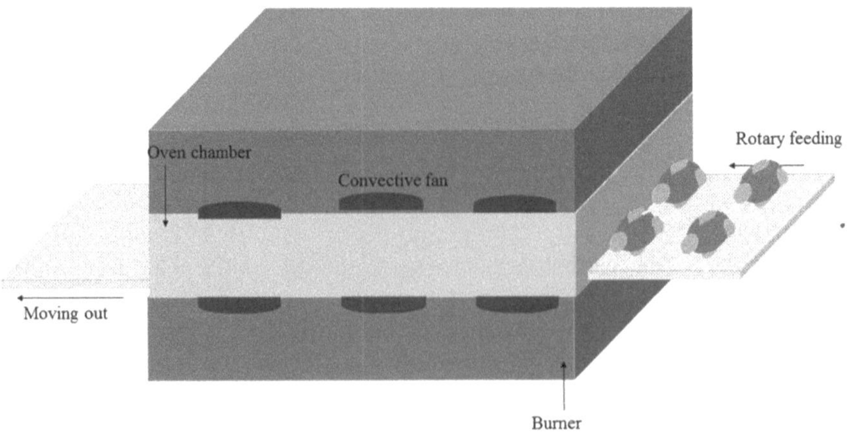


**Fig. 6.11** Continuous convection roasting oven.

consumption, hot air recirculation should be implemented to recycle the hot air into the system. Furthermore, the oven door and walls should be well insulated to reduce heat loss. The main operating parameters are the roasting temperature (95–180°C), the speed of the conveyor (around 200 mm/min), and the thickness of the food (4–10 mm). During roasting, the roasting temperature in each zone can be adjusted individually to obtain a good color, flavor and taste for the cooked product.

Roasting is completed once the food has passed through the transition zone, roasting is finished. Then the roasted food passes through the cooling zone that allows cooler air from outside to enter the oven to remove heat from the roasted foods. Eventually the surface temperature of the roasted foods should be cooled to ambient temperature before being sent to a packing machine. A continuous convection roasting oven is good for roasting peanuts, cashew nuts, walnuts, hazelnuts, sunflower seeds, pumpkin seeds, and almonds among other products.

In addition to single conveyor roasters or single belt roasting ovens, a continuous roasting oven can be designed as a multilayer conveyor roasting oven. The multilayers allow a greater amount of food to be placed in the oven. The main advantages of this design are to save the space as the length of the oven can be shortened and the energy efficiency should be improved. For example, the energy consumption of LPG and gas may be 5 to 12 kg/h and 5 to 15 $Nm^3$/h, respectively, and around 94 kW for electric power (Servval, 2020).

## 6.5 Novel baking and roasting ovens

To improve baking quality and efficiency, other oven types have been developed using novel technology such as infrared, microwave, air impingement, and combined techniques.

### 6.5.1 *Infrared oven*

Infrared which is a rapid heating technique can be applied in the oven chamber for baking, to increase the heating rate of the radiant heat. Instead of the radiant heat being transferred from hot surfaces, it is generated from a halogen lamp that produces near-infrared. Near infrared wavelengths have a small penetration depth, compared with microwaves, although both are electromagnetic waves. Heat is generated and accumulated on the cooking surface, resulting in a dry and crispy crust (Fig. 6.12). The food product for infrared baking should be thin and flat with a broad surface area such as pizza.

A far-infrared oven can be used to supply radiant heat for roasting. This type of oven is composed of a heat source generally on the top to irradiate the food in the chamber. The distance between the heat source and the food should not be more than 8 mm and the roasting time should not be longer than 10 min. The shorter roasting time is a result of the fast heating rate compared with a convection oven. Although the temperature range (23.0–159.5°C) is much lower than for superheated steam roasting (22.2–232.2°C), color development in food cooked using a far-infrared oven is much

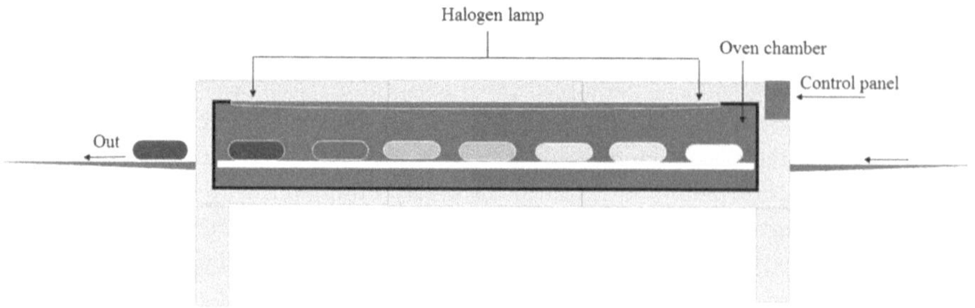


**Fig. 6.12** Continuous infrared oven.

faster. The carbonization phase in far-infrared oven roasting begins after 8 min, instead of after 12 min for superheated steam roasting. The lack of oxygen during roasting is not a factor in far-infrared oven roasting; thus, the kinetic rate constant of the L* is much higher than for convection-type roasting using superheated steam as the heating medium (Matsuda et al., 2013).

For an infrared oven with single-side roasting, the roasting time is divided into 2 periods, with the first for one side of the food and the second after flipping the food to cook the other side. The roasting time for each side can be the range of 20 to 150 s (Braeckman et al., 2009). In addition to having a rapid processing time, infrared as a heat source in the roasting oven avoids benzo[*a*]pyrene (BaP) formation with roasted pork belly and this could be health benefit (Park et al., 2017).

## 6.5.2 Microwave oven

A microwave baking oven might be a good option to increase the heating rate for large-sized products. In the oven, a magnetron is used to generate microwaves that can penetrate deep into the food being cooked that should have a high moisture so its high dielectric properties result in high energy absorption and its subsequent effective conversion to heat (Fig. 6.13). The inner layer of the food tends to be rapidly heated and completely baked whereas the outer later is slowly heated and so does not form a substantial dry crust. Thus, to obtain proper crust formation, the microwaves should be applied together with convective heat or irradiated heat to reduce

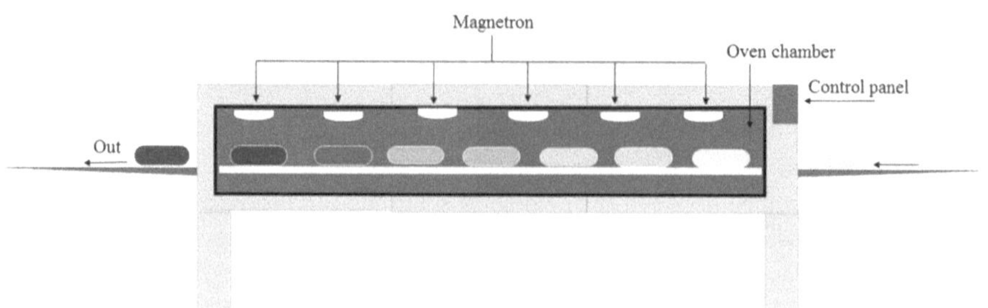


**Fig. 6.13** Continuous microwave oven.

water condensation on the surface and increase the surface temperature for a non-enzymatic browning reaction to occur. In addition to shortening the baking time, microwave baking can produce bakery products with an increased content of slowly digested starch. For example, with gluten-free bread made from rice flour, microwave baking could affect the starch-protein interaction and thereby reduce the glycemic index (GI) from around 90 (conventional hot air baking) to below 70. With a GI below 75, microwave-baked bread can be categorized as a medium GI food (Therdthai et al., 2016). However, retrogradation seems to be enhanced by microwave baking as the crystallinity of the starch in gluten-free bread was significantly changed during storage for 3 d at 4°C (Srikanlaya et al., 2017) and this may be disadvantage of microwave baking that may concern some users.

A continuous roasted microwave oven is designed to for roasting foods such as beans, nuts, and coffee on a mesh conveyor that moves forward into the microwave oven. The microwaves operate at a frequency of 2450 Hz. Magnetrons are installed along the length of the roasting oven to provide microwave energy for cooking the food and heat can be generated in the food rapidly. The length of the oven and the speed of conveyor must be properly controlled to ensure properly roasting. In addition, microwave heating was reported as a potential method to reduce carcinogenic compounds to 65 ng/g PAH from 177 ng/g PAH (charcoal heating of chicken). A similar trend was also observed with beef (Farhadian et al., 2011). The advantages of a continuous roasted microwave oven include: continuous production (first-in, first-out), safe operation, clean (no pollution), zone operation, time and energy saving, and a sterilization effect.

### 6.5.3 Impingement oven

An impingement oven is designed to speed up the heating rate in a convective oven by increasing the air velocity, as a high air velocity can increase the heat transfer coefficient and thereby the heating rate of the oven. The heat transfer coefficient in an impingement oven is approximately 150 to 250 $W/m^2K$ which is much higher than for a conventional hot air oven (approximately 100 $W/m^2K$) (Shevade et al., 2019). Thus, the rapid cooking is the main advantage of this design. A possible drawback is the high weight loss due to the high air flow velocity that enhances rapid moisture transfer from the cooking surface to the oven chamber. Generally, the air velocity is controlled in the range of 10 to 15 m/s. However, the air is not circulated throughout the oven as is usual for a convective oven. Instead, the air is actually forced to flow through nozzles to increase its pressure and subsequently its velocity. The forced air can flow directly to the cooking surface which can then progressively heat the inner layer (Fig. 6.14). Model simulation indicated that impingement could reduce the energy required by 12% to 16% compared with hot air baking (Li et al., 2013).

An impingement oven can be designed for batch operation or continuous operation. Food with a large surface area exposed to the forced air (such as pizza) is suited to this oven design. However, air impingement was recently applied to muffins. This was advantageous not only for the speed of cooking, but also for the reduction in the fat content in the muffin formula while still retaining the same height and volume

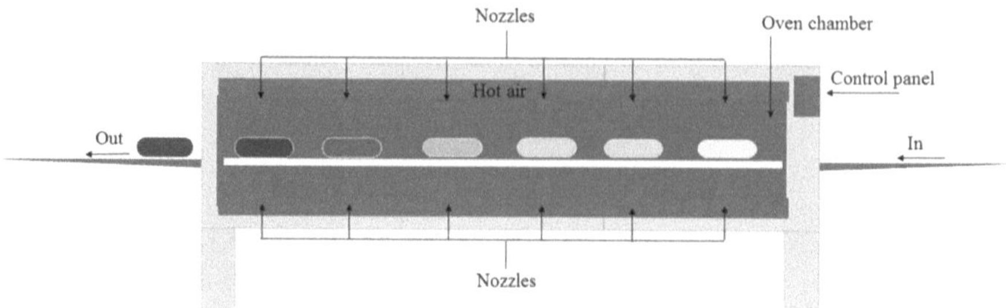


**Fig. 6.14** Air impingement oven.

as the regular muffin (Carrascal et al., 2019). Thus, an impingement baking oven has the potential to reduce the use of fat in bakery products.

### 6.5.4 Hybrid oven

Hybrid baking oven is designed to improve baking performance and save energy. For example, cooking crackers and biscuits requires a high level of heating from the beginning. Thus, the hybrid baking oven is designed to provide direct radiant heat in the beginning zone while the later zone uses indirect radiant heat. The direct radiant heat is applied for one-third of the cooking time and consumes 50% of the total energy for the whole baking process, whereas the remaining two-thirds of the baking time use the other 50% of total energy to provide indirect radiant heat. For small, flat food with some sugar, a lower heat is required to prevent burning. Thus, the hybrid baking oven is suitable for these semisweet products with radiant heat in the beginning zone for one-third of the baking time and convective heat for the remaining period (Davidson, 2018).

Microwaves are often used together with convective heating to obtain better baking quality as the microwaves can generate heat rapidly and save on baking time. This makes fresh baking and food service possible. The convective heating provides the dry heat from the outer layer to the inner layer. With the assistance of the convective heating, proper crust formation becomes possible in a microwave oven. Similarly, infrared is sometimes applied together with microwave baking to develop a dry, crispy brown crust on the bakery product. As infrared is categorized as a fast heating method, not only does it ensure a suitable crust following microwave baking, it also saves on baking time. Air impingement is another heating method that makes it possible to create a golden-brown crust quickly in microwave baking. Therefore, hot air-assisted microwave baking, infrared-microwave baking, and impingement-microwave oven are hybrid methods that require a shorter baking time (Fig. 6.15). The combination between infrared and air impingement has been also studied. However, the muffin from this hybrid oven was humid and elastic. In addition, the oily feeling was not improved, compared with a muffin produced using a baking oven with only air impingement (Carrascal et al., 2019).

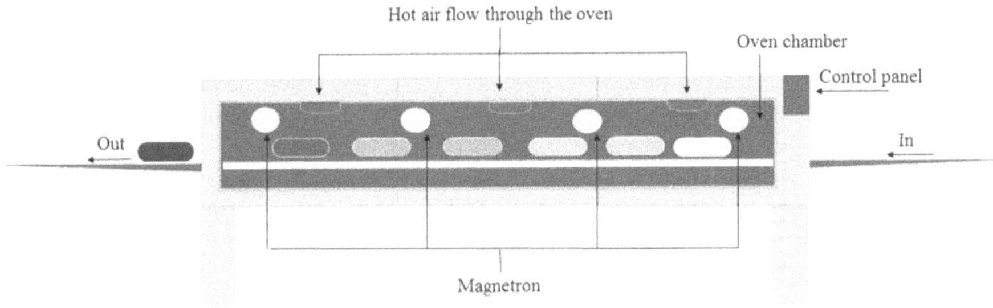


**Fig. 6.15** Hybrid oven.

## 6.6 Conclusion

Baking and roasting ovens have been invented and developed since ancient time. Initially, wood burning was used to generate heat for baking and roasting. Later, the energy source changed to gas, fuel oil, and electromagnetic waves. The heating modes cover conductive, convective, and radiative heating. Oven capacity was initially small, focusing on household applications as baked and roasted products became popular. Then, oven capacity expanded to accommodate mass production. For small-scale production, the oven was usually designed for batch operation (charcoal, convective, deck, and rack types). For industrial-scale production, the preference was for continuous ovens (tunnel and traveling-tray types). In addition, some novel ovens have been invented to reduce processing time such as the microwave, infrared, impingement, and combi types of ovens, with each type having its unique characteristics. Selecting the proper oven to suit the application is key to achieving the best quality of baked and roasted products.

## References

Anonymous, 2020. A history of the bread oven. https://www.mot.be/en/opzoeken/bakovens/geschiedenis/geschiedenis-van-de-bakoven. (Accessed 22 November 2020).

Berk, Z., 2018. Frying, baking, and roasting. In: Food Process Engineering and Technology, third ed. Academic Press, ISBN: 978-0-12-812018-7, pp. 583–590 (742 pp., Chapter 24).

Bernal, M.A., 2016. A Brief History of Baking. https://maryannbernal.blogspot.com/2016/10/a-brief-history-of-baking.html. (Accessed 22 November 2020).

Braeckman, L., Ronsse, F., Hidalgo, P.C., Pieters, H.J., 2009. Influence of combined IR-grilling and hot air cooking conditions on moisture and fat content, texture and color attributes of meat patties. J. Food Eng. 93, 437–443.

Carrascal, A., Rasines, L., Ríos, Y., Rioja, P., Rodríguez, R., Alvarez-Sabatel, S., 2019. Development of reduced-fat muffins by the application of jet-impingement microwave (JIM) technology. J. Food Eng. 262, 131–141.

Charrington-Hollins, S., 2020. Life Through the Baker's Window. https://www.dawnfoods.com/uk/campaigns/seren-charrington-hollins—100years-of-bakery/life-through-the-bakers-window-over-the-last-century. (Accessed 22 November 2020).

Chiang, C.-F., Hsu, K.-C., Cho, C.-Y., Tsai, T.-Y., Hsu, C.-H., Yang, D.-J., 2020. Comparison and establishment of appropriate methods to determine EU priority PAHs in charcoal-grilled chicken drumsticks with different treatments and their dietary risk assessments. Food Chem. Toxicol. 142, 111400.

Chung, S.Y., Yettella, R.R., Kim, J.S., Kwon, K., Kim, M.C., Min, D.B., 2011. Effects of grilling and roasting on the levels of polycyclic aromatic hydrocarbons in beef and pork. Food Chem. 129, 1420–1426.

Davidson, I., 2018. Baking oven. In: Biscuit, Cookie and Cracker Production. Academic Press, ISBN: 978-0-12-815579-0, pp. 75–91, https://doi.org/10.1016/B978-0-12-815579-0.00009-X. 244 pp., Chapter 9.

Farhadian, A., Jinap, S., Hanifah, H.N., Zaidul, I.S., 2011. Effects of meat preheating and wrapping on the levels of polycyclic aromatic hydrocarbons in charcoal-grilled meat. Food Chem. 124, 141–146.

Güneş, D., Kükrer, E., Yıldırım, C., 2020. Experimental and computational design improvement assessment of a household toast grill. Therm. Sci. Eng. Prog. 19, 100609.

Hurwood, J., 2015. A Brief History of Ovens through the Ages. https://www.canstarblue.co.nz/appliances/ovens/a-brief-history-of-ovens/. (Accessed 22 November 2020).

International Energy Agency (IEA), 2020. Appliances and equipment. https://www.iea.org/fuels-and-technologies/appliances-equipment. (Accessed 22 November 2020).

Lee, J.-G., Kim, S.-Y., Moon, J.-S., Kim, S.-H., Kang, D.-H., 2016. Effects of grilling procedures on levels of polycyclic aromatic hydrocarbons in grilled meats. Food Chem. 199, 632–638.

Li, X.-D., Alamir, M., Witrant, E., Della-Valle, G., Rouaud, O., Boillereaux, L., Josset, C., 2013. Further investigations on energy saving by jet impingement in bread baking process. In: Proceedings of the 5th IFAC Symposium on System Structure and Control, Grenoble, France, February 4–6, 2013.

Matsuda, H., Llave, Y., Fukuoka, M., Sakai, N., 2013. Color changes in fish during grilling – influences of heat transfer and heating medium on browning color. J. Food Eng. 116, 130–137.

Park, K.-C., Pyo, H., Kim, W., Yoon, K.S., 2017. Effects of cooking methods and tea marinades on the formation of benzo[a]pyrene in grilled pork belly (Samgyeopsal). Meat Sci. 129, 1–8.

Rocca-Poliméni, R., Vilet, Z.N., Roux, S., Jean-Luc Bailleul, Broyart, B., 2019. Continuous measurement of contact heat flux during minced meat grilling. J. Food Eng. 242, 163–171.

Servval, 2020. Multilayer Conveyor Roasting Machines. https://www.sevvalmakine.com/en/7/srx90-nuts-roasting-machine-2.html. (Accessed 22 November 2020).

Shevade, S., Rahman, M., Guldiken, R., 2019. Optimization of turbulent air jet impingement for energy efficient commercial cooking. Energy Procedia 160, 691–698.

Srikanlaya, C., Therdthai, N., Ritthiruangdej, P., Zhou, W., 2017. Effect of butter content and baking condition on characteristics of the gluten-free dough and bread. Int. J. Food Sci. Technol. 52, 1904–1913.

Sundararajan, N., Ndife, M., Basel, R., Green, S., 1999. Comparison of sensory properties of hamburgers cooked by conventional and carcinogen reducing 'safe grill' equipment. Meat Sci. 51, 289–295.

Szterk, A., Jesionkowska, K., 2015. Influence of the cold storage time of raw beef meat and grilling parameters on sensory quality and content of heterocyclic aromatic amines. LWT Food Sci. Technol. 61, 299–308.

Therdthai, N., Tanvarakom, T., Ritthiruangdej, P., Zhou, W., 2016. Effect of microwave assisted baking on quality of rice flour bread. J. Food Qual. 39, 245–254.

Wall, K.R., Kerth, C.R., Miller, R.K., Alvarado, C., 2019. Grilling temperature effects on tenderness, juiciness, flavour and volatile aroma compounds of aged ribeye, strip loin, and top sirloin steaks. Meat Sci. 150, 141–148.

# Section C

# Thawing operations

# Fundamentals of thawing processes for frozen foods

7

*Nusrat Jan, Gousia Gani, Omar Bashir, Tawheed Amin, Nazrana Rafique Wani, Abida Jabeen, and Taha Mukhtar*
Division of Food Science and Technology, SKUAST-Kashmir, Srinagar, India

## 7.1 Introduction

Converting frozen water into liquid is essentially referred to as thawing of food products. It entails heat transfer to a frozen commodity by which the ice that developed within the food commodity during the freezing process is melted (Archer et al., 2008). The phase change involving ice-to-water conversion needs a great deal of energy and typically occurs in foods at glide temperature. Since energy is required in a large quantity to accomplish this change in phase, the thawing products at this glide temperature often requires extended amounts of time, as shown in Fig. 7.1 (Haugland, 2002).

Thawing is totally opposite to freezing. In freezing, heat is extracted from food product, whereas in thawing heat is supplied to the product. Thawing, however, is a very complicated operation and its management is laborious regarding thawing food items, and if the quality of material is to be retained as much as possible, great care should be taken. Theoretically, thawing is the reverse of the freezing process; they are not only distinct in heating and cooling process and phase change direction, but also in variations of internal temperature and in time required (Min, 2001). Similar to freezing, thawing must be performed quickly in order to preserve product quality; however, it must not be so fast that it has a detrimental effect on the commodity (Archer et al., 2008).

Freezing is widely used in the food industry to improve the shelf life of food items and to maintain their quality during storage and transport. Due to its effectiveness in preserving product quality, food storage by freezing has been employed for thousands of years (Persson and Londahl, 1993). In general, the quality of frozen food is determined by the cycles of freezing and thawing. All frozen food items, except ice cream, must be thawed or tempered above their freezing temperature for further processing or direct consumption. However, the thermal conductivity of water (0.6 W/mK) is slightly lower as compared to ice (2.25 W/mK), and so thawing food requires a comparatively longer period of time than freezing. During thawing, the surface layer of the frozen food thaws first, which creates a film of water that acts as a thermal insulator and slows down the thawing phase. Compared to freezing, the larger thawing period will adversely influence frozen foods (Park et al., 2006). The energy costs rise as the time of thawing of foods increases, and in addition, microbial growth on the food

High-Temperature Processing of Food Products. https://doi.org/10.1016/B978-0-12-818618-3.00006-9

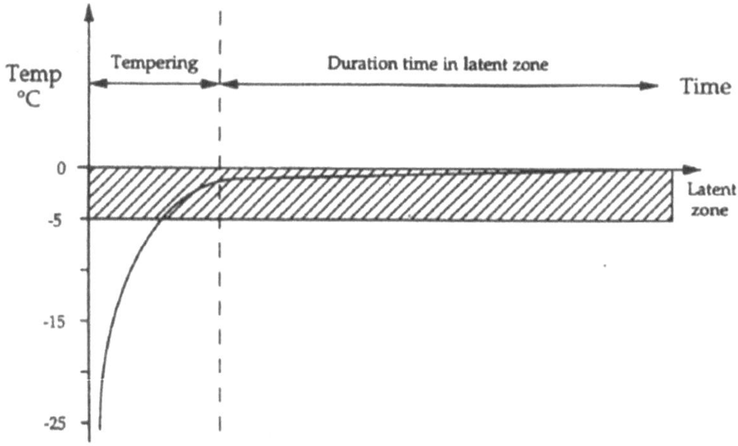


**Fig. 7.1** Glide temperature range of food thawing.

surface increases, nutritional significance due to solubilization of vitamins decreases and proteins get leached, thereby resulting in the production of waste water high in biological and chemical oxygen demands (Icier et al., 2010; Roberts et al., 1998).

There are many reasons why it is more complex to thaw than to freeze. First of all, as the thawing process continues, the thermal resistance of the commodity increases. In freezing, most of the product's outer layers begin to freeze and create a conduction surface through which the heat escapes, which is the opposite of what occurs in thawing. Secondly, thawing cannot be performed at any temperature. That is, in order not to affect product quality, the difference between the original temperature of the frozen product and the temperature of the thawing medium can only occur at a certain interval. This is particularly necessary in operations where surface heat is applied to the product.

Meat and meat products have an exceptional place in human nutrition. Because of their natural structure, meat and meat products compose an appropriate environment for the reproduction. Since frozen meat is widely used in the meat industry, considerable attention must be given to the thawing process. The water fraction of meat is primarily influenced by the freezing and thawing process. Since the water is present in intermuscular and intramuscular fibers of the meat, compartments are created in the tissue, which complicates the process. The homeostasis of the complex meat system gets disturbed when the concentration of the residual solutes (lipids, carbohydrates, proteins, minerals, and vitamins) rises as water freezes (Lawrie, 1998). The characteristics of the cell membrane are affected due to the changes in the internal environment of muscle fibers, thereby affecting the meat quality (Fellows, 2000). In particular, problems occur with large frozen blocks (e.g., 25 kg) in the industrial-scale thawing of frozen meat, resulting in non-uniform thawing (Fellows, 2000). The food industry has been increasingly dependent on the use of thawed meat in meat processing recently. Therefore, it is extremely necessary to reduce all the changes encountered by meat by rapid thawing in order to obtain meat with the technological attributes closest to refrigerated meat (Gambuteanu et al., 2013). Various researchers have demonstrated techniques in recent years that allow for rapid thawing. Rapid thawing at low temperatures is preferable in order to reduce the changes in nutrient content and quality during the thawing operation (Zhu et al., 2004).

The methodologies and technologies employed for thawing and freezing processes are necessary to preserve the quality of products. The primary objective of freezing in terms of preservation and customer preferences is the conservation of fresh qualities such as color, texture, and juiciness (Alizadeh et al., 2007). In addition, it is essential that during thawing, the surface of the product must not get too hot, as this may speed-up spoilage. Thus, once the product is thawed, it should be stored chilled or immediately processed (Archer et al., 2008).

## 7.2 Quality evaluation of thawed food

### 7.2.1 Texture

An essential aspect of a products' acceptability to consumers is texture. In the case of meat products, texture is very important as it affects the juiciness of meat. In this respect, many studies showed that the freezing process as well as the meat aging rate before and after freezing could contribute to changes in meat texture after thawing. Shanks et al. (2002) demonstrated variations between the shear force of unfrozen chilled aged meat and that of frozen stored meat for 2 months, constituting a lower shear force for the latter. These findings were believed to be a consequence of the weakening of muscle cells due to the formation of ice in intra-cellular spaces during freezing, which contributed to a decrease in the shear force in thawed and frozen products.

Dransfield (1994) found that the freezing rate has a pivotal role in meat aging; rapid freezing enhanced the rate of aging by three times compared to chilled beef, unlike slow commercial freezing. These findings were justified in this situation by the appearance of cellular lesions, thereby increasing the rate of aging (Vieira et al., 2009).

### 7.2.2 Drip loss

A very important aspect in meat industry—especially from a financial point of view—is the drip loss after thawing. This is the reason why it is important to recognize the variables that cause drip loss after thawing. The influence of the rate of freezing on drip loss was shown by Ngapo et al. (1999). It has been shown that the drip loss was the same for refrigerated meat as for fast-frozen pork (for 12–120 min). However, the drip loss was slightly greater in slowly frozen meat (for 240–900 min) than for refrigerated meat. Sakata et al. (1995) stated that no connection was observed between freezing temperature ($-20°C$ and $-80°C$) and drip loss and no major changes had been recorded with respect to freezing temperature. Petrovic et al. (1993) conducted a study of freezing velocity and it was shown that major deteriorations of fibers and microfibers occurred in both cases (slow freezing and quick freezing), reduced solubility of myofibrillar proteins, and substantial loss of thawing.

The freezing process of fishery products is generally considered more critical than thawing for drip loss, which is known to be related with the formation of large intracellular ice crystals, causing greater destruction to the walls of individual cells. During frozen storage, and particularly long frozen storage periods, slow freezing rates, and high and fluctuating temperatures are considered to be correlated with the development of large ice crystals (Archer et al., 2008). In minimizing drip loss, it must be taken into account that very quick thawing is associated with the increase in drip loss, likely due to reduced drip reabsorption time. Water reabsorption is a slow process that can take numerous hours, especially for seafood, whose muscle structure, after it has been frozen, has less reabsorption capacity. However, long thawing periods for the majority of fish species are unlikely to be practically advantageous (Archer et al., 2008). As a consequence, the utilization of quick freezing techniques, good temperature management in all phases of handling and processing, and well-controlled frozen storage conditions is certainly the best way to reduce thawing drip loss in fishery products (Archer et al., 2008).

### *7.2.3 Microbial growth*

The amount of viable microbes present in meat does not tend to be decreased by either freezing or thawing. Microbial spoilage is efficiently eliminated during freezing, however, as the microbes become inactive. Unfortunately, during thawing, they resume their activity (Londahl and Nilaaon, 1993). Since thawing is considered to be a less uniform and far slower process than freezing, microbial growth will be subject to more favorable temperature conditions in certain areas of the meat. Vieira et al. (2009) observed that beef previously aged for 3 to 10 days and frozen for up to 90 days did not spoil due to microbial growth. However, they reported a rise in the levels of psychrotrophic bacteria during the 90 day frozen storage that were probably favored by the thawing process over other bacteria (48 h at 4°C in a cooler). Greer and Murray (1991) found that in frozen/thawed pork, the lag phase of bacterial growth was shorter than for fresh meat, but that the time to develop spoilage odors was not affected.

## 7.3 Conventional/traditional techniques of thawing

There are number of techniques by which the process of thawing can be accomplished. The conventional/traditional thawing methods include air thawing, water thawing, and vacuum thawing. However, these techniques are slow, inconvenient, and expensive.

### *7.3.1 Still air thawing*

The widely used method of conventional thawing is still air thawing because its implementation is easy. However, the process and the respective ambient temperature under which still air thawing is carried out affect the raw material quality. A boundary layer around the product is formed as they are laid out to thaw, resulting in the product thawing with cooler air temperatures near to its surface. Due to this, the overall thawing process becomes longer, as the heat transfer from the product to the environment is considerably slowed down. This method involves little or no cost to thaw food products, which is one of the reasons why this method is still used (Archer et al., 2008). However, this method has some limitations, viz., it involves high drip loss, is generally practicable only on a small scale, suffers accelerated bacterial growth due to slower thawing, and thawing is slow compared to other methods (Archer et al., 2008).

### 7.3.2 Forced air/blast thawing

Thawing products using forced air/blast thawing involves stimulation of the air around them. Unlike still air thawing, where a boundary layer is formed, here, no formation of boundary layer takes place, due to the circulation of air, which isolates the frozen product from warmer air. Due to this, the overall efficiency of the process is increased, as the convection coefficient increases from around 15–30 to over 100 $W/m^2$ K (Archer et al., 2008). This method has some drawbacks, viz., danger of surface drying/overheating, products might need tempering, and hard to acquire uniform thawing (Archer et al., 2008).

### 7.3.3 Water thawing

The easiest and cheapest way of thawing products of all types is thawing in warm water, provided that clean water in an ample quality should be available. Thawing of products can be done by spraying or by immersion, or by their combination. The immersion of frozen products in water used by commercial companies is the most common practice of thawing methods, aiming to deliver high quality products by thawing large quantities of produce. However, storing blocks in a tank overnight or in highly sophisticated tanks with water filtering and temperature control systems largely depends on the way the process is carried out. The issue with this implementation of the water immersing method is also the fact that since there is little recirculation of the water and essentially no filtration, bacteria from guts and blood finds it easy to go from one product to another and contaminate the whole tank. The danger of contamination is always there, however, a thawing medium with lower temperatures would limit the growth of microorganisms under such conditions (Archer et al., 2008).

### 7.3.4 Vacuum thawing

Vacuum thawing is a process which involves heat transfer through the surface, but cannot be considered as a conventional procedure due to its nature. With conventional methods, there is always a risk of uneven thawing of products. However, this method is very efficient and relatively hygienic for small products with high surface area to volume ratio. By adjusting pressure, temperature is controlled within the chamber; however, controlling transfer of heat into the product is quite a difficult task.

The major drawbacks of this method are unsuitability for large batches or quantities of food products and non-uniform thawing (Archer et al., 2008).

## 7.4 Novel techniques of thawing

To avoid problems associated with long thawing times and changes in quality characteristics, rapid thawing techniques are preferred (Table 7.1) such as high-voltage electrostatic field (HVEF) thawing, ultrasound-assisted (UA) thawing, ohmic heating (OH) thawing, high pressure (HP) thawing, microwave thawing (MW), and radiofrequency (RF) thawing (Mousakhani-Ganjeh et al., 2015, 2016a) (Fig. 7.2).

**Table 7.1** Applications of novel techniques in food.

| Parameters | Food material | Effect | References |
|---|---|---|---|
| High pressure (HP) thawing | Fish, chicken breast | Thawing time decreases, thawing loss decreases, and relaxation time increases | Zhu et al. (2014); Li et al. (2014) |
| Ultrasound-assisted (UA) thawing | Livestock meat | No significant differences in microbiological, chemical, and textural attributes | Gambuteanu and Alexe (2013) |
| Ohmic heating (OH) thawing | Beef meat | Electrical conductivity values decrease effectively as the fat content of frozen minced beef meat increases | Cevik and Icier (2018) |
| Radiofrequency (RF) thawing | Lean beef meat | Heating uniformity improves and thawing time decreases | Bedane et al. (2017) |
| Microwave (MW) thawing | Pork meat | Thawing time reduces | Kondratowicz et al. (2008) |
| High-voltage electric field (HVEF) thawing | Tuna fish | Oxidized quicker at high electrostatic filed intensities; the increasing prevalence of thiobarbituric acid-reactive substance is accompanied by an increase in voltage and reduction of electrode gap | Mousakhani-Ganjeh et al. (2016b) |

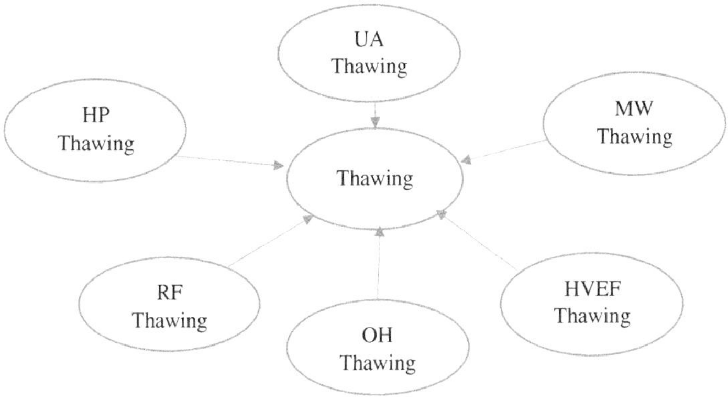


**Fig. 7.2** Novel techniques of thawing.

### 7.4.1 High pressure (HP) thawing

A modern application of thawing in the food industry is HP thawing (Fig. 7.3). Although HP thawing is well known, it has not received significant attention compared to HP freezing, several studies has recently shown that HP thawing can maintain the quality of food and decrease the necessary thawing time (Makita, 1992; Zhao et al., 1998), demonstrating its importance for the food sector. Makita (1992) observed that at atmospheric pressure, HP thawing of frozen meat only needed one-third of the time, but yielded sensory characteristics equivalent to those of traditional thawed items. The improvement in the texture of konnyaku (konjac glucomannan gel) was recorded by Teramoto and Fuchigami (2000) when frozen konnyaku was thawed at 200–400 MPa. HP thawing in frozen tofu was also more efficacious in texture enhancement than atmospheric pressure thawing. The drip loss of beef was too small to observe during HP thawing and there were no adverse effects in terms of cooking loss, penetration force, or color of thawed beef (Zhao et al., 1998). Microbial growth was inhibited by HP treatments of 100 MPa (Hite, 1899). The pressure, however, reduced salt bonds and hydrophobic interactions, which could lead to denaturation of proteins and degradation of texture (Fernandez-Martın et al., 2000; Wen et al., 2015). Various studies have shown that water-holding capacity is also enhanced by HP thawing (Li et al., 2014; Okamoto and Suzuki, 2002).

It is of high importance to mention that in certain cases such as fish fillets, the organoleptic attributes of fish thawed with a HP of 200 MPa do not vary compared to fish thawed at 15°C in water and that the loss of thawing is minimal (Schubring et al., 2003). Moreover, the microbiological benefit of HP thawing is also provided by Otero and Sanz (2003), namely that HP and low temperature may have a synergetic impact on microorganism lethality. Research has also proved that HP thawing retains the characteristics of the commodity and decreases thawing time, indicating its great importance for the industry.

The rate of thawing varies with respect to heat conduction, as pressure is evenly distributed throughout the sample (Kalichevsky et al., 1995). Zhao et al. (1998) found that the degree of pressure and the duration of treatment inevitably affect the rate of

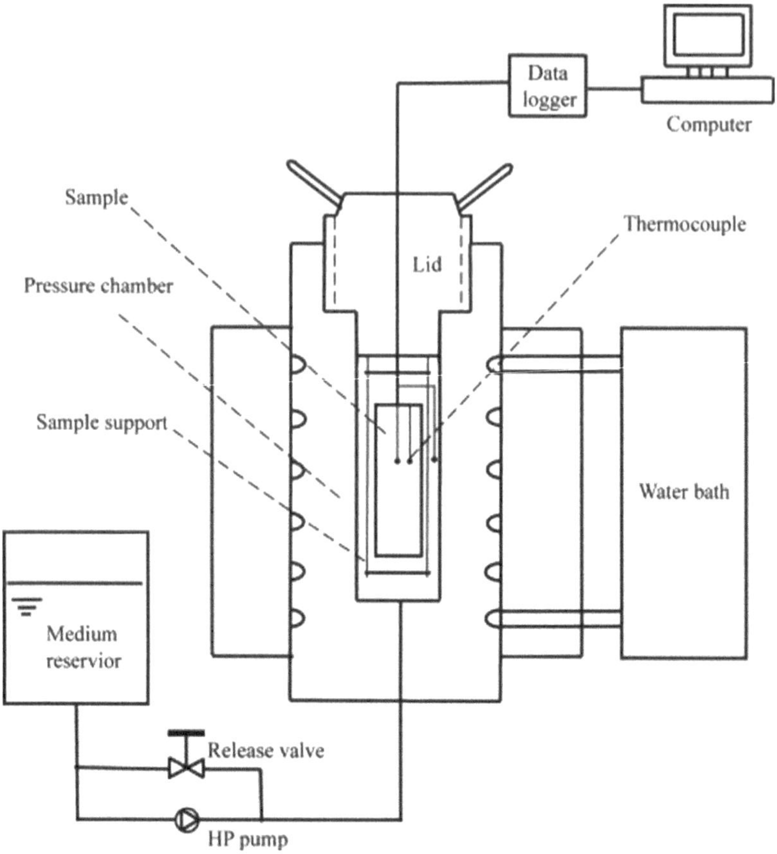


**Fig. 7.3** Schematic representation of high pressure thawing (Zhu et al., 2014).

thawing and the quality of the product, while the properties of the product, including initial temperature and size, did not influence the rate of thawing, suggesting that thawing a larger volume of product at HP is beneficial. HP thawing is primarily limited by high costs, as is HP freezing, pressure-induced protein denaturation, and meat decoloration (Kalichevsky et al., 1995; Mertens and Deplace, 1993). Research on fundamental data governing the process of HP thawing and its optimization are, therefore, critical for its commercial application.

## *7.4.2 Ultrasound-assisted (UA) thawing*

In the food sector, ultrasonic processing has emerged in recent decades as an innovative and promising technology (Fig. 7.4A). During the UA thawing process, the front of the phase change approaches faster than in conventional thawing (Fig. 7.4B). However, some drawbacks related to ultrasound, like high power consumption, localized heating, and poor penetration, are hampering its wider application (Li and Sun, 2002).

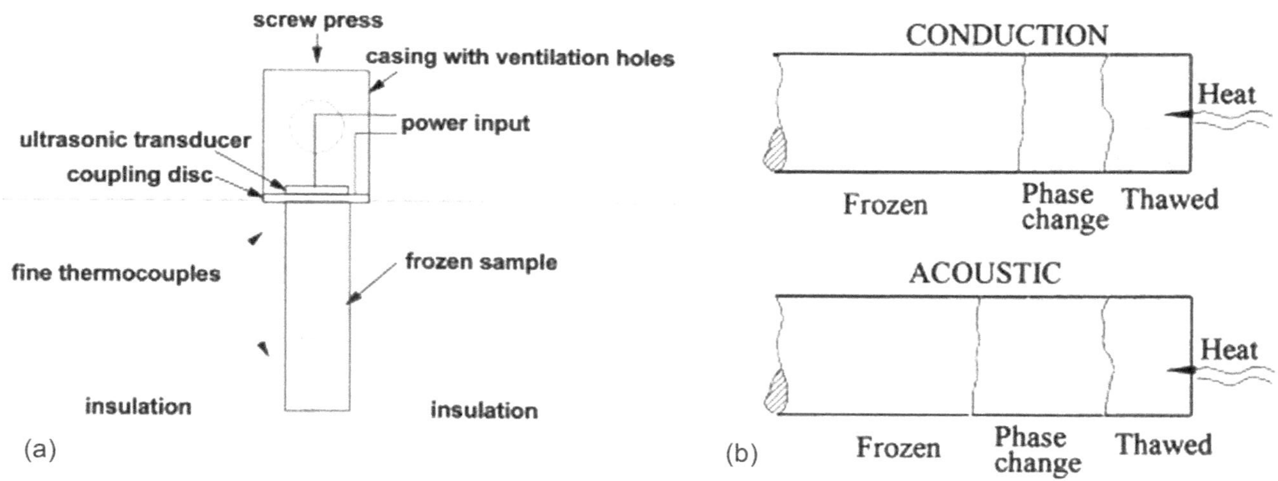


**Fig. 7.4** (A) Schematic representation of UA thawing (Miles et al., 1999) and (B) effect of UA thawing on the phase change region (Li and Sun, 2002).

Several research studies have shown that the application of ultrasound for rapid meat thawing has proved very beneficial. A decrease in thawing time was obtained in this regard by Kissam (1984) and Miles et al. (1999). A patent filed by Kissam in 1984 showed that the thawing phase of fish blocks immersed in water was shortened due to the application of ultrasound at low power and frequencies. Thermal transfer was stimulated by the employment of 60 W power and 1500 Hz frequency to reduce the time kept at phase change temperatures from −5°C to −1°C by 82% (Kissam, 1984). Miles et al. (1999) recorded that, with intensity and frequency, the speed of heating in thawed regions increases and when transmission is parallel to muscle fiber, it is higher. It has been found that when combining 500 kHz frequency and 0.5 W/cm$^2$ intensities, thawing takes place without extreme surface heating. In addition, several research papers have concentrated on the employment of ultrasound in the meat industry.

The influence of ultrasound on the physical characteristics of chilled meat has been thoroughly investigated, but the findings published are inconsistent (Jayasooriya et al., 2007; Got et al., 1999; Lyng et al., 1997; Stadnik et al., 2008). Lyng et al. (1997) assessed the ultrasonic methods on the muscles of longissimus, biceps femoris, and semitendinosus, and observed that there were no major effects on the collagen solubility, beef tenderness, and proteolysis of myofibrillar proteins by employing ultrasonic water baths with frequencies of 30 to 47 kHz and intensities of 0.29 to 0.62 Wcm$^2$. On the other hand, Jayasooriya et al. (2007) reported that the ultrasound treatment significantly reduced hardness, cook and total loss of beef muscle. Furthermore, Cheng et al. (2014) stated that the thawing time of frozen edamames can be significantly decreased by UA thawing.

### 7.4.3 Ohmic heating (OH) thawing

Heat is created immediately within the food when an electric current with high electrical resistance passes through it, thereby raising the temperature of the food product (Fu and Hsieh, 1999). This heating technique is known as electro-heating or ohmic heating (Fig. 7.5). The ohmic heating application receives special attention in the food industry for pasteurization and aseptic processing of particulate foods as it produces effectively reliable results compared to microwave heating; this is because almost all of the energy reaches the food as heat and ohmic heating has no limit to depth of penetration. In addition, ohmic heating has several benefits including high rate of heating, volumetric heating, high energy conversion efficiency, etc., over conventional/traditional heating (Reznick, 1996; Fellows, 2000). In 1920, the first industrial ohmic heating applications for thermal processing were introduced. The method has been licensed for blanching of vegetables, continuous pasteurization of milk, and direct electric heating of sausages. Although high electricity costs and the rise in alternative maintenance thermal processes, including the UHT process, have resulted in a reduction in the use of ohmic heating, it has been further developed for the single objective of thawing food (Jaeger et al., 2016).

An innovative technique is the use of ohmic heating to thaw frozen food. Ohtsuki (1991, 1993) patented an ohmic thawing process where frozen foods positioned with

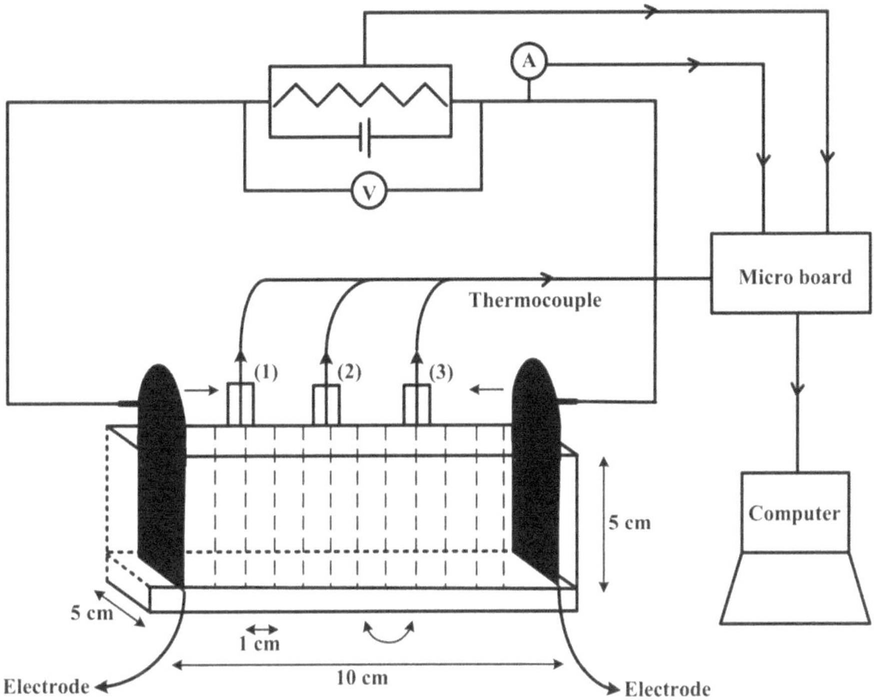


**Fig. 7.5** Schematic representation of ohmic heating thawing (Cai et al., 2019).

negative electrons were introduced into a high voltage electrostatic field. Frozen food products can be thawed easily in the temperature range of −3 to 3°C using this process. For example, under the same temperature conditions, the time of thawing for frozen beef, tuna, and eggs was dramatically reduced to 1/4–1/3. The effects of electrostatic thawing on the texture and amount of cell damage were examined by Fuchigami et al. (1994). For frozen carrots, transmission electronic microscopy and light microscopy showed that ohmic heating thawing avoided drip loss, cell damage, and softening. The employment of an alternating electric field would be more effective because electrolysis would be caused by an electrostatic field and thus costly electrodes would be needed. Yun et al. (1998) investigated OH thawing with 60–210 V (AC) at 60 Hz–60 kHz frequencies of frozen meat chucks combined with traditional water immersion thawing. It was observed that frequency changes did not affect the time of thawing significantly, and when lower voltages were applied, OH thawed samples showed increased water holding capacity and reduced drip loss.

The immersion ohmic heating thawing application significantly reduced the thawing period of frozen tuna fish cubes, as was investigated by Fattahi and Zamindar (2019). Furthermore, in OH treatment, thawing loss and complete loss were substantially lower than in conventional methods. There was no major impact of OH treatment on the water-holding capacity of samples. In their research, Seyhun et al. (2013) showed that the nutrient and flavor retention in potato puree improved with shorter thawing time. Duygu and Umit (2015) conducted research to determine the applications and effects of the ohmic heating system in the thawing of meat and its quality. The study, however, revealed that shorter thawing time and less weight loss were shown by frozen meat thawed by ohmic heating, whereas other methods showed increased pH value and decreased water activity (aw) and moisture values. Miao et al. (2007) assessed that OH thawing showed excellent results with a concentration rate of 0.2% for frozen surimi with stronger gels, which is 12 times greater than the traditional thawing rate. A higher concentration of electrode solution would indicate an increase in the thawing rate linearly in the OH thawing period. Chicken (Sarang et al., 2008), pork (Mckenna et al., 2006), and fish are among the other studied commodities ( Jin et al., 2015). Ohmic heating technology shows the ability to provide high quality thawed food products. However, very little research has been performed to date on OH thawing.

### 7.4.4 Microwave (MW) thawing

The specific characteristics of microwaves that penetrate and generate heat deep inside the food materials (Tong et al., 1993) allow them to accelerate thawing. Microwave heating is most appropriate for heating small food products rapidly, but it has been observed to be less reliable for heating larger food products (Taher and Farid, 2001). This may be because of the runaway heating issue (which occurs with ice melting because water heats quicker owing to its high dielectric loss factor), resulting in non-uniform heating and finite penetration depth in comparison to radiofrequency heating (Buffler, 1993). The run-away heating has limited the application of microwave thawing to food systems. The major cause of runaway heating is the preferential absorption of microwaves by liquid water. So in this case, food products undergo risk of excess water loss and thermally chemical deterioration also takes place (Rosenberg and Bogl, 1987; Virtanen et al., 1997). Tong et al. (1993) demonstrated that the unique feature of microwaves of penetrating food and producing heat within it makes them capable of accelerating thawing. For defrosting frozen food products, microwave ovens are used, as this method allows smaller thawing time and limited processing space for thawing. It also decreases microbial problems, drip loss, and chemical degradation (Meisel, 1973; Rosenberg and Bogl, 1987; Virtanen et al., 1997). Tong et al. (1993) engineered a microwave oven with a temperature controller and variable continuous power in order to maintain the optimal temperature gradient within a model food system. Compared to conventional thawing at atmospheric temperature, the thawing time is shortened by a factor of seven using this apparatus in suitable conditions. A diagram of such microwave thawing is presented in Fig. 7.6.

In microwave thawing, the rate of thawing of frozen samples varies, depending on the material characteristics, dimensions, and magnitude and frequency of electromagnetic radiation (Pangrle et al., 1991). The thawing process is complicated by factors including thermal characteristics that differ with irregular shapes, temperature, and food heterogeneity (Taoukis et al., 1987). Techniques have been developed to tackle the difficulty of predicting the rate of thawing and optimize the thawing rate and use microwaves effectively. Numerical solutions based on methods of finite elements or

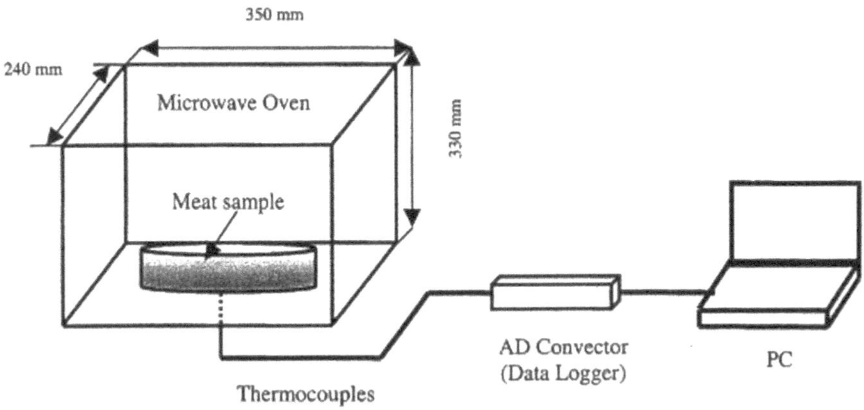


**Fig. 7.6** Schematic representation of microwave thawing (Taher and Farid, 2001).

finite difference will estimate the actual process more realistically (Taoukis et al., 1987). Using the efficient heat capacity process, Basak and Ayappa (1997) examined microwave thawing of tylose slabs and obtained the liquid volume fractions, temperature, and microwave power by finite element. Using the finite elements process, Pangrle et al. (1991) engineered microwave thawing of pure water and 0.1 M NaCl cylinders and showed that there may be a two-phase mushy area and, at the center of the cylinder, an additional thawing front may develop. Moreover, Taoukis et al. (1987) used modified isotherm migration method which treats the problem as the propagation of a phase front to solve a mathematical model of microwave thawing of homogeneous food products, model and experimental resultsfor thawing a lean beef cylinder heated at low power using 2450 MHz frequency were compared well.

Kondratowicz et al. (2008) noticed that thawing loss in the case of microwave thawing is lower than in the case of conventional thawing only for the pork stored frozen for a short period of time, but not for a longer period of time. Apart from these observations, a comparative analysis of thawing methods was carried out by Xia et al. (2012) and it was observed that of all the traditional thawing techniques tested, the most damage was observed when microwave was employed for thawing of pork. More than this, microwave thawing had a negative effect on the quality of thawed pork, increasing cooking loss, cutting force, carbonyl content, thus the results related to the qualities of the meat thawed by this method being unsatisfactory.

### *7.4.5 Radiofrequency (RF) thawing*

Radiofrequency heating was originally proposed as a method for medical treatment processes in 1895. It later came to be used in the food processing industry for blanching, followed by cooking and dehydration (Proctor and Goldblith, 1948; Moyer and Stotz, 1947). In the 1960s, a number of efforts were made to use this method in order to thaw frozen foods. A diagram of radiofrequency thawing is presented in Fig. 7.7.

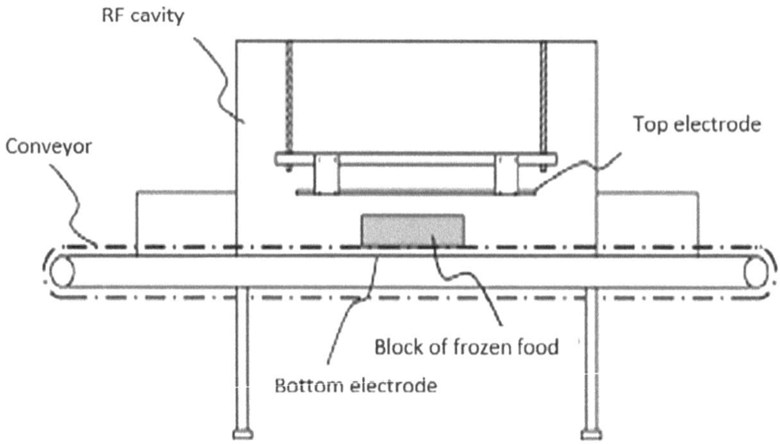


**Fig. 7.7** Schematic representation of radiofrequency thawing (Palazoglu and Miran, 2017).

Radiofrequency heating is a technique of bulk heating that offers quick and volumetric generation of heat within food matrices. Usually, 1–100 MHz electromagnetic waves penetrate food and interact with atoms, ions, and molecules in order to produce internal heat ( Jiao et al., 2018). While dipole rotation is a significant contributor to the larger frequencies used in MW applications—2450 MHz (home appliances) and 915 MHz (industrial)—ionic displacement in RF applications is more prominent at the lower frequencies (13.56 and 27.12 MHz) ( Jones, 1992). Higher penetration depth is observed in the case of RF due to the longer wavelength ($\approx$22 and 11 min free space, respectively) in comparison to MW ($\approx$0.327 and 0.122 min free space, respectively), and runaway heating are extensively reduced inside the sample with the aid of lesser frequency applications (Uyar et al., 2015). Therefore, during a process on an industrial scale, the RF method could be agreed to have the ability to increase uniformity in heating, taking into account the bigger sizes of frozen goods in an industrial context (Taher and Farid, 2001). In the food industry, RF is used extensively, particularly in the processes of drying (Wang et al., 2014), sterilization (Schlisselberg et al., 2013), baking (Kocadagli et al., 2012), and thawing (Basaran-Akgul et al., 2008), owing of its capacity to heat food volumetrically. Wen (2015) used devices for food thawing employing radiofrequencies of approximately 1 to 300 MHz. According to Jiao et al. (2014), the process of thawing is accelerated when dissociated ions travel with alternating electric fields in frozen foods and thus produce heat through the friction process between molecules.

A self-constructed RF heater (27.12 MHz, 500 W) was used by Farag et al. (2008) for tempering beef blocks (4 kg) from −20°C to −3.6°C. The findings revealed that the time taken by conventional methods was reduced by RF tempering from 50 h to 35 min. The findings also indicated that the improvement in the homogeneity of sample composition could improve the uniformity of thawing. In order to determine the quality parameters of beef defrosted by RF and traditional methods, further study was conducted. Compared to air-thawed products, the findings revealed a substantial reduction in micronutrient and drip losses in RF-defrosted meat (Farag et al., 2008). Uyar et al. (2015) also developed a mathematical model on the basis of the experimental results of Farag et al. (2008) and validated the model with experimental data using various meat block sizes. They stated that sample size affected power absorption and thawing times within the food systems. However, owing to non-uniformity of the distribution of temperature in the food system and potential methods of reducing the higher temperature encountered on the corners and surfaces of the product, a variety of issues remain to be addressed. Llave et al. (2014) investigated the dielectric characteristics in the tempering temperature range of frozen tuna and determined the effectiveness of a frozen tuna block tempered by RF. The findings revealed a three-fold reduction in time of thawing in comparison to conventional thawing method and that the best heating uniformity is obtained by using an electrode size equal to the sample surface.

The majority of previous RF thawing studies have focused on batch processes, whereas in the food industry, continuous treatment methods are needed for the benefits of handling large quantities of products and minimizing handling costs. The very common techniques such as thawing and tempering of frozen items include bigger sized frozen food items to be processed at one time, particularly in the meat products processing industries, and therefore, require quick and continuous processing techniques. This remains an open field of research aimed at extending the application of continuous processes to practical technologies for industrial processing. Houben et al. (1991) showed that sausage products had a smooth surface and good appearance when heated in a continuous RF system and also did not show release of moisture or fat. For post-harvest treatments of insect control in chestnuts (Hou et al., 2014), walnuts (Wang et al., 2007a,b), lentils (Wang et al., 2010; Jiao et al., 2012), and rice (Zhou and Wang, 2016a,b), an industrial RF heating device was used. In continuous RF processing, the uniformity of heating was affected by physical parameters including speed of belt conveyor and, consequently, the residence time.

RF thawing/tempering methods are now widely used in the industry, but are limited to fish and meat in standard (or roughly regular) shapes. In order to solve the issues of thawing irregular-shaped-food, further research must be performed, which may expand the RF thawing application to a large extent.

### 7.4.6 High-voltage electric field (HVEF) thawing

A relatively new, non-thermal approach is HVEF thawing (Dalvi-Isfahan et al., 2016), and several researchers have investigated it in pork (He et al., 2013, 2014), chicken (Hsieh et al., 2010), tuna fish (Mousakhani-Ganjeh et al., 2015, 2016a), and apple tissue (Wiktor et al., 2015). The benefits generally include decreased time of thawing, food quality maintenance, inhibition of microbial growth, and reduced energy consumption. Under HVEF, the time of thawing for frozen chicken was 2/3 the time taken to thaw meat using a common refrigerator when the thawing temperature was set at −3°C (Hsieh et al., 2010). A diagram of HVEF is presented in Fig. 7.8.

The treatment with HVEF substantially reduced the time of thawing of frozen pork tenderloin meat and reduced the time of thawing to 2/3 of the control (He et al., 2013).

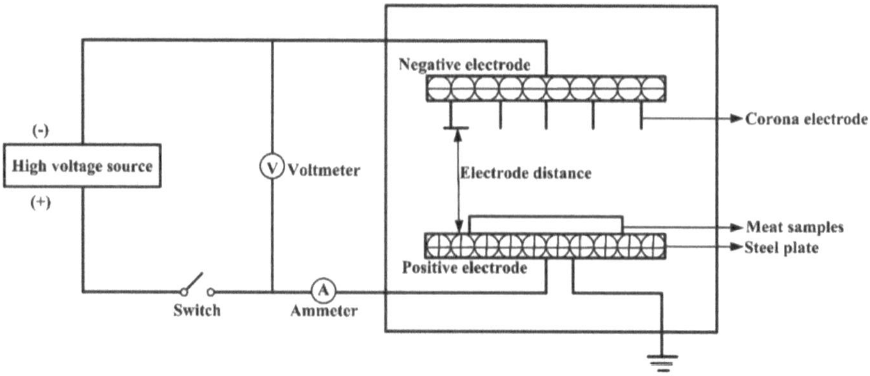


**Fig. 7.8** Schematic representation of HVEF thawing (Cai et al., 2019).

Similarly, HVEF greatly increased the thawing rate in the case of tuna fish (Mousakhani-Ganjeh et al., 2015). Deng et al. (2017) stated that the thawing rate of frozen tofu can be improved by the HVEF process. With the strengthening of the applied voltage, the thawing rate increases. He et al. (2016) recorded that the thawing period can be decreased by 50% at most in comparison to the conventional air-thawing process. Without affecting meat quality, the HVEF process decreased the total microbial counts in thawed frozen meat by 0.5–1 log CFU/g, and decreased volatile basic nitrogen production during storage (He et al., 2013). In order to determine the relationships and factors that influence HVEF treatment, parameters including electric field strength, voltage, and distance were studied (He et al., 2014; Mousakhani-Ganjeh et al., 2016a). Mousakhani-Ganjeh et al. (2016a) stated that, due to generation of ozone, high-voltage electric field could enhance the susceptibility of tuna fish to lipid oxidation. Compared to the other thawing strategies, the energy usage of HVEF thawing was comparatively much lower (He et al., 2014; Mousakhani-Ganjeh et al., 2016a). These experiments were conducted under high-voltage direct current electric field.

## 7.5 Limitations and challenges

There has been a rapid growth in advanced thawing technologies. To achieve some of the optimal parameters for various foods, complex thawing processes have been developed. However, no single thawing technology appears to be sufficient for all food types and various sizes of frozen food materials. For vegetables and fruits that quickly lose water, soften, and deteriorate, thermal methods, including ohmic heating or microwave thawing are not sufficient for thawing. For small-sized foods, using HVEF or microwave is not appropriate for thawing. In the microwave, the thawing rate is so high for such materials that overheating is not easily managed. Moreover, the machinery used for HVEF thawing is usually so massive that small-scale food thawing operations are difficult. Because of the limited penetration depth of microwave thawing, it is observed that the method is not a suitable technique for deep food thawing. In order to protect the quality of commodities in microwave thawing, an acceptable depth for the frozen food is therefore necessary. There are also some detrimental effects on the quality of frozen food from the thawing treatment. Nevertheless, the drawbacks of space requirement for large equipment, runaway heating, the operational issues, high energy usage, and high costs are gradually being solved. Several studies have been published recently on the prevention of runaway heating and non-uniformities in ohmic heating thawing, microwave thawing, and RF thawing.

"Cold surface heating" techniques and the combination of thawing techniques have been applied to OH thawing technique in order to prevent overheating (Pain and Muller, 2014). In order to increase the uniform temperature distribution, air jet impingement or infrared non-thermal thawing and conventional thawing can be combined with microwave (Datta and Ni, 2002). Simulation models are employed to study the effects of a corona wind during HVEF thawing on the uneven distribution of charge and temperature variance (He et al., 2014). However, in order to find more convenient and complete thawing techniques, more effort is still required. Moreover, in the peer-reviewed literature, the economic viability of different technologies has not been widely evaluated at the industrial level.

## 7.6 Conclusion

Thawing is a difficult process, which entails heat transfer, and has a potential for a number of changes, including physical and chemical, that can have a detrimental effect on product quality. New techniques are necessary in terms of energy saving or improvement of quality. The novel techniques of ultrasound-assisted thawing, high pressure thawing, microwave thawing, radiofrequency thawing, ohmic heating thawing, and high-voltage electric field thawing can reduce the time necessary for thawing, thus decreasing drip loss and enhancing product quality. Further research is needed to demonstrate more practical applications of novel techniques.

## References


Alizadeh, E., Chapleau, N., de Lamballerie, M., LeBail, A., 2007. Effects of freezing and thawing processes on the quality of Atlantic salmon (*Salmo salar*) fillets. J. Food. Sci. 72, E279–E284.

Archer, M., Edmonds, M., George, M., 2008. Seafood Thawing. Seafish, Edinburgh.

Basak, T., Ayappa, K.G., 1997. Analysis of microwave thawing of slabs with effective heat capacity method. AICHE J. 43 (7), 1662–1674.

Basaran-Akgul, N., Basaran, P., Rasco, B., 2008. Effect of temperature (−5 to 130°C) and fiber direction on the dielectric properties of beef semitendinosus at radio frequency and microwave frequencies. J. Food Sci. 73 (6), E243–E249.

Bedane, T.F., Chen, L., Marra, F., Wang, S., 2017. Experimental study of radio frequency (RF) thawing of foods with movement on conveyor belt. J. Food Eng. 201, 17–25.

Buffler, C.R., 1993. Microwave Thawing of Foods (Ph.D. thesis). Cornell University, Ithaca.

Cai, L., Cao, M., Regenstein, J., Cao, A., 2019. Recent advances in food thawing technologies. Compr. Rev. Food Sci. Food Saf. 18 (4), 953–970.

Cevik, M., Icier, F., 2018. Effects of voltage gradient and fat content on changes of electrical conductivity of frozen minced beef meat during ohmic thawing. J. Food Process Eng. 41 (4), e12675.
Cheng, X.F., Zhang, M., Adhikari, B., 2014. Effects of ultrasound-assisted thawing on the quality of edamames [*Glycine max* (L.) Merrill] frozen using different freezing methods. Food Sci. Biotechnol. 23, 1095–1102.
Dalvi-Isfahan, M., Hamdami, N., Le-Bail, A., Xanthakis, E., 2016. The principles of high voltage electric field and its application in food processing: a review. Food Res. Int. 89 (1), 48–62.
Datta, A.K., Ni, H., 2002. Infrared and hot-air assisted microwave heating of foods for control of surface moisture. J. Food Eng. 51 (4), 355–364.
Deng, S., Gao, Z., Xu, J., Wang, G., Bai, Y., Ding, C., 2017. The thawing characteristics of frozen tofu under high voltage alternating electric field. J. Food Qual. 2017, 3914074.
Dransfield, E., 1994. Optimization of tenderization, ageing and tenderness. J. Meat Sci. 36, 105–121.
Duygu, B., Umit, G., 2015. Application of ohmic heating system in meat thawing. Procedia Soc. Behav. Sci. 195, 2822–2828.
Farag, K.W., Lyng, J.G., Morgan, D.J., Cronin, D.A., 2008. A comparison of conventional and radio frequency tempering of beef meats: effects on product temperature distribution. Meat Sci. 80, 488–495.
Fattahi, S., Zamindar, N., 2019. Effect of immersion ohmic heating on thawing rate and properties of frozen tuna fish. Food Sci. Technol. Int. 26, 1–9.
Fellows, P., 2000. Food Processing Technology – Principles and Practice, second ed. Ellis Horwood, Chichester, UK, pp. 369–380.
Fernandez-Martın, F., Otero, L., Solas, M.T., Sanz, P., 2000. Protein denaturation and structural damage during high-pressure-shift freezing of porcine and bovine muscle. J. Food Sci. 65 (6), 1002–1008.
Fu, W.R., Hsieh, C.C., 1999. Simulation and verification of two dimensional ohmic heating in static system. J. Food Sci. 64 (6), 946–949.
Fuchigami, M., Hyakuumoto, N., Miyazaki, K., Nomura, T., Sasaki, J., 1994. Texture and histological structure of carrots frozen at a programmed rate and thawed in an electrostatic field. J. Food Sci. 59 (6), 1162–1167.
Gambuteanu, C., Alexe, P., 2013. Effects of ultrasound assisted thawing on microbiological, chemical and technological properties of unpackaged pork longissimusdorsi. Ann. Univ. Dunarea Jos Galati Fascicle VI Food Technol. 37 (1), 98–107.
Gambuteanu, C., Borda, D., Alexe, P., 2013. The effect of freezing and thawing on technological properties of meat: review. J. Agroaliment. Process. Technol. 19 (1), 88–93.
Got, F., Culioli, J., Berge, P., Vignon, X., Astruc, T., Quideau, J.M., Lethiecq, M., 1999. Effects of high-intensity high-frequency ultrasound on ageing rate, ultrastructure and some physico-chemical properties of beef. Meat Sci. 51, 35–42.
Greer, G.G., Murray, A.C., 1991. Freezing effects on quality, bacteriology and retail case life of pork. J. Food Sci. 56, 891–894.
Haugland, A., 2002. Industrial Thawing of Fish – To Improve Quality, Yield and Capacity. Norwegian University of Science and Technology, Trondheim.
He, X., Jia, G., Tatsumi, E., Liu, H., 2016. Effect of corona wind, current, electric field and energy consumption on the reduction of the thawing time during the high-voltage electrostaticfield (HVEF) treatment process. Innov. Food Sci. Emerg. Technol. 34, 135–140.
He, X., Liu, R., Nirasawa, S., Zheng, D., Liu, H., 2013. Effect of high voltage electrostatic field treatment on thawing characteristics and post-thawing quality of frozen pork tenderloin meat. J. Food Eng. 115 (2), 245–250.

He, X., Liu, R., Tatsumi, E., Nirasawa, S., Liu, H., 2014. Factors affecting the thawing characteristics and energy consumption of frozen pork tenderloin meat using high-voltage electrostatic field. Innov. Food Sci. Emerg. Technol. 22, 110–115.

Hite, B.H., 1899. The Effect of Pressure in the Preservation of Milk. 54 Bulletin of the West Virginia Agricultural Experiment Station of Morgantown, pp. 15–35.

Hou, L., Ling, B., Wang, S., 2014. Development of thermal treatment protocol for disinfesting chestnuts using radio frequency energy. Postharvest Biol. Technol. 98, 65–71.

Houben, J., Schoenmakers, L., Van Putten, E., Van Roon, P., Krol, B., 1991. Radio-frequency pasteurization of sausage emulsions as a continuous process. J. Microw. Power Electromagn. Energy. 26 (4), 202–205.

Hsieh, C.W., Lai, C.H., Ho, W.J., Huang, S.C., Ko, W.C., 2010. Effect of thawing and cold storage on frozen chicken thigh meat quality by high-voltage electrostatic field. J. Food Sci. 75 (4), M193–M197.

Icier, F., Izzetoglu, G.T., Bozkurt, H., Ober, A., 2010. Effects of ohmic thawing on histological and textural properties of beef cuts. J. Food Eng. 99 (3), 360–365.

Jaeger, H., Roth, A., Toepfl, S., Holzhauser, T., 2016. Opinion on the use of ohmic heating for the treatment of foods. Trends Food Sci. Technol. 55.

Jayasooriya, S.D., Torley, P.J., D'Arcy, B.R., Bhandari, B.R., 2007. Effect of high power ultrasound and ageing on the physical properties of bovine semitendinosus and longissimus muscles. Meat Sci. 75, 628–639.

Jiao, S., Johnson, J.A., Tang, J., Wang, S., 2012. Industrial-scale radio frequency treatments for insect control in lentils. J. Stored Prod. Res. 48, 143–148.

Jiao, S., Luan, D., Tang, J., 2014. Principles of radio-frequency and microwave heating. In: Awuah, G.B., Ramaswamy, H.S., Tang, J. (Eds.), Radio-Frequency Heating in Food Processing: Principles and Applications. Taylor & Francis Group, Boca Raton, FL, pp. 1–2.

Jiao, Y., Tang, J., Wang, Y., Koral, T.L., 2018. Radio-frequency applications for food processing and safety. Annu. Rev. Food Sci. Technol. 9 (6), 1–6.

Jin, Y., Cheng, Y.D., Fukuoka, M., Sakai, N., 2015. Electrical conductivity of yellowtail (*Seriola quinqueradiata*) fillets during ohmic heating. Food Bioprocess Technol. 8 (9), 1904–1913.

Jones, P.I., 1992. Dielectric heating for food processing. J. Nutr. Food Sci. 2, 14–15.

Kalichevsky, M.T., Knorr, D., Lillford, P.J., 1995. Potential food applications of high-pressure effects on ice-water transitions. Trends Food Sci. Technol. 6, 253–258.

Kissam, A.D., 1984. Acoustic Thawing of Frozen Food. United States Patent No. 4464 401. filed 22 April 1982.

Kocadagli, T., Palazoglu, T.K., Gokmen, V., 2012. Mitigation of acrylamide formation in cookies by using Maillard reaction products as recipe modifier in a combined partial conventional baking and radio frequency post-baking process. Eur. Food Res. Technol. 235 (4), 711–717.

Kondratowicz, J., Chwastowska-Siwiecka, I., Burczyk, E., 2008. Technological properties of pork thawed in the atmospheric air or in the microwave oven as determined during a six-month deep-freeze storage. Anim. Sci. Paper Rep. 26 (3), 175–181.

Lawrie, R.A., 1998. Meat Science, sixth ed. CRC Press Publishers Ltd., Abington, England, pp. 143–244.

Li, B., Sun, D.W., 2002. Novel methods for rapid freezing and thawing of foods – a review. J. Food Eng. 54, 175–182.

Li, W., Wang, P., Xu, X., Xing, T., Zhou, G., 2014. Use of low-field nuclear magnetic resonance to characterize water properties in frozen chicken breasts thawed under high pressure. Eur. Food Res. Technol. 239 (2), 183–188.

Llave, Y., Terada, Y., Fukuoka, M., Sakai, N., 2014. Dielectric properties of frozen tuna and analysis of defrosting using a radio-frequency system at low frequencies. J. Food Eng. 139, 1–9.

Londahl, G., Nilaaon, T., 1993. Storage of frozen foods. In: Caballero, B. (Ed.), Encyclopaedia of Food Science and Nutrition, second ed. Academic Press, Oxford, pp. 2732–2735.

Lyng, J.G., Allen, P., McKenna, & B. M., 1997. The influence of high intensity ultrasound baths on aspects of beef tenderness. J. Muscle Foods 8, 237–249.

Makita, T., 1992. Applications of high pressure and thermophysical properties of water to biotechnology. Fluid Phase Equilib. 76, 87–95.

Mckenna, B.M., Lyng, J., Brunton, N., Shirsat, N., 2006. Advances in radio frequency and ohmic heating of meats. J. Food Eng. 77 (2), 215–229.

Meisel, N., 1973. Microwave applications to food processing and food systems in Europe. J. Microwave Power 8 (2), 143–146.

Mertens, B., Deplace, G., 1993. Engineering aspects of high pressure technology in the food industry. Food Technol. 47 (6), 164–169.

Miao, Y., Chen, J.Y., Noguchi, A., 2007. Studies on the ohmic thawing of frozen surimi. Food Sci. Technol. Res. 13 (4), 296–300.

Miles, C.A., Morley, M.J., Rendell, M., 1999. High power ultrasonic thawing of frozen foods. J. Food Eng. 39, 151–159.

Min, Y.Y., 2001. Review on thawing technology of frozen foods. Food Sci. 22, 87–90.

Mousakhani-Ganjeh, A., Hamdami, N., Soltanizadeh, N., 2015. Impact of high voltage electric field thawing on the quality of frozen tuna fish (*Thunnus albacares*). J. Food Eng. 156, 39–44.

Mousakhani-Ganjeh, A., Hamdami, N., Soltanizadeh, N., 2016a. Effect of high voltage electrostatic field thawing on the lipid oxidation of frozen tuna fish (*Thunnus albacares*). Innov. Food Sci. Emerg. Technol. 36, 42–47.

Mousakhani-Ganjeh, A., Hamdami, N., Soltanizadeh, N., 2016b. Thawing of frozen tuna fish (*Thunnus albacares*) using still air method combined with a high voltage electrostatic field. J. Food Eng. 169, 149–154.

Moyer, J.C., Stotz, E., 1947. The blanching of vegetables by electronics. Food Technol. 1, 252–257.

Ngapo, T.M., Babare, I.H., Reynolds, J., Mawson, R.F., 1999. Freezing and thawing rate effects on drip loss from samples of pork. Meat Sci. 53, 149–158.

Ohtsuki, T., 1991. Process for Thawing Foodstuffs. US Patent 5,034,236.

Ohtsuki, T., 1993. Process for Thawing Foodstuffs. European Patent 0,409,430.

Okamoto, A., Suzuki, A., 2002. Effects of high hydrostatic pressure-thawing on pork meat. Prog. Biotechnol. 19, 571–576.

Otero, L., Sanz, P.D., 2003. High pressure-assisted and high pressure-induced thawing: two different processes. J. Food Sci. 68, 2523–2528.

Pain, J.P., Muller, F.L., 2014. Modeling basics as applied to ohmic heating of liquid and wall cooling. In: Ramaswamy, H.S., Marcotte, M., Sastry, S., Abdelrahim, K. (Eds.), Ohmic heating in Food Processing. CRC Press, Taylor & Francis Group, Boca Raton, FL, pp. 199–211.

Palazoglu, T.K., Miran, W., 2017. Experimental comparison of microwave and radio frequency tempering of frozen block of shrimp. Innov. Food Sci. Emerg. Technol. 41, 292–300.

Pangrle, B.J., Ayappa, K.G., Davis, H.T., Davis, E.A., Gordon, J., 1991. Microwave thawing of cylinders. AICHE J. 37 (12), 1789–1800.

Park, S.H., Ryu, H.S., Hong, G.P., Min, S.G., 2006. Physical properties of frozen pork thawed by high pressure assisted thawing process. Food Sci. Technol. Int. 12 (4), 347–352.

Persson, P.O., Londahl, G., 1993. Freezing technology. In: Mallett, C.P. (Ed.), Frozen Food Technology. Blackie Academic & Professional, Glasgow, UK, pp. 20–58.
Petrovic, L., Grujic, R., Petrovic, M., 1993. Investigation of the physico-chemical properties of beef M. Longissimusdorsi frozen at different freezing rates. Meat Sci. 33, 319–331.
Proctor, B.E., Goldblith, S.A., 1948. Radar energy for rapid food cooking and blanching, and its effect on vitamin content. Food Technol. 2, 95–104.
Reznick, D., 1996. Ohmic heating of fluid foods. Food Technol. 50, 250–251.
Roberts, J.S., Balaban, M.O., Zimmerman, R., Luzuriaga, D., 1998. Design and testing of prototype ohmic thawing unit. Comput. Electron. Agric. 19 (2), 211–222.
Rosenberg, U., Bogl, W., 1987. Microwave thawing, drying, and baking in the food industry. Food Technol., 85–91.
Sakata, R., Oshida, T., Morita, H., Nagata, Y., 1995. Physico-chemical and processing quality of porcine M. Longissimusdorsi frozen at different temperatures. Meat Sci. 39, 277–284.
Sarang, S., Sastry, S.K., Knipe, L., 2008. Electrical conductivity of fruits and meats during ohmic heating. J. Food Eng. 87 (3), 351–356.
Schlisselberg, D.B., Kler, E., Kalily, E., Kisluk, G., Karniel, O., Yaron, S., 2013. Inactivation of food borne pathogens in ground beef by cooking with highly controlled radio frequency energy. Int. J. Food Microbiol. 160, 219–226.
Schubring, R., Meyer, C., Schluter, O., Boguslawski, S., 2003. Impact of high pressure assisted thawing on the quality of fillets from various species. Innovative Food Sci. Emerg. Technol. 4 (3), 257–267.
Seyhun, N., Ramaswamy, H.S., Zhu, S., Sumnu, G., Sahin, S., 2013. Ohmic tempering of frozen potato puree. Food Bioprocess Technol. 6 (11), 3200–3205.
Shanks, B.C., Wulf, D.M., Maddock, R.J., 2002. Technical note: the effect of freezing on Warner-Bratzler shear force values of beef longissimus steaks across several post-mortem aging periods. J. Anim. Sci. 80, 2122–2125.
Stadnik, J., Dolatowski, Z.J., Baranowska, H.M., 2008. Effect of ultrasound treatment on water holding properties and microstructure of beef (m. semi membranosus) during ageing. LWT Food Sci. Technol. 41, 2151–2158.
Taher, B.J., Farid, M.M., 2001. Cyclic microwave thawing of frozen meat: experimental and theoretical investigation. Chem. Eng. Process. 40, 379–389.
Taoukis, P., Davis, E.A., Davis, H.T., Gordon, J., Takmon, Y., 1987. Mathematical modelling of microwave thawing by the modified isotherm migration method. J. Food Sci. 52 (2), 455–463.
Teramoto, A., Fuchigami, M., 2000. Changes in temperature texture and structure of konnyaku (konjacglucomannan gel) during high pressure-freezing. J. Food Sci. 65 (3), 491–497.
Tong, C.H., Lentz, R.R., Lund, D.B., 1993. A microwave oven with variable continuous power and a feedback temperature controller. Biotechnol. Prog. 9, 488–496.
Uyar, R., Bedane, T.F., Erdogdu, F., Palazoglu, T.K., Farag, K.W., Marra, F., 2015. Radio-frequency thawing of food products e a computational study. J. Food Eng. 146, 163–171.
Vieira, C., Diaz, M.Y., Martinez, B., Garcia-Cachan, M.D., 2009. Effect of frozen storage conditions (temperature and length of storage) on microbial and sensory quality of rustic crossbred beef at different stages of aging. Meat Sci. 83, 398–404.
Virtanen, A.J., Goedeken, D.L., Tong, C.H., 1997. Microwave assisted thawing of model frozen foods using feed-back temperature control and surface cooling. J. Food Sci. 62 (1), 150–154.
Wang, S., Monzon, M., Johnson, J.A., Mitcham, E.J., Tang, J., 2007a. Industrial-scale radio frequency treatments for insect control in walnuts: I. Heating uniformity and energy efficiency. Postharvest Biol. Technol. 45 (2), 240–246.

Wang, S., Monzon, M., Johnson, J.A., Mitcham, E.J., Tang, J., 2007b. Industrial-scale radio frequency treatments for insect control in walnuts: II. Insect mortality and product quality. Postharvest Biol. Technol. 45 (2), 247–253.

Wang, S., Tiwari, G., Jiao, S., Johnson, J.A., Tang, J., 2010. Developing postharvest disinfestation treatments for legumes using radio frequency energy. Biosyst. Eng. 105, 341–349.

Wang, Y.Y., Zhang, L., Gao, M.X., Tang, J., Wang, S.J., 2014. Pilot-scale radio frequency drying of macadamia nuts: heating and drying uniformity. Dry. Technol. 32 (9), 1052–1059.

Wen, G.Y., 2015. Solutions of electromagnetic field problems. In: Wen, G.Y. (Ed.), Foundations for Radio Frequency Engineering. World Scientific, Hackensack, NJ, pp. 1–3.

Wen, X., Hu, R., Zhao, J.H., Peng, Y., Ni, Y.Y., 2015. Evaluation of the effects of different thawing methods on texture, colour and ascorbic acid retention of frozen hami melon (Cucumismelo var. Saccharinus). Int. J. Food Sci. Technol. 50 (5), 1116–1122.

Wiktor, A., Schulz, M., Voigt, E., Witrowa-Rajchert, D., Knorr, D., 2015. The effect of pulsed electric field treatment on immersion freezing, thawing and selected properties of apple tissue. J. Food Eng. 146, 8–16.

Xia, X., Kong, B., Liu, J., Diao, X., Liu, Q., 2012. Influence of different thawing methods on physicochemical changes and protein oxidation of porcine longissimus muscle. Food Sci. Technol. 46, 280–286.

Yun, C.G., Lee, D.H., Park, J.Y., 1998. Ohmic thawing of a frozen meat chunk. J. Food Sci. Technol. 30 (4), 842–847.

Zhao, Y., Flores, R., Olson, D., 1998. High hydrostatic pressure effects on rapid thawing of frozen beef. J. Food Sci. 63 (2), 272–275.

Zhou, L., Wang, S., 2016a. Industrial-scale radio frequency treatments to control Sitophilusoryzae in rough, brown and milled rice. J. Stored Prod. Res. 68, 9–18.

Zhou, L., Wang, S., 2016b. Verification of radio frequency heating uniformity and Sitpphilusoryzae control in rough, brown and milled rice. J. Stored Prod. Res. 65, 40–47.

Zhu, S., Ramaswamy, H.S., Simpson, B.K., 2004. Effect of high pressure versus conventional thawing on color, drip loss and texture of Atlantic salmon frozen by different methods. LWT Food Sci. Technol. 37 (3), 291–299.

Zhu, S.M., Su, G.M., He, J.S., Ramaswamy, H.S., Bail, A.L., Yu, Y., 2014. Water phase transition under pressure and its application in high pressure thawing of agar gel and fish. J. Food Eng. 125 (1), 1–6.

# Thawing equipment for the food industry

8

*Elham Ansarifar*[a], *Sara Hedayati*[b], *and Seid Mahdi Jafari*[c]
[a]Social Determinants of Health Research Center, Department of Public Health, School of Health, Birjand University of Medical Sciences, Birjand, Iran, [b]Nutrition Research Center, School of Nutrition and Food Sciences, Shiraz University of Medical Sciences, Shiraz, Iran, [c]Faculty of Food Science and Technology, Gorgan University of Agricultural Sciences and Natural Resources, Gorgan, Iran

## 8.1 Introduction

From ancient times, freezing has always been a common method for the long-term food preservation, owing to its ability to retain the quality and extend the shelf-life of foods (Backi, 2022). Quality of frozen food products is firmly associated with freezing and thawing conditions. The formation of tiny ice crystals during freezing lowers the drip loss and tissue damage during thawing. Thawing is more complicated than freezing, behind which there are many reasons. In the first place, the thermal conductivity of water is lower than ice, and the thawing rate decreases as defrosting keeps on. This is in contrast with what occurs during freezing, where the outer layers of the product start to freeze and create a conductive surface for the heat to transfer through. In the second place, thawing cannot be done at all temperatures; i.e., the difference between the initial temperature of the frozen product and that of the thawing medium could only be within a specific interval to prevent the damage of food products (Bozzato et al., 2021; Xu et al., 2021). This is particularly critical in processes where the product is exposed to heat through its surface (Cai et al., 2020; Backi, 2022).

Frozen foods are typically stored at −18 °C; as a result, tempering or thawing is required for further. At the same cooking or processing time, the physical, chemical, and microbiological changes taking place during long periods of tempering or thawing lead to undesirable changes in the quality of the product. Consequently, the selection of techniques that reduce the duration of tempering or thawing and, therefore, extending the food's quality retention period and safety are vital for the frozen food industry (Xu et al., 2021). Thawing can occur by two different concepts, which differ in how the frozen food is exposed to heat (Cai et al., 2019a). Heat transfer through the frozen food surface is the first and most common thawing method (Uyar et al., 2015). This can be carried out by conduction, convection, and radiation (Cai et al., 2019a). Heat generation inside the product is the other thawing technique, which is a more uniform process (Cai et al., 2019a; Singla and Sit, 2021; Svendsen et al., 2022). These innovative techniques can reduce the processing time, thus lowering drip loss and enhancing the quality of the product (Dalvi-Isfahan et al., 2016; Backi, 2018; Cai et al., 2020;

High-Temperature Processing of Food Products. https://doi.org/10.1016/B978-0-12-818618-3.00010-0

Bhargava et al., 2021; Llave and Erdogdu, 2022). The thawing method plays a crucial role in determining the product final quality. This chapter aims to discuss different types of the thawing equipment by gathering information from different documents and reporting the *pros* and *cons* of each thawing technique.

## 8.2 Air thawing

### 8.2.1 Theory

The simplest method of thawing food is to place it in a room with still air at ambient or chilled (0–4 °C) temperatures (Shen et al., 2020). Air thawing occurs due to heat transfer by convection, whether natural or forced. In air thawing methods, heat is transferred to the frozen product via conduction during contact with boundary layer of static air (Backi, 2018). The efficiency of heat transfer is dependent on the temperature difference between the air and the product, the air velocity, the relative humidity (RH), shape, and surface area of the product to be thawed (Peng et al., 2021; Xu et al., 2021). Air thawing increases the risk of lipid oxidation and undesirable surface drying.

### 8.2.2 Still air thawing

This method involves the thawing of the product by static and nonmoving air in a cold room (Fig. 8.1). Still air thawing takes a long time due to the slow rate of heat exchange, there may be a significant drip loss, and the result may not always be homogeneous (Mousakhani-Ganjeh et al., 2016). Blocks of thin frozen products (<10 cm) may be thawed at room temperature for up to 24 h without the surface become too dry. However, natural convection is not recommended for thicker materials due to prolonged thawing, which can deteriorate surface layers before thawing the inner layers of the product. Still air thawing is only feasible on small-scale since much space is needed, processes is uncontrolled, and the duration of thawing often interferes with processing cycles (Cai et al., 2019a, 2020).

**Fig. 8.1** Still air thawing device.
From https://www.Sairem.com/ the different industrial food hawing technologies.

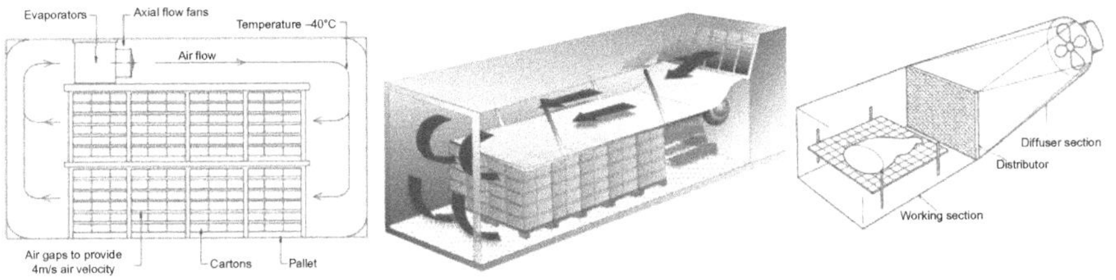


**Fig. 8.2** Schematic representation of the Blast air thawing.
From https://klingecorp.com/thawing units/.

### 8.2.3 *Blast air thawing*

Industrial air thawing systems use moving air as a thawing medium. Air blast thawing produces a greater heat exchange rate than still air thawing. Moving air increases the heat transfer rate, as a result thawing is faster and more controlled than natural convection (Choi et al., 2017). There are both batch and continuous defroster in the industry, but the batch type is more common because it is more cost-effective. In batch thawing, the product is placed on trays, stacked on trolleys and take into the thawing chamber with uniform air circulation. Therefore, trays and trolleys must be spaced apart for optimal airflow (Backi, 2022; Svendsen et al., 2022). A thawing chamber should allow the direction of airflow to be adjusted to thaw the product uniformly. Thus, a reversible fan equipped with heated water nozzles is required (Backi, 2022). Lipid oxidation of food generally decreased compared to the still air thawing method due to the shorter thawing time. Nonetheless, there is a slight risk of surface drying (Aidani et al., 2014; Cai et al., 2019a, 2020). Fig. 8.2 illustrates a blast air thawing system.

### 8.2.4 *Tumbling air thawing*

This system uses a drum that rotates while its walls are slightly heated (Fig. 8.3). Cold steam under vacuum may occasionally be used to improve the process. Compared to defrosting in a cold room, this method is quicker, as it takes only hours rather than days. However, this process is not continuous, so it is not suitable for processing large quantities of food and is not suitable for delicate products such as fruits and vegetables (Cai et al., 2019a).

### 8.2.5 *Impingement air thawing*

Impingement thawing refers to applying heated fluid at high speeds (approximately 10–50 m/s) to the surface of the product. The high air speed efficiently removes the food boundary layer, consequently increases the heat transfer (Kerry, 2011). As shown in Fig. 8.4 impingement is a thermal processing system that has been successfully used in the food industry to reduce the thawing time (Bozzato et al., 2021). Anderson and Singh (2006a, b) used air impingement technology for thawing frozen foods. For this

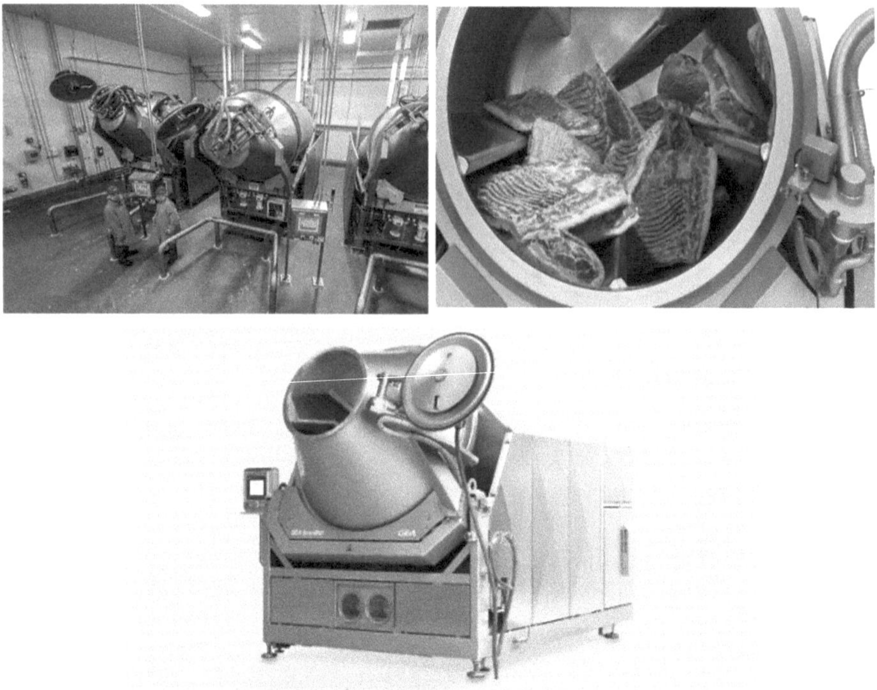

**Fig. 8.3** Schematic representation of the tumbling air thawing.
From https://www.sairem.com/the-different-industrial-food-thawing-technologies.

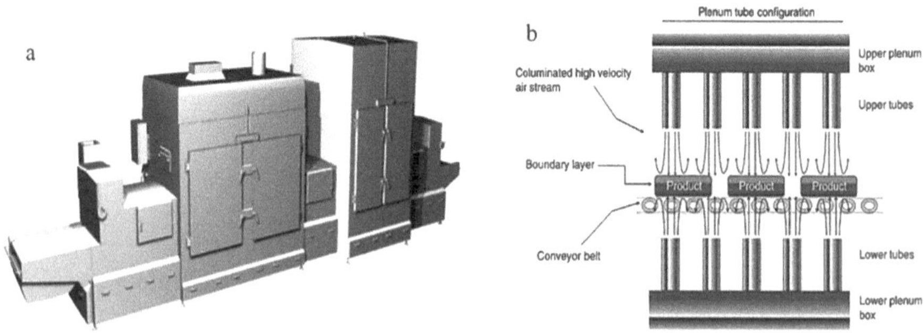


**Fig. 8.4** Schematic representation of the continuous impingement air heating (A) detail of top and bottom impingement jet tube array (B).
From Kerry, J.F., 2011. Effects of novel thermal processing technologies on the sensory quality of meat and meat products. In: Processed Meats: Improving Safety, Nutrition and Quality. Woodhead Publishing Limited. https://doi.org/10.1533/9780857092946.3.617.

purpose, a model air impingement for thawing frozen food was developed and tested experimentally. They used a single impingement nozzle with a round, unconfined airflow. The samples thawed in less than 3 h under a single impingement jet, which was more than four times faster than thawing without impingement. It was suggested that air impingement could be used to increase the thawing rate of frozen foods without increasing the temperature at which microbial growth is potentially problematic (Anderson and Singh, 2006a, b).

Bozzato et al. (2021) evaluated the effect of air impingement (AI) for thawing chicken fingers. They designed and assembled a pilot AI thawing prototype (Fig. 8.5). The thawing chamber contained a blower and a chiller to provide air circulation and controlling the air temperature, respectively. The plenum with nozzles (diameter: 1.2 cm, and the distance between the nozzles: 7.2 cm) was placed above the thawing chamber to control the uniform airflow velocity. According to the results, AI thawing significantly decreased the thawing time (up to 85%) compared to conventional refrigerated thawing. Moreover, by reducing chicken fingers temperature during thawing, the structures of meat tissue and proteins were better preserved (Bozzato et al., 2021).

## 8.2.6 Two-stage air thawing

Many researchers have proposed two-stage air thawing as a method to shorten the thawing process ( James and James, 2002; Lee et al., 2021). First, a high air temperature is maintained until the surface of the frozen product reaches a certain temperature and then the air temperature is reduced to below 10 °C, until the thawing process is complete. As a result, bacterial growth is controlled due to low temperature and short thawing time.

Lee et al. (2021) examined the effect of two-stage air thawing on reducing thawing time and maintaining the freshness of frozen pork. The two-stage air thawing process was carried out by modifying the thawing temperature of frozen products to the transition zone borderline during the thawing process, such that the samples were initially

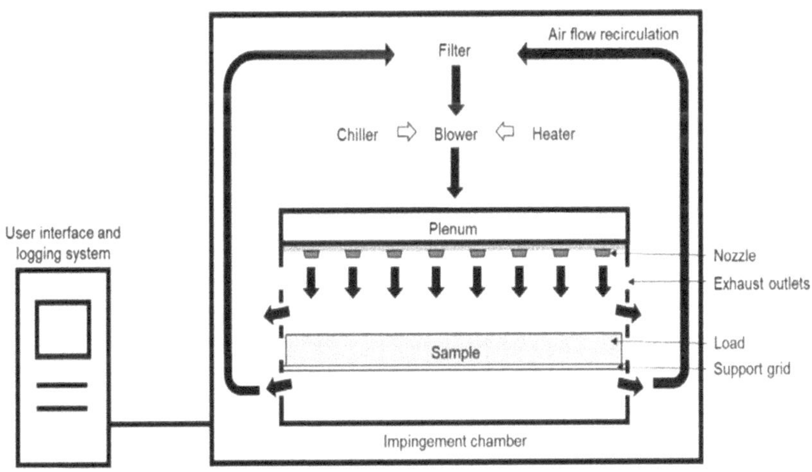


**Fig. 8.5** Diagram of the air impingement equipment.
From Bozzato, A., et al., 2021. Air impingement to reduce thawing time of chicken fingers for food service. J. Food Process. Preserv. 45 (11), 1–10. https://doi.org/10.1111/jfpp.15962.

thawed at 25 °C. After the core temperature of the sample reached −5 °C, the samples were chilled to −1 °C or 2 °C. Due to a large amount of heat provided at the higher thawing temperature during the first stage, the two-stage air thawing process considerably decreased the total thawing time. Therefore, this method can be regarded as a promising option for shortening the thawing time and preserving meat quality (Lee et al., 2021).

## 8.3 Water thawing

### 8.3.1 Theory

Water thawing (immersion thawing) is a process widely used in food industry, during which a product is immersed in water (Ersoy et al., 2008; Xu et al., 2021). This method is faster than thawing in a cold room, because the coefficients of heat transfer obtained in water are considerably more significant than those in the air; thawing times of frozen food are effectively reduced by water (Ersoy et al., 2008). Ice has a slightly lower density than liquid water and the product remains completely submerged. Consequently, the risk of drying the surface of food products and lipid oxidation is reduced and leads to higher food quality compared with air-based thawing methods (Wang et al., 2021). Using water for thawing is limited because packaged or boxed items must be removed from their packaging prior to thawing (unless they are shrink-wrapped or vacuum-sealed). The blocks of boned-out cuts or entire fish, break apart and become dispersed in the thawing tank, resulting in handling difficulties (Cai et al., 2020; Xu et al., 2021). Also, the lower the temperature differences between medium and product, the longer the thawing time. Furthermore, if the thawing medium is not changed frequently, microbial populations will increase and cause contamination of the thawed product and reduced quality.

### 8.3.2 Equipment

Immersion thawing systems may be batch or continuous. An immersion batch thawing system consists of a rectangular tank filled with water and a product to be thawed. Fig. 8.6 illustrates an immersion batch thawing machine designed to thaw frozen fish blocks. The immersion thawing process takes about 1 h at 7 °C. As a result of convection, the water temperature remains constant throughout the thawing operation, ensuring uniform defrosting of the raw material. The design is simple and ensures high reliability.

Continuous immersion thawing is another type of thawing operation, which exists in a several shapes and sizes (Fig. 8.6). There is a large cylindrical tank filled with water and divided into compartments where thawing products are placed. By rotating the compartments around the cylinder's rotational axis, the products are entirely thawed after one turn.

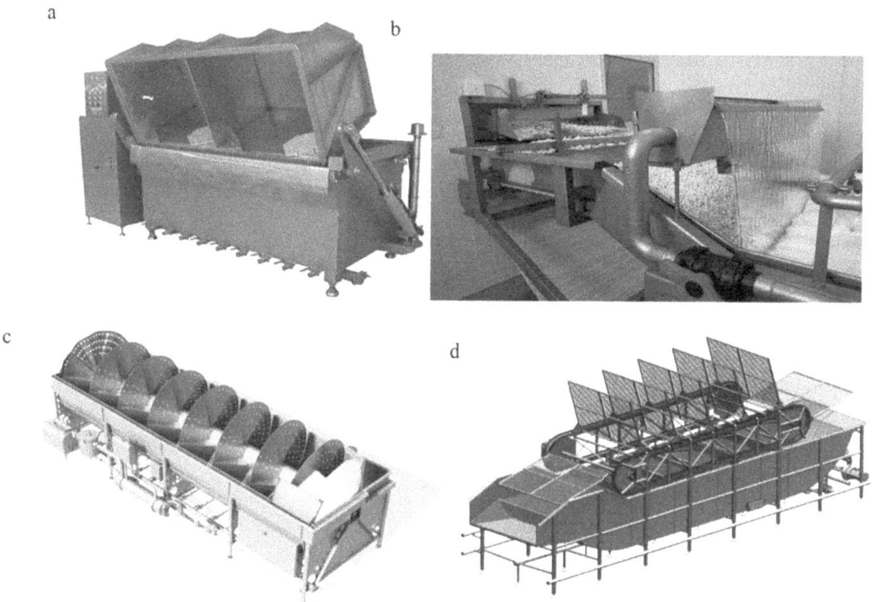


**Fig. 8.6** Batch Immersion thawing machine (A and B); continues immersion thawing machine (C and D).
From (A) https://www.directindustry.com/prod/cabinplants/product-94695-2132457.html.
(B) https://www.sairem.com/the-different-industrial-food-thawing-technologies/.
(C and D) https://www.directindustry.com/prod/uni-food-technic/product-209863-2130923.html.

## 8.4 Vacuum thawing

### 8.4.1 Theory

A vacuum thawing system works on the basis that the decreased pressure lowers the boiling point of water. In the vacuum chamber, water vapor condenses on the product's surface while it releases heat from condensation (Watanabe and Ando, 2021; Xu et al., 2021). Purposes of this method are to decrease the condensation temperature of the steam and to raise the rate of heat transport by the removal of the air molecules. Vacuum thawing occurs at a higher rate than water thawing, with less pollution and water absorption (Watanabe and Ando, 2021).

### 8.4.2 Equipment

Vacuum thawing machine is commonly used for thawing beef, meat, poultry, and seafood. Blocks of the frozen product are stacked on trolleys and wheeled into an airtight chamber, which is then evacuated by a liquid ring pump. A diagram of a commercial vacuum thawing unit can be seen in Fig. 8.7. Water vapor is produced by heating a small quantity of water in a shallow tank at the bottom of the chamber. Typically, steam or hot water is used for heating, but sometimes a heat exchanger or electricity

**Fig. 8.7** Vacuum thawing machine for aquatic product and seafood.
From http://www.frperc.com/FRPERC.com/Thawing and tempering.html.

is used to achieve a controlled temperature (usually 18 °C). As the vapor is distilled from the water in the tank, it condenses completely onto the frozen product and ensures that the product is not contaminated with bacteria in the tank. Because the walls of the chamber are completely dry, there is no odor generated, making the whole process hygienic and requiring little cleaning. In the presence of air, condensation is slowed down, but without the air, the rate of condensation and heat transfer increases. In addition, removing the air allows the water vapor to penetrate small spaces between blocks of product. This can be a great advantage in some food products; for example, shrimp blocks of 65 mm thick can be thawed in 78 min.

## 8.5 Ultrasound thawing

### *8.5.1 Theory*

Ultrasonic waves are like sound waves but have frequencies higher than 16 kHz and cannot be heard by a human. Ultrasound is divided into two categories, namely low-intensity ultrasound ($<1\,W\,cm^{-2}$) and high-intensity ultrasound ($10–1000\,W\,cm^{-2}$) (Singla and Sit, 2021). The former is utilized as a nondestructive analytical technique to evaluate the structure, composition, and flow rate of food products, and the latter is applied at higher frequencies (up to 2.5 MHz) for the physical destruction of tissues, emulsification, cleaning of equipment, and promotion of chemical reactions (e.g., oxidation) (Natarajan and Ponnusamy, 2020; Khadhraoui et al., 2021). Ultrasonic waves generate a force when colliding with the surface of a substance. A compression wave is generated when the force is vertical to the surface of food material. On the other hand, a shearing wave is generated when the force is parallel to the surface. Both kinds of waves are attenuated when they pass through the substance (Natarajan and Ponnusamy, 2020). Ultrasound generates extremely fast localized alterations in temperature and pressure, bringing about cavitation (bubble creation in liquid foods), disruption of cell membranes, localized heating, and generation of free radical, all of which produce destructive impacts on microorganisms (Fig. 8.8) (Singla and Sit, 2021).

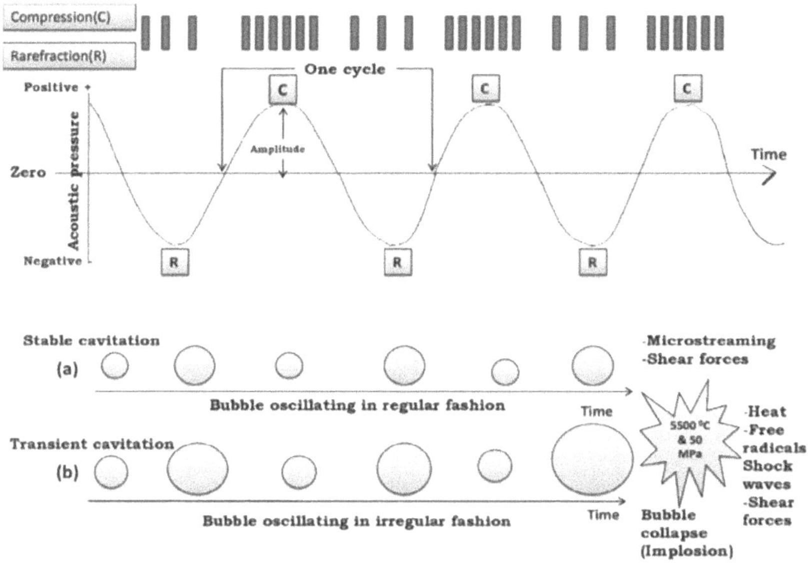


**Fig. 8.8** Ultrasonic cavitation.
From Singla, M., Sit, N., 2021. Application of ultrasound in combination with other technologies in food processing: a review. Ultrason. Sonochem. Elsevier B.V., 73, 105506. https://doi.org/10.1016/j.ultsonch.2021.105506.

Ultrasound is generated with a transducer that converts electrical pulses into acoustic energy at the required intensity (Bhargava et al., 2021). There are two kinds of transducers that are mainly employed for the generation of ultrasound, magnetostrictive and piezoelectric (Fig. 8.9). A magnetostrictive transducer works as an electroacoustic transducer to generate ultrasonic waves. Piezoelectric transducers interconvert acoustic and electrical energies.

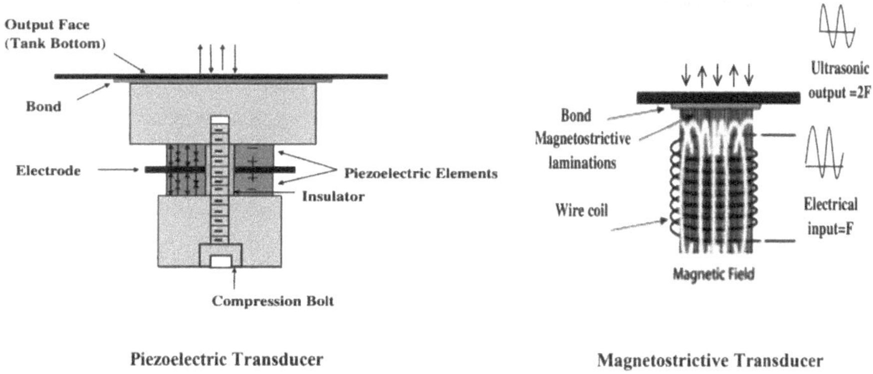


**Fig. 8.9** Different ultrasound generators.
From Bhargava, N., et al., 2021. Advances in application of ultrasound in food processing: a review. Ultrason. Sonochem. Elsevier, 70 (June 2020), 105293. https://doi.org/10.1016/j.ultsonch.2020.105293.

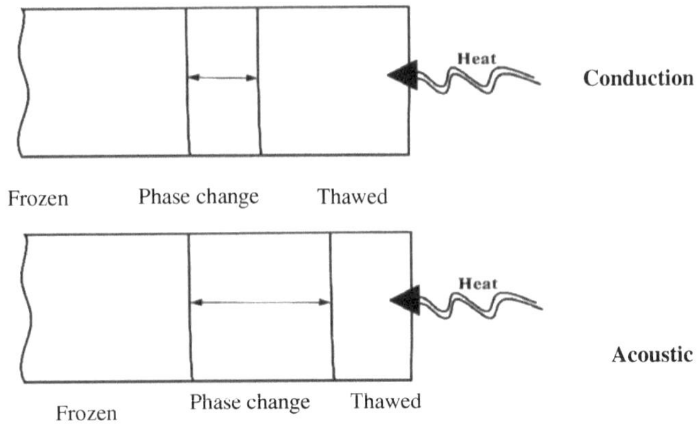


**Fig. 8.10** Comparison of ultrasound thawing with conventional thawing on the region of phase shift.
From Khadhraoui, B., et al., 2021. Review of ultrasound combinations with hybrid and innovative techniques for extraction and processing of food and natural products. Ultrason. Sonochem. Elsevier B.V., 76, 105625. https://doi.org/10.1016/j.ultsonch.2021.105625.

As ultrasonic waves pass through a frozen food, a fraction of their energy converts into heat to contribute to the thawing process. The surface temperature of the frozen slab will quickly rise. Nevertheless, the rise in the temperature of the slab core is slower. As the temperature exceeds the initial freezing point, the frozen part in the vicinity of the frozen/thawed boundary is capable of absorbing more energy to accelerate the thawing process (Fig. 8.10). During thawing, the physical impacts of ultrasound can raise the water temperature and generate extremely fast jets that will turn, causing the bubbles to collapse asymmetrically. This phenomenon enhances heat transfer (Chemat et al., 2011; Khadhraoui et al., 2021).

## 8.5.2 Equipment

The use of ultrasonic is a growing technique for thawing frozen foods in order to accelerate the thawing rate and to retain their quality (Singla and Sit, 2021). Several studies have examined the effects of ultrasound-assisted thawing on food quality characteristics. Liu et al. (2019) studied the effect of ultrasound-assisted thawing on the sensory and nutritional properties of frozen mango pulp. In this study, it was shown that ultrasound processing at optimal conditions could be an alternative method to conventional thawing and lead to improved thawing efficiency and nutritional value (Liu et al., 2019). In general, ultrasound is not widely used in thawing due to its limited penetration, localized heating, and high power requirements. Nevertheless, this method can still be used in combination with conventional thawing in order to speed up the thawing process (Qiu et al., 2020; Khadhraoui et al., 2021). Hong et al. (2014) demonstrated the effect immersion thawing in water or brine combined with ultrasound treatment on pork quality characteristics. After removing frozen meat samples from the package, they were connected to a 100 MV mobile corder and located into a glass

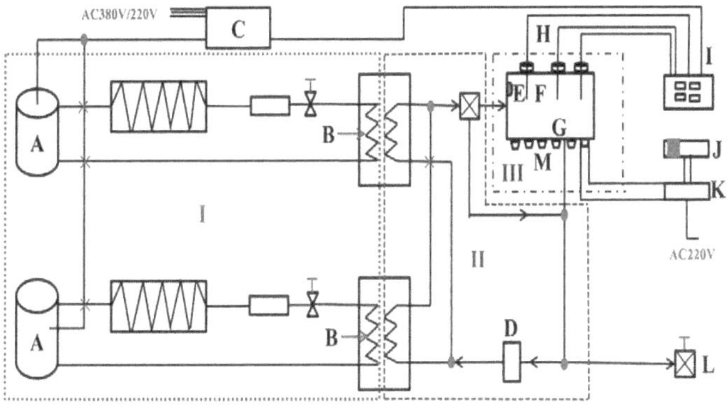


**Fig. 8.11** Diagram of ultrasound-assisted thawing device.
From Guo, Z., et al., 2021. Ultrasound-assisted thawing of frozen white yak meat: effects on thawing rate, meat quality, nutrients, and microstructure. Ultrason. Sonochem. Elsevier, 70 (1), 105345. https://doi.org/10.1016/j.ultsonch.2020.105345.

tray containing distilled water or brine (2% NaCl, w/v). A sonicator was used to generate ultrasound waves at a 40 kHz frequency and a 150 W power level. The ultrasound treatment in brine reduced cooking losses and shear force compared to the control (Hong et al., 2014).

Guo et al. (2021) explored the impact of ultrasound-assisted thawing (UAT) on the quality of white yak meat. A diagram of UAT apparatus is displayed in Fig. 8.11. This device included the following components: A, compressor; B, plate heat exchanger; C, control panel; D, pump; E, coolant circulation inlet; F, ultrasonic thawing tank; G, coolant circulation outlet; H, thermocouple; I, temperature measuring system; J, ultrasonic controller; K, ultrasonic generator; L, air-vent; M, ultrasonic transducer. As a result of UAT treatment, thawing time was decreased by 30%–64% compared to control. The use of an appropriate ultrasonic power can minimize the thawing loss of meat, avoid mineral and water-soluble vitamin degradation, and result in reduced cutting force. In conclusion, UAT improves thawing efficiency and reduces the undesirable changes caused by thawing in the meat industry (Guo et al., 2021).

Gambuteanu and Alexe (2015) assessed the impact of ultrasound-assisted thawing on texture and technical characteristics of pork meat. The thawing was carried out using an ultrasound generator, a transducer, and a water bath (Fig. 8.12). Ultrasound is transmitted through a liquid medium from the generator (1) through a plate transducer (2) located above the water bath (3). As the transducer is located above the water bath, ultrasound can be transmitted uniformly throughout the entire sample, thereby thawing the sample uniformly. (5) The samples were placed on a grating (4) to avoid contact with the bottom of the vessel. According to the results, ultrasound treatment at suitable parameters (25 kHz; 0.6 W/$cm^2$) considerably decreased the thawing time of the pork samples by 87% without affecting their textural or technological characteristics. Another study by the same researchers showed that ultrasonic thawing of pork meat using low intensities (25 kHz; 0.2; 0.4 W/$cm^2$) had no significant impact on chemical, microbiological, or texture attributes (Gambuteanu and Alexe, 2015).

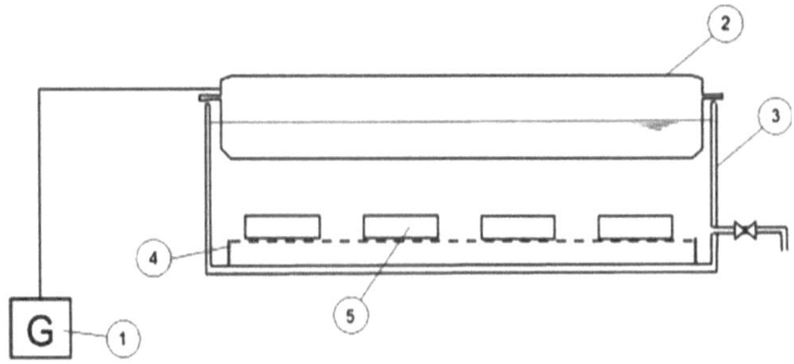


**Fig. 8.12** Diagram of ultrasound-assisted thawing device. (1) Generator; (2) plate transducer; (3) water bath; (4) grating; and (5) meat samples.
From Gambuteanu, C., Alexe, P., 2015. Comparison of thawing assisted by low-intensity ultrasound on technological properties of pork longissimus dorsi muscle. J. Food Sci. Technol. 52 (4), 2130–2138. https://doi.org/10.1007/s13197-013-1204-7.

Cai et al. (2019b) evaluated the effect of ultrasonic combined with far-infrared thawing (UFT) or microwave thawing (UMT) on properties of *Sciaenops ocellatus*. UMT device is presented in Fig. 8.13; the ultrasonic instrument was placed in the bottom of the tomographic ark. A microwave generator was attached to the separation layer of the tomographic ark with high-temperature rope, so the microwave transmitter outlet faces a heat-resistant plastic basin. Samples were placed in the basin until the core temperature reached 4 °C. Results showed that UMT and UIT

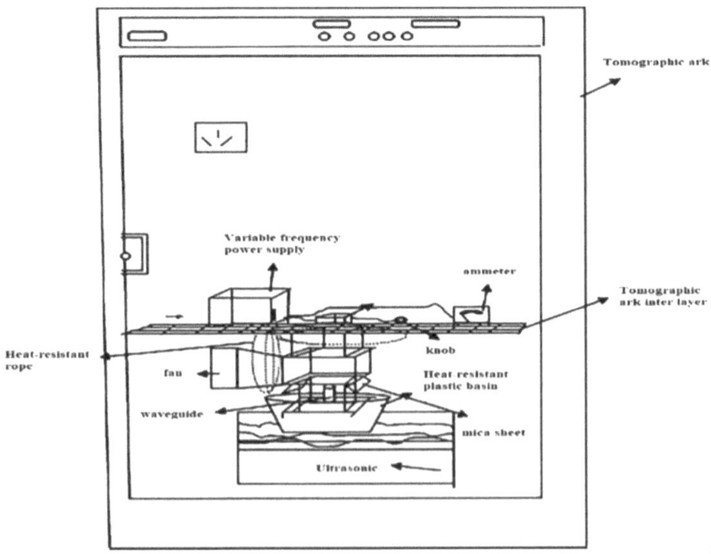


**Fig. 8.13** Diagram of ultrasonic and microwave combined.
From Cai, L., et al., 2018. Ultrasound or microwave vacuum thawing of red seabream (Pagrus major) fillets. Ultrason. Sonochem. Elsevier, 47 (March), 122–132. https://doi.org/10.1016/j.ultsonch.2018.05.001.

**Fig. 8.14** Ultrasonic microwave thawing equipment.
From https://chinathaw.com/ultrasonic-microwave-thawing-equipment/.

preserved muscle fiber structure and protein stability more than any other thawing method. Therefore, UMT and UIT minimized protein denaturation and water migration. Another study by these authors concluded that ultrasonic vacuum thawing (UVT) and microwave vacuum thawing (MVT) maintained actin thermostability and kept tertiary structure better than other thawing technologies (Cai et al., 2018, 2019b).

Chinathaw Company produces industrial ultrasonic microwave thawing device. This equipment is shown in Fig. 8.14. The device consists of body (1) and a freezing zone (2). There is a thawing cabinet (21) and a refrigerating cabinet (22) in the freezing area (2). The thawing cabinet is equipped with ultrasonic thawing devices (3), which will be thawed inside the cabinet. Because food is usually placed on the bottom surface of the cabinet, the ultrasonic thawing device is placed outside this surface to maximize thawing efficiency. Ultrasonic microwave thawing equipment uses ultrasonic vibrations to break the ice crystals and increase the rate of thawing.

Kissam et al. (1982) examined the effects of simultaneous ultrasound and water thawing on frozen pacific cod blocks. Cod blocks were thawed in a circulated water bath at 18°C while simultaneously exposed to acoustic energy at 1500 Hz. The ceramic transducer was used to make light contact with the frozen block. A motorized belt moved the block so that the 239 $cm^2$ transducer was distributed over the 1500 $cm^2$ block surface. The combined use of ultrasound and water thawing reduced thawing time by 71% compared to water thawing alone (Kissam et al., 1982).

## 8.6 High-pressure thawing

### *8.6.1 Theory*

High-pressure processing (HPP), also called high hydrostatic pressure (HHP) or ultra-high pressure (UHP), can inactivate both spoilage microorganisms and pathogens by the application of pressure instead of heat (Elamin et al., 2015). To the best of our knowledge, high pressures merely influence noncovalent bonds (ionic, hydrophobic, and hydrogen bonds) and do not affect the covalent ones. This inhibits microbial activity without significantly impacting upon the texture-contributing food molecules, vitamins, flavoring agents, and pigments of the food (Okamoto and Suzuki, 2002). High-pressure thawing is based on the premise that the freezing temperature of water drops with increasing pressure (up to 210 MPa). The freezing temperature of water at ambient pressure is 0 °C, while at 210 MPa, it is around −21 °C. Takai et al. (1991) were the first researchers who applied UHP-assisted thawing. They realized that this technique could be utilized at temperatures lower than 5 °C and shorten the process duration (Boonsumrej et al., 2007). UHP reduces the specific heat capacity and the melting point of ice, giving rise to its thermal conductivity. The lower melting point elevates the temperature gap between the phase change front and the heating medium. Consequently, the thawing rate rises which can be ascribed to the dramatic difference temperature between thawing medium and the frozen food under high pressure (Rouillé et al., 2002; Zhu et al., 2014). High-pressure systems have benefits, like (1) the process could be run at room, or even low temperatures and (2) nutrients or natural colors and flavors are not severely damaged, which lead to the production of high-quality products (Ganessingh et al., 2018). The homogenous exertion of high pressures on all components of food material, irrespective of its shape and size, is an essential advantage over other processing techniques, as the entire product is treated homogenously. HPP has no "heating" or "cooling" periods while comprising fast pressurization/depressurization cycles. This shortness processing duration relative to thermal processing, HPP, could be applied together with other kinds of processing methods, hence increasing the number of unit operations for food manufacturers, resulting in the creation of novel processes and products (Elamin et al., 2015; Verma et al., 2020).

## 8.6.2 Equipment

The essential parts of a HPP system are pressure transmitting fluid, pressurization chamber, supporting heating–cooling components, and material handling unit (Fig. 8.15) (Verma et al., 2020). During HPP, the food packaged in flexible containers/packs is put in the holding basket, and the basket is placed in the reaction chamber. HHP is generated by a pump or piston that compresses a pressure-transmitting fluid and uniformly distributes pressure in the food matrix (Elamin et al., 2015). Materials are handled using automatic equipment in in-container processing, similar to what is utilized for loading/unloading batch retorts. Bulk handling is more accessible needing pumps, and valves (Verma et al., 2020).

There are three types of industrial high-pressure system: batch, continuous, and semicontinuous. Batch processing is suitable for both liquid and solid products, whereas continuous and semicontinuous processing are only suitable for pumpable products and liquids (Elamin et al., 2015).

*Batch*: A batch pressure system includes a cylindrical pressure vessel, closures for the ends, a fastening device, and an intensifying pump (Fig. 8.16). The pressure vessel is filled with liquid (generally water), and food is vacuum-packed into containers to

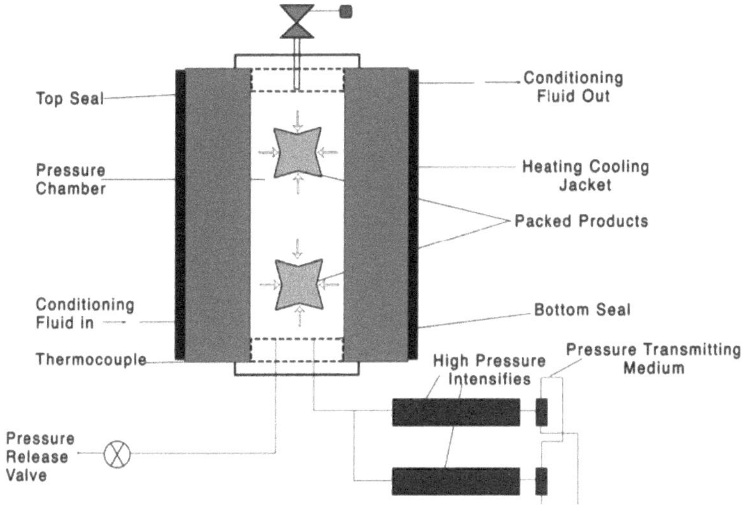


**Fig. 8.15** Schematic representation of high-pressure processing equipment.
From Ganessingh, V., Sahibdeen, R., Maharaj, R., 2018. An evaluation of the impact of novel processing technologies on the phytochemical composition of fruits and vegetables. In: Phytochemicals—Source of Antioxidants and Role in Disease Prevention, (November 2018). https://doi.org/10.5772/intechopen.77730.

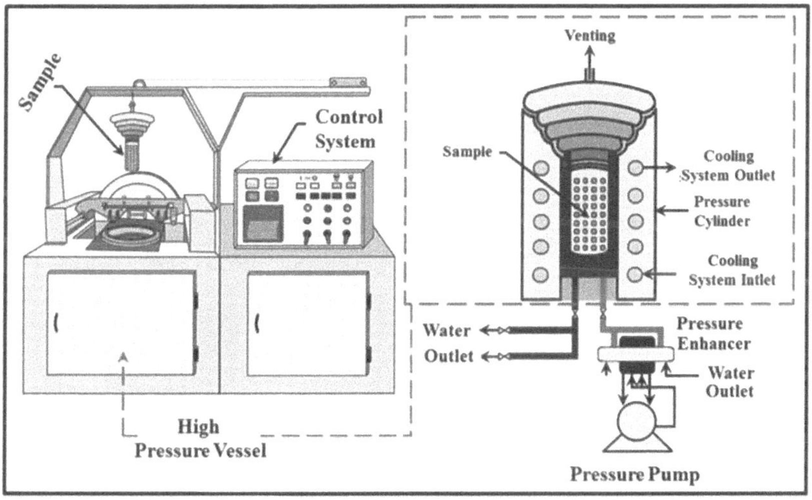


**Fig. 8.16** Schematic representation of a batch setup of high-pressure processing.
From Elamin, W.M., et al., 2015. High pressure processing technology and equipment evolution: a review. J. Eng. Sci. Technol. Rev. 8 (5), 75–83. https://doi.org/10.25103/jestr.085.11.

prevent mixing. The fluid that transmits pressure is pumped into the vessel from the storage tank and the inlet valves are closed when reaching the required pressure. The pressure in the vessel is released when the product has been held at 600 MPa for the required time (3–5 min) (Elamin et al., 2015; Verma et al., 2020).

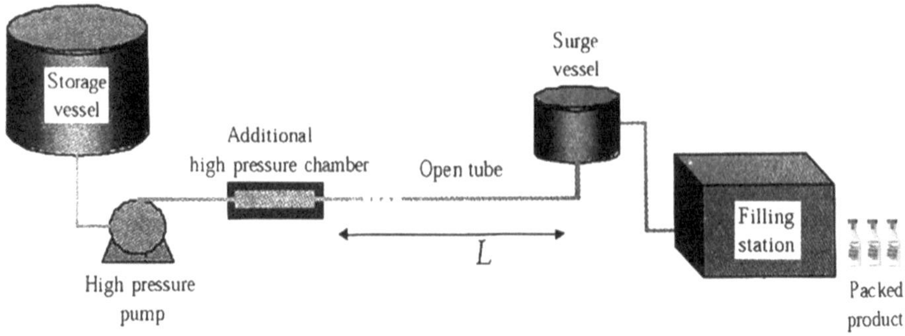


**Fig. 8.17** Schematic of the continuous high-pressure processing system.
From Elamin, W.M., et al., 2015. High pressure processing technology and equipment evolution: a review. J. Eng. Sci. Technol. Rev. 8 (5), 75–83. https://doi.org/10.25103/jestr.085.11.

*Continuous*: The continuous processing method is shown in Fig. 8.17. Products are passed through open-end tubes that are pressurized at 100 MPa or higher using high-pressure intensifiers. After depressurization, the liquid is transferred into a sterile tank for clean filling, thus preventing extreme shear and heat. As a result of high shear forces, friction heating, and other flow phenomena, this method has been replaced by semicontinuous operations (Elamin et al., 2015; Verma et al., 2020).

*Semicontinuous*: Semicontinuous systems, as shown in Fig. 8.18 (Lelieveld and Hoogland, 2016), typically consist of multiple pressure vessels, a low-pressure pump for filling the vessels, a high-pressure transfer pump, sterile tanks, and valves to control the process. Free-moving pistons are used separate the pressurizing fluid from the product in each pressure vessel. A product is pumped into the vessel, and then the inlet valve is locked, allowing pressurizing fluid to be compressed into the vessel and moving the piston to compress the liquid food. The vessel is depressurized after a certain

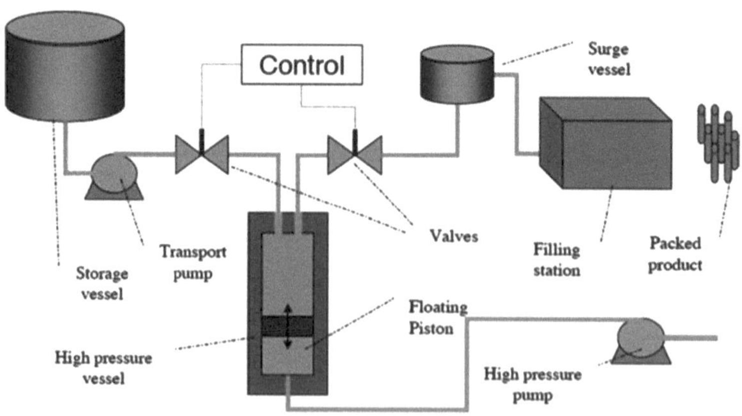


**Fig. 8.18** Schematic of the semicontinuous high-pressure processing system.
From Elamin, W.M., et al., 2015. High pressure processing technology and equipment evolution: a review. J. Eng. Sci. Technol. Rev. 8 (5), 75–83. https://doi.org/10.25103/jestr.085.11.

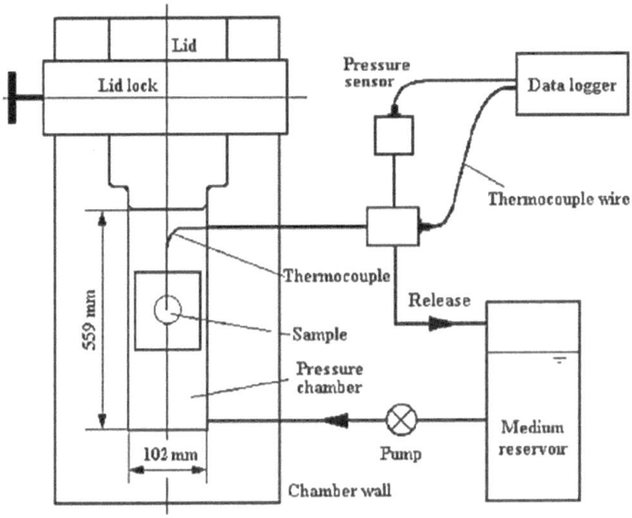


**Fig. 8.19** Schematic description of the high-pressure equipment.
From Zhu, S., Ramaswamy, H.S., Simpson, B.K., 2004. Effect of high-pressure versus conventional thawing on color, drip loss and texture of Atlantic salmon frozen by different methods. LWT Food Sci. Technol. 37 (3), 291–299. https://doi.org/10.1016/j.lwt.2003.09.004.

time. As the high-pressure is released, the food products are decompressed and the piston is returned to its original position (Cheftel et al., 2000; Elamin et al., 2015).

Zhu et al. (2004, 2014) assessed the effect of high-pressure thawing (HPT) on frozen agar gel and Atlantic salmon. HPT tests were conducted in a cold isostatic equipment with a capacity of 4.5L (Fig. 8.19). The system used distilled water containing mineral oils as the medium for pressure transmission. The HPT treatment was performed at three pressure levels (100, 150, and 200MPa). The results revealed that HPT significantly accelerated the thawing process due to the depressed ice melting temperature and HPT should operate below 150MPa to minimize salmon discoloration and toughness (Zhu et al., 2004, 2014).

## 8.7 Dielectric thawing

### 8.7.1 Theory

Dielectric thawing is a collective term representing all types of thawing techniques employing high-frequency waves (radio frequency (RF) and microwave (MW)) for thawing foods (Backi, 2018). In the electromagnetic spectrum, RF waves typically lie in the range of 3kHz to 300MHz, while MWs mainly vary from 300MHz to 300GHz, and infrared light waves range between 300GHz and 385THz. At the same time, merely a small number of frequencies in these ranges are permitted to be used in scientific, industrial, and medical applications (13.56, 27.12, and 40.68MHz for the RF waves and 915MHz, 2450MHz, 5.8GHz, and 24.124GHz for the MWs)

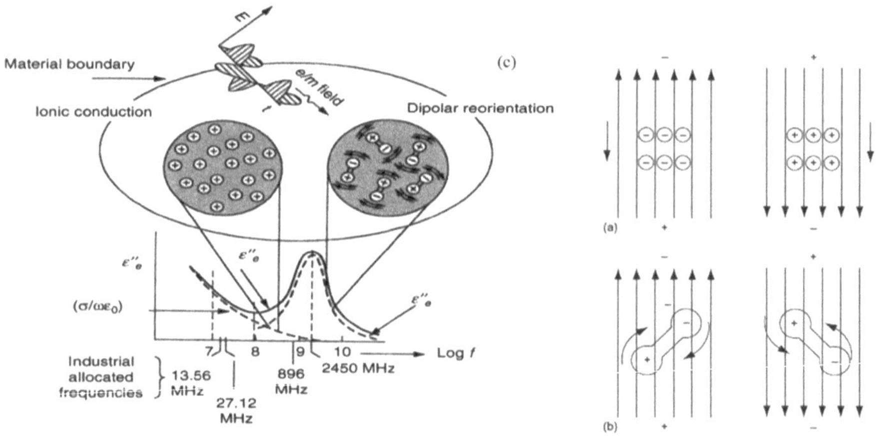


**Fig. 8.20** Schematic representation of (A) ionic polarization and (B) dipolar rotation in alternating MW field (C) mechanisms of dipolar and ionic loss and their contributions to dielectric properties.
From Llave, Y., Sakai, N., 2018. Dielectric Defrosting of Frozen Foods, Food Processing for Increased Quality and Consumption. Elsevier Inc. https://doi.org/10.1016/B978-0-12-811447-6.00013-8.

(Llave and Sakai, 2018). Both MWs and RF waves are classified under the non-ionizing type of electromagnetic radiation and could be explained to time-dependent magnetic and electric fields that move in the space based on wave-like patterns (Uyar et al., 2015). Unlike conventional techniques, dielectric heating, also called volumetric heating, can produce heat inside the product. During dielectric heating, energy is absorbed into food products, mainly by two mechanisms: ionic conduction and dipolar relaxation, also known as dipolar reorientation (Fig. 8.20) (Uyar et al., 2015; Llave and Sakai, 2018). Water is mostly the main compound liable for dielectric heating. Owing to dipolar nature, water molecules try to orient in the direction of electric fields as they alternate at extremely high frequencies. The molecules begin to rotate in an oscillating electric field and collide with one another, and the elevation of the kinetic energy results in the production of heat. The rotation of the molecules of water generates heat ( James et al., 2017; Llave and Sakai, 2018; Guzik et al., 2021). In addition to the movement of the ions, once food products are subjected to RF waves, polar molecules begin the dipole rotation in the direction of the oscillating electromagnetic field, which contributes to the conversion of the electromagnetic energy into heat, particularly in frozen foods (Llave and Erdogdu, 2022). After that, the thawing rate is increased because of the phase change of ice to water, which increases the dielectric properties (DP). Consequently, the electromagnetic energy is more absorbed and converted into heat. As the polarization indicates its influence, the separated ions begin to rerotate in the direction of the oscillating electric field. Therefore, heat is produced by the friction between the molecules (Sumnu and Sahin, 2005; Uyar et al., 2015; Llave and Sakai, 2018). In conclusion, dielectric heating using MWs in higher frequencies primarily continues by dipole reorientation. This is in contrast with ionic conduction, the main mechanism in heating with RF waves with lower frequencies (Backi, 2018; Llave and Sakai, 2018). In conventional RF heating, including the systems of parallel plate electrodes, the electromagnetic field is typically developed in a directional way between the plate electrodes (Fig. 8.20), whereas the MW electromagnetic field may diffuse into the foodstuff from various directions considering magnetron structure, cavity design, and waveguides location (Chen et al., 2021; Guzik et al., 2021).

In conventional RF heating, including the systems of parallel plate electrodes, the electromagnetic field is typically developed in one direction between the plate electrodes, whereas the MW electromagnetic field may diffuse into the foodstuff from various directions considering magnetron structure, cavity design, and waveguides location (Krishnamurthy et al., 2008; Tang et al., 2021). Llave and Erdogdu (2022) have defined this phenomenon on the corners as the magnet-like effect. Therefore, compared with the nonuniform and complex field generated in MW devices, RF waves usually deliver an uncomplicated and more homogenous electromagnetic field (Llave and Erdogdu, 2022). RF controls temperature increase during the process more efficiently, owing to the homogeneity in the distribution and penetration depth of temperature. The longer wavelength of RF implies it can penetrate more than MW and thus might be more appropriate for large food products, whereas MW energy is better absorbed and as a result, it has higher rates of heating (Rastogi, 2012; Llave and Sakai, 2018).

### 8.7.2 Microwave thawing (MW)

A microwave system is composed of magnetron, step-up transformer, oven cavity, waveguide, and applicator (Fig. 8.21). The magnetron converts electric energy into microwave energy at 4000 V. The microwaves produced by magnetron travel through a waveguide to the cavity. As the microwave applicator in a household microwave oven, a rectangular cavity represents a multimode resonator (Sumnu and Sahin, 2005; Guzik et al., 2021).

The MW applicators are grouped into multimode, single-mode, and traveling applicators (Fig. 8.22). The multimode applicator is used in both domestic and industrial MW ovens. It consists of a rectangular metal box in which the frozen food is held. Waveguides transmit microwave energy to the applicator, which is reflected multiple times by the cavity walls to create a distributed electric field within the cavity and within the product. This produces stable modes of propagation of the electric field at multiple points, hence it is called multimode applicator (Swain and James, 2005; Llave and Sakai, 2018; Guzik et al., 2021). A single-mode cavity operates most simply and is physically the smallest. The single-mode cavity allows only one mode of propagation within a defined area. Samples are located in an electrically conductive rectangular or cylindrical cavity at predetermined locations where a concentrated electric field occurs. The single-mode applicator consists of a feeding waveguide, a tuning plunger, and a small microwave resonator (Guzik et al., 2021).

The microwave thawing has many advantages over conventional systems, such as high thawing rate, minimal drip loss, low space requirements, and high flexibility.

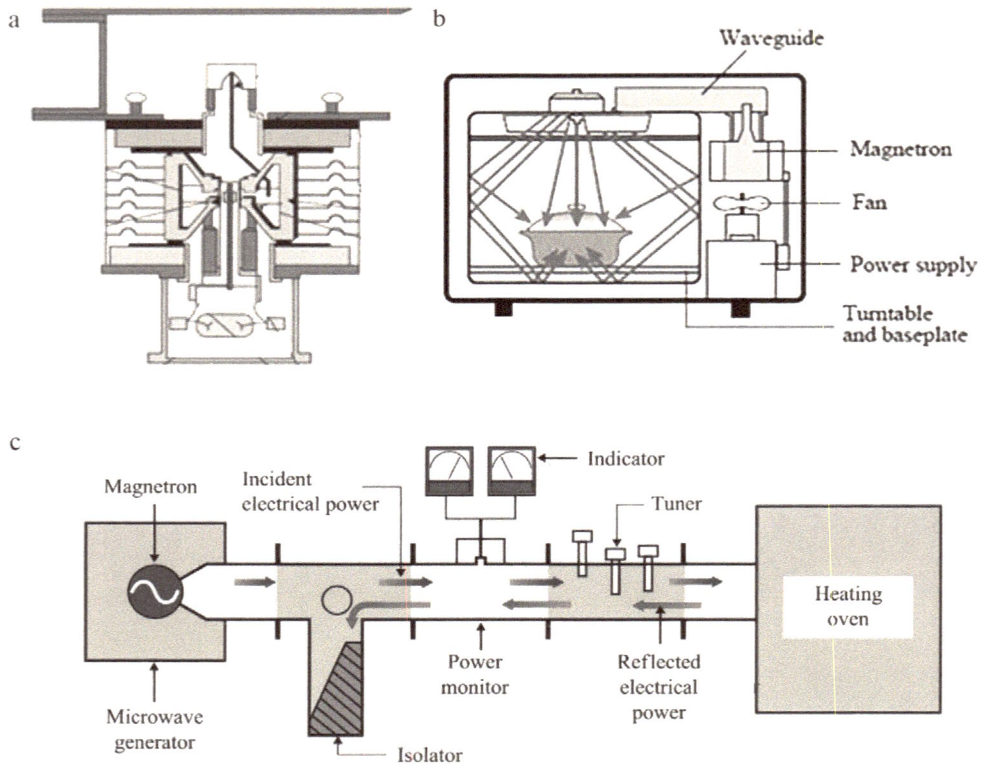


**Fig. 8.21** Schematic (A) components of a MW heating system, (B) components of a domestic turntable MW heating oven, and (C) configuration of a MW heating device (conveyor type).
From Llave, Y., Sakai, N., 2018. Dielectric Defrosting of Frozen Foods, Food Processing for Increased Quality and Consumption. Elsevier Inc. https://doi.org/10.1016/B978-0-12-811447-6.00013-8.

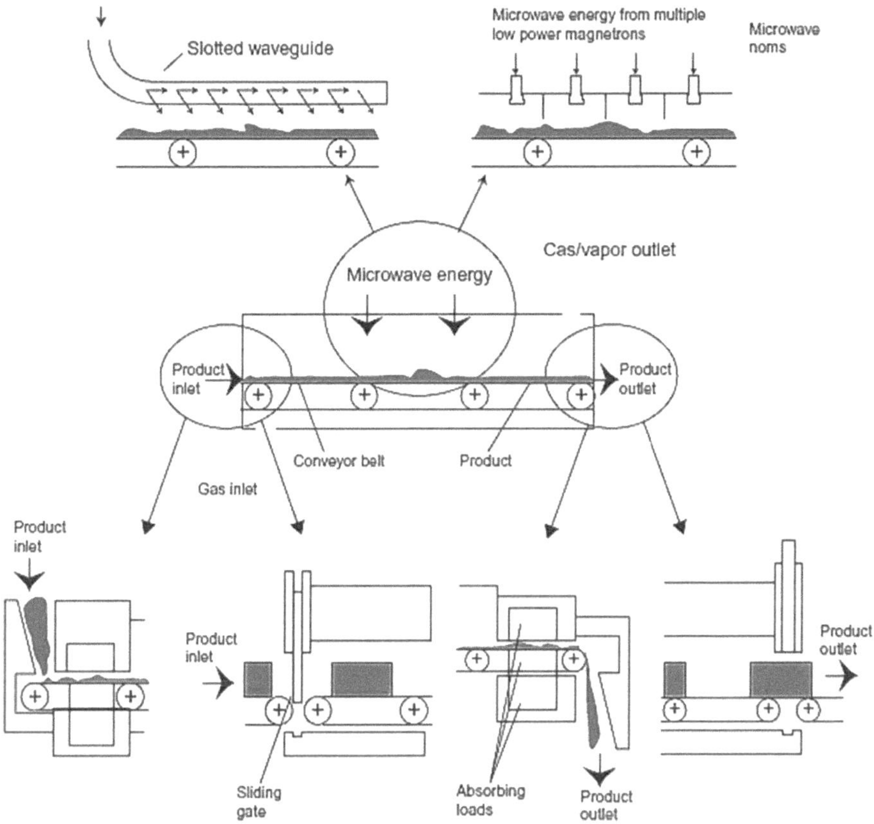


**Fig. 8.22** Types of magnetrons and parts of conveyorized MW systems.
From Umesh Hebbar. H., Rastogi, K., 2012. Microwave heating of fluid foods. In: Novel Thermal and Non-Thermal Technologies for Fluid Foods, pp. 369–409 (Chapter 12).

However, the high cost of initial investments, process optimization, operation costs, and increased workforce and energy requirements are currently limitations faced by industrial microwave applications (Wray and Ramaswamy, 2015). Industrial systems can be batch or continuous with different sizes ranging from small to large continuous tunnels. The MIP4 mini-batch microwave oven (Ferrite's Industrial Microwave Systems, Nashua, NH, USA) is ideal for low-to-mid volume foodservice operations that need to defrost moderate quantities of food in a small space (Fig. 8.23). It is excellent for meat, poultry, fish, fruits, and bakery products. The microwave oven powered by a 40kW magnetron operating at 915MHz and can defrost up to 680kg/h raw frozen products.

Microwave Tunnel Tempering System (MIP8) (Ferrite's Industrial Microwave Systems, Nashua, NH, USA) is used for mid-to-high volume food processing (Fig. 8.24). MIP8 is fully compliant with EU emissions standards. The conveyor ends are provided with unique vestibules to prevent microwave energy from leaking into the environment. The use of a microwave tunnel improves product quality and reduces the drip loss.

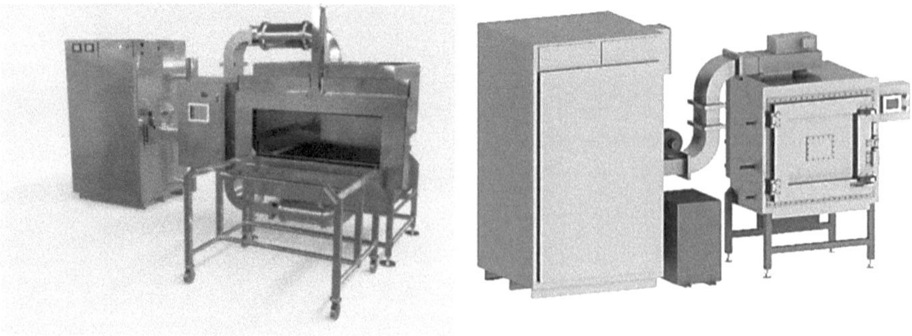

**Fig. 8.23** The MIP4 mini-batch microwave oven.
From https://ferriteinc.com/industrial-microwave-systems/mini-batch-microwave-tempering-oven/.

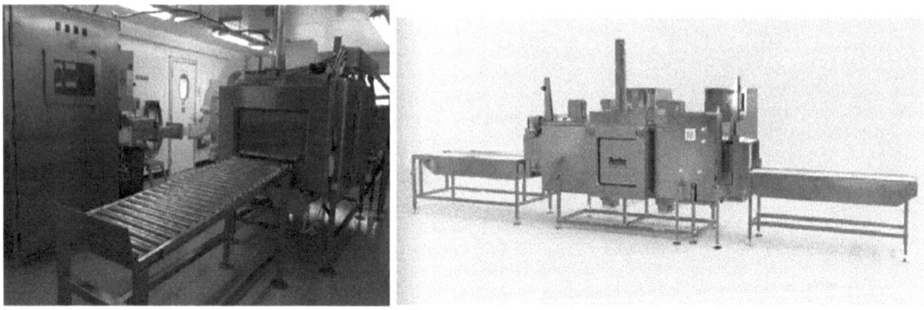

**Fig. 8.24** Tunnel Microwave System (MIP8).
From https://ferriteinc.com/industrial-microwave-systems/TunnelMicrowaveSystem (MIP8).

Microwave Tempering Tunnel Oven System (MIP12) (Ferrite's Industrial Microwave Systems, Nashua, NH, USA) is a continuous tempering microwave system used in industrial food processing (Fig. 8.25). MIP12 has the capacity to temper up to 7700 kg of frozen product per hour by connecting four 75 kW generators to each 2.5 m tunnel. Temperature control enables quick tempering of seafood, frozen meat, poultry, fruits, and bakery items. This instrument reduces the production time from days to minutes.

Zhu et al. (2020) evaluated different thawing methods on physicochemical changes and protein denaturation in pork. All microwave-based thawing methods avoided local overheating. Microwave combined with air convection (MAT) considerably decreased protein denaturation and the formation of carcinogenic compounds. According to the findings of this study, MAT successfully reduced the thawing time and preserved the meat quality (Zhu et al., 2020). In order to improve the effectiveness of MWs, researchers have recommended combining them with other methods (MW with vacuum or hot air) and applying them at the end of the process. Nguyen et al. (2013) designed and constructed a continuous microwave-ohmic heater. The chamber was equipped with two dual magnetrons capable of delivering 0.9 kW each and two

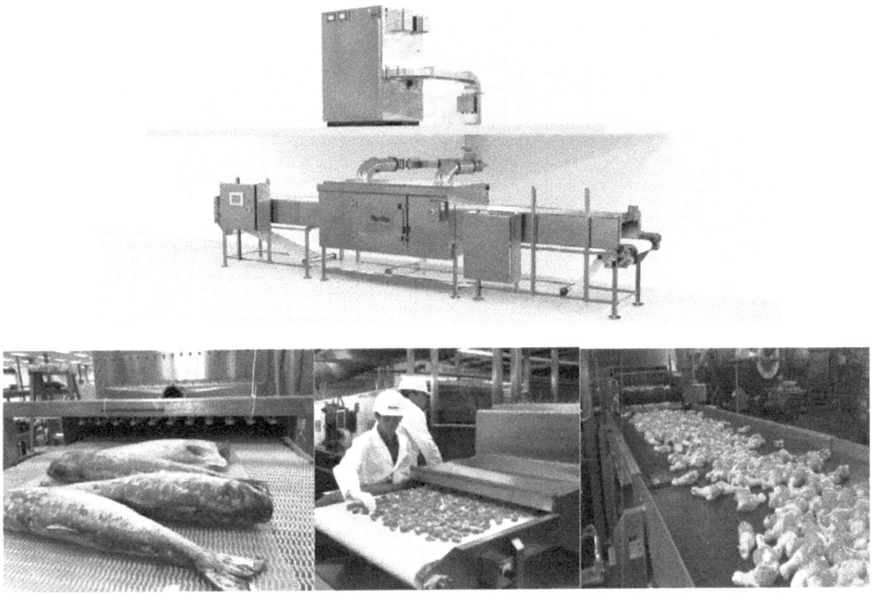

Fig. 8.25 Microwave Tempering Tunnel Oven System (MIP12).
From https://ferriteinc.com/industrial-microwave-systems/tempering-tunnel-oven.

transmission ports connected to the waveguide built with nickel-coated brass (Fig. 8.26). A polytetrafluoroethylene (PTFE) tube was placed in the cavity and used as an applicator. As an ohmic heating chamber, the applicator consisted of two ring-shaped titanium electrodes that were connected to the power supply by lead wires. The findings exhibited that in microwave, ohmic and microwave-ohmic heating, maximum temperature differences between solids and liquids were approximately 8.1, 8.0, and 0.5 °C, respectively. As a result, the combination heating had a higher energy efficiency than microwave heating, and the energy conversion increased by 12.8% (Nguyen et al., 2013).

In another study, Lee et al. (2015) developed a multicylinder ohmic and microwave combined heater to minimize thermal drop back when heating particulate foods. In this study, ohmic and microwave power sources were used to create multicylindrical heating chamber to concentrate electric field strengths (Fig. 8.27). The chamber composed of two ports connected to waveguide on each end an ohmic tube was mounted. Two titanium electrodes were placed at either end of the ohmic heating tube. In order to generate an alternating pulsed waveform, electrodes were connected to an integrated gate bipolar transistor power supply. This study showed that the highest thermal differences were 7.1 °C (microwave), 11.9 °C (ohmic), and 3.08 °C (combination heater). The study claims that the combined heating method can reduce over-processing of multiphase foods caused by changes in electrical conductivity in food components (Lee et al., 2015).

Watanabe and Ando (2021) examined the effects of air, microwave (MW), and vacuum microwave (VMW) thawing on characteristics of frozen apples, also the

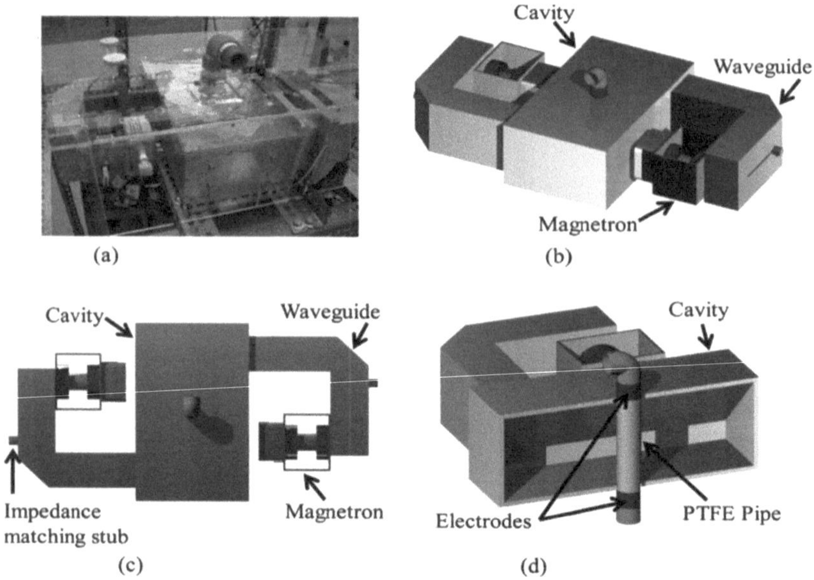


**Fig. 8.26** Schematic diagram of the microwave-Ohmic heater. (A) Combination heater, (B) 3D view, (C) upper view, and (D) internal view.
From Nguyen, L.T., et al., 2013. Exploring the heating patterns of multiphase foods in a continuous flow, simultaneous microwave and ohmic combination heater. J. Food Eng. Elsevier Ltd, 116 (1), 65–71. https://doi.org/10.1016/j.jfoodeng.2012.11.011.

uniformity of heating during MW and VMW thawing. Fig. 8.28A, shows a schematic for a vacuum microwave device. A microwave reactor contained a glass desiccator with three samples. A vacuum pump maintained a pressure of 5 kPa inside the desiccator during heating. The microwave at 2450 MHz was irradiated with a 300 W output power. In this study, it was demonstrated that VMW thawing inhibited local and overheating of the sample because thawing was achieved quickly thus it suggests an effective thawing technique for various foods (Fig. 8.28B) (Watanabe and Ando, 2021).

In the 1980s, a microwave/vacuum thawing system was developed (Swain and James, 2005). It consisted of a cylindrical vacuum chamber and a microwave oven (Fig. 8.29). Microwaves were introduced into the chamber through waveguides near the top of the plant at its front and rear. During thawing, a large circular disc was rotated within the chamber to create an evenly distributed microwave field (Orsat et al., 2017). The microwave-vacuum thawing method provides better physicochemical properties in thawed food compared to other thawing methods (conventional, microwave, ultrasound or far-infrared) (Rastogi, 2012). Yagi and Shibata (2002) submitted a patent on using microwave under vacuum to thaw frozen foods. Sublimation at the surface and a relatively low microwave power led to uniform product temperatures and low weight losses (Rastogi, 2012; Cai et al., 2019a).

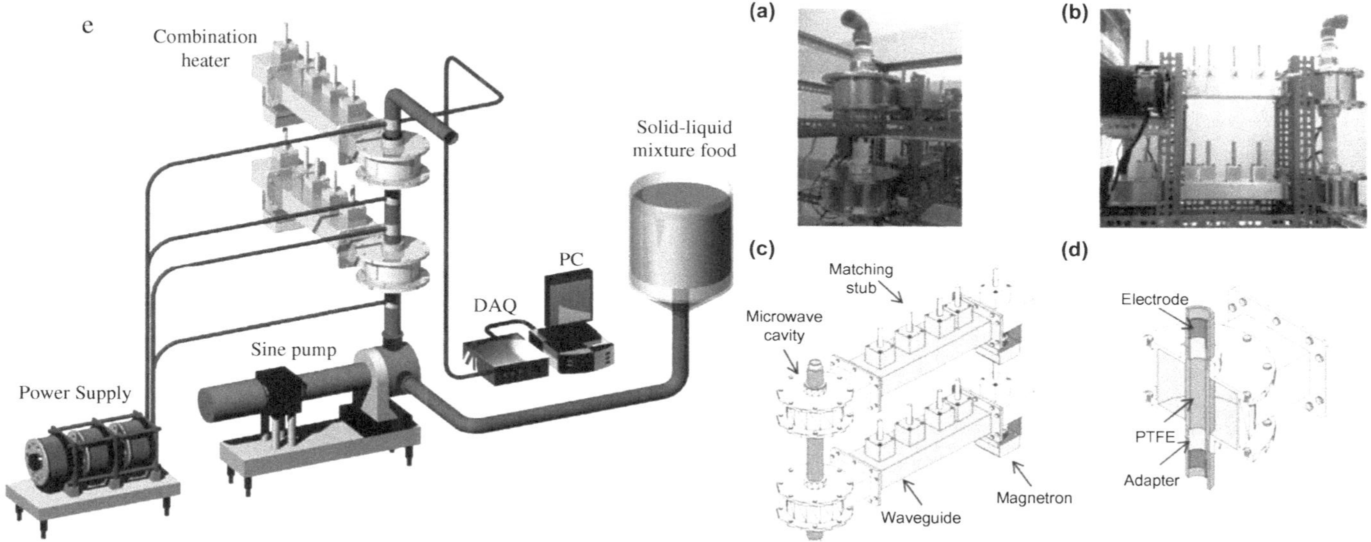


**Fig. 8.27** Schematic diagram of microwave-Ohmic heater: (A) front view, (B) side view, (C) 3D schematic, (D) cross-sectional view of the chamber, and (E) a schematic of the experimental set-up.
From Lee, S.H., et al., 2015. Development of a dual cylindrical microwave and ohmic combination heater for minimization of thermal lags in the processing of particulate foods. LWT. Elsevier Ltd, 63 (2), 1220–1228. https://doi.org/10.1016/j.lwt.2015.04.042.

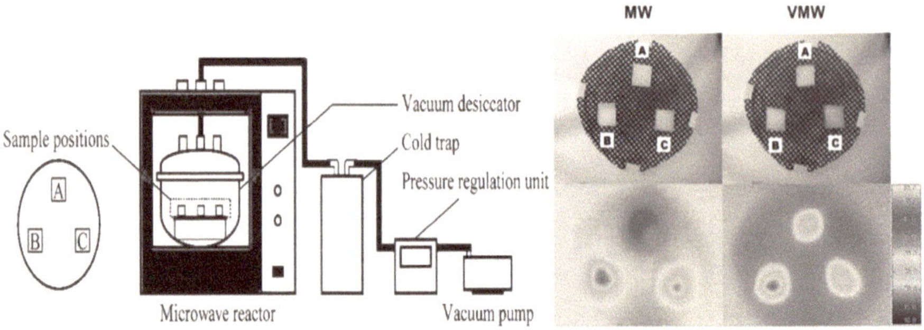


**Fig. 8.28** Schematic diagram of the vacuum-microwave heater apparatus (A) thermographs of thawed samples (B).
From Watanabe, T., Ando, Y., 2021. Evaluation of heating uniformity and quality attributes during vacuum microwave thawing of frozen apples. LWT. Elsevier Ltd, 150 (June), 111997. https://doi.org/10.1016/j.lwt.2021.111997.

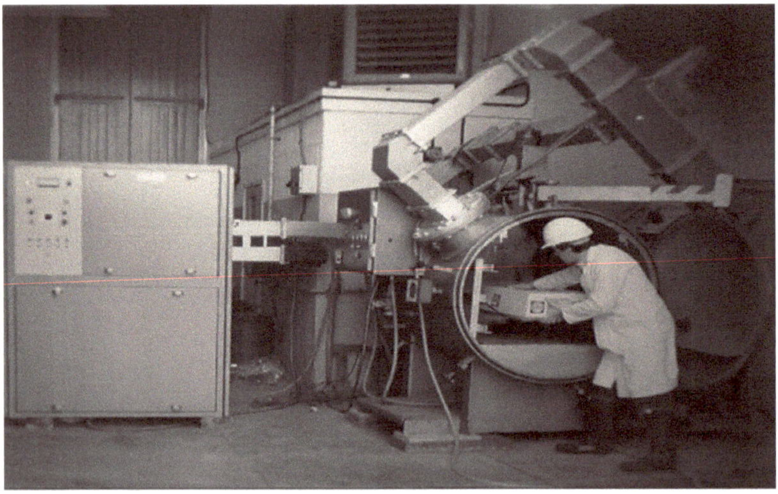

**Fig. 8.29** A hybrid microwave-vacuum thawing system.
From Swain, M., James, S., 2005. Thawing and Tempering Using Microwave Processing, The Microwave Processing of Foods. Woodhead Publishing Limited. https://doi.org/10.1533/9781845690212.2.174.

## 8.7.3 Radio frequency thawing (RF)

The RF process uses two electrodes, one as the ground to store the energy and the other as the potential applied electrode (Tang et al., 2021). The electromagnetic energy is absorbed by the product that is placed between two electrodes and then the energy is converted into heat by bipolar rotation and ionic conduction (Fig. 8.30; Huang et al., 2016).

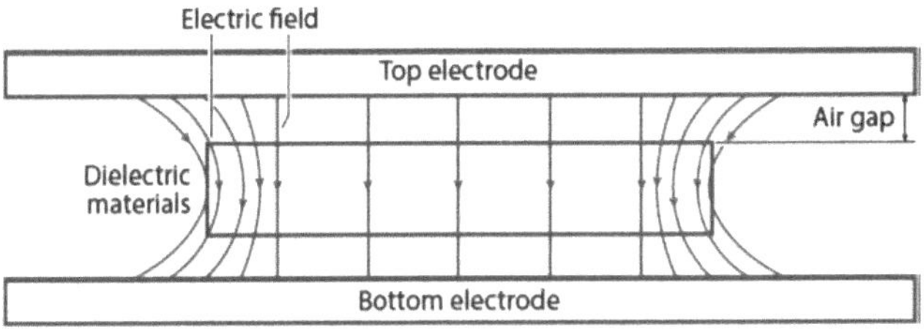


**Fig. 8.30** The electric field strength between two parallel plate electrodes with the slab of a dielectric sample placed in the centroid of the cavity.
From Huang, Z., Marra, F., Wang, S., 2016. A novel strategy for improving radio frequency heating uniformity of dry food products using computational modeling. Innov. Food Sci. Emerg. Technol. Elsevier Ltd, 34, 100–111. https://doi.org/10.1016/j.ifset.2016.01.005.

RF systems are classified into two categories: the self-oscillator RF system (also called open circuit RF heating system) and the 50-X system (Llave and Sakai, 2018). A self-oscillator RF system is generally used in many industrial applications because it is low-cost, flexible, and simple. The 50-X system is comparatively new and has demonstrated some advantages over the other method, such as: setting a fixed frequency, controlling power, and being compatible with requirements.

As shown in Fig. 8.31A, the self-oscillator RF system consists of a high-voltage transformer, rectifier, oscillator, tuned circuit, coupling and matching system, and applicator. AC power is transformed to a high voltage and changed into DC power using a rectifier. The oscillator generates radiofrequency energy through a thermionic tube that drives the resonant circuit. After the RF signal is transmitted to the product load, it generates heat by dielectric loss (Llave and Sakai, 2018; Llave and Erdogdu, 2022). The 50-X RF system consists of a crystal-controlled oscillator, solid-state amplifier, automatic impedance matching network, and applicator (Fig. 8.31B).

Bedane et al. (2017) evaluated the thawing of frozen beef meat blocks using pilot-scale RF under the two conditions batch (placing sample between electrodes) and continuous (transferring sample between electrodes along with the RF system) conditions. The RF equipment includes an adjusted electrode at the top and conveyor belt (speed between 1 and 60 m/h) (Fig. 8.32). In batch thawing, the amount of heat generation inside the frozen food is influenced by the gap between electrodes and the time of heating. Besides the gap between electrodes, the speed of conveyor belt and time play a significant role in continuous thawing. In this study, the potential of continuous and bath RF thawing of beef was compared based on heating uniformity and temperature–time histories. The results exhibited that the temperature distribution of the sample was between −0.2 and 28.4 °C and −1 to 12.8 °C and average temperature of $-0.2 \pm 1.5$ °C and $0.4 \pm 0.7$ °C after 17 min batch and continuous process, respectively. Additionally, continuous thawing of frozen blocks improved the heating uniformity throughout the food product with the minimized difference between hot and cold spots. In this study, RF thawing improves heat distribution and temperature uniformity, as well as meat quality (Bedane et al., 2017).

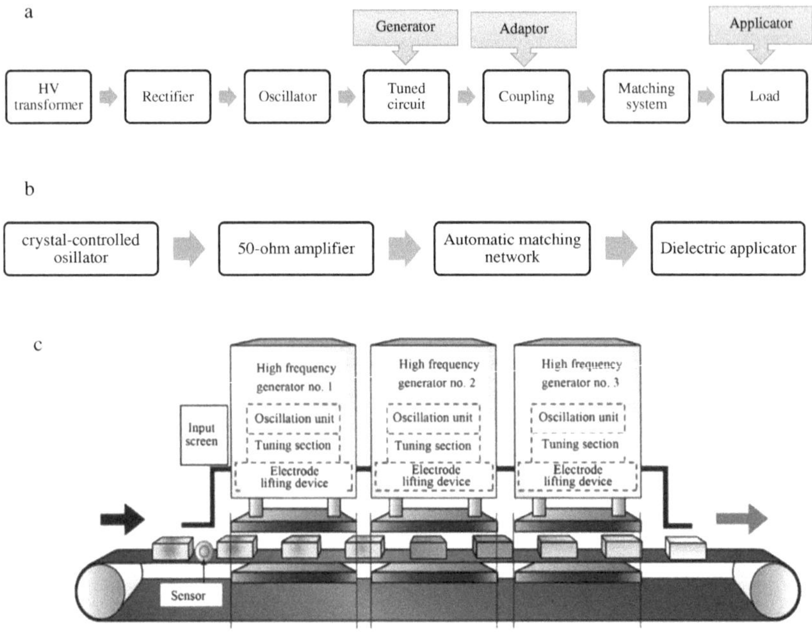


**Fig. 8.31** Scheme of (A) a self-oscillator RF system, (B) a 50-X RF system, and (C) configuration of a RF continuous defroster.
From Llave, Y., Sakai, N., 2018. Dielectric Defrosting of Frozen Foods, Food Processing for Increased Quality and Consumption. Elsevier Inc. https://doi.org/10.1016/B978-0-12-811447-6.00013-8.

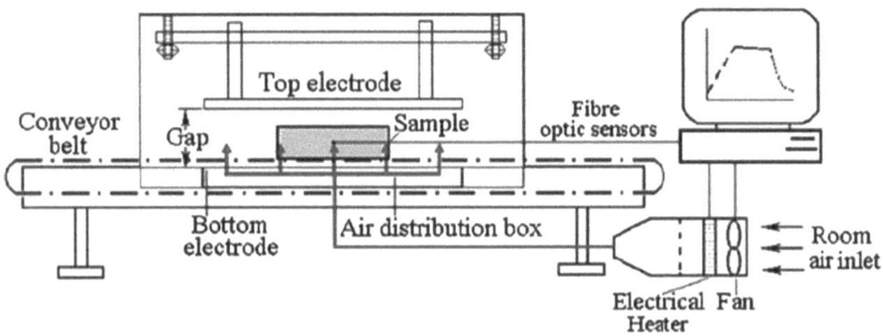


**Fig. 8.32** Schematic view of RF equipment with conveyor belt and electrodes.
From Bedane, T., Marra, F., Wang, S., 2017. Performance comparison between batch and continuous thawing of food products assisted by radio frequency heating. Chem. Eng. Trans. 57, 2017–2022. https://doi.org/10.3303/CET1757337.

Tang et al. (2021) developed a RF thawing chamber for rapid food thawing. The chamber consisted of an aluminum shell, a plastic tray, and two metal cylinders with metal disks located on top and bottom of the chamber. The resonant frequency could be adjusted based on the volumes of frozen product to achieve maximum energy transmission. As results, the field was uniformly distributed inside the frozen product and the thawing time was decreased from approximately 8 h (maximum) to about 4 min (Tang et al., 2021). Koray Palazoğlu and Miran (2017) explored the effects of microwave (MW) and radiofrequency (RF) methods on the rate and uniformity of frozen shrimp tempering. The MW tempering exhibited a shorter tempering time (4–10 min) than RF tempering (7–11 min). Fig. 8.33 showed the thermal images of shrimp block after tempering measured. At both power settings of microwave tempering, local surface overheating was observed. In contrast, radiofrequency tempering results in a uniform temperature distribution without overheating at the surface (Koray Palazoğlu and Miran, 2017).

## 8.8 Electrostatic field (EF) thawing

Electrostatics field (EF) is a relatively new technology used in food preservation that can be used in a wide range of food products ( Jafari and Saien, 2021). Electric field (EF) is used for several applications such as extraction, drying, enzyme inactivation and thawing. There are two types of EF induced thawing methods, high-voltage electric field (HVEF) and low-voltage electric field (LVEF) (Xie et al., 2021). Due to high costs and complex conditions of using high voltage (safety issues), the use of a low voltage electrostatic field (LVEF, $\leq$2500 V) is more energy-efficient, safe, and applicable (Hu et al., 2021; Jafari and Saien, 2021). A HVEF thawing setup comprises a plate electrode and wire or needle electrodes placed on a specific air gap. Frozen food is located on the plate between the two conductive electrodes. Next, a high-voltage source is connected to the electrodes, and the voltage is increased to the point where no discharge can occur (Dalvi-Isfahan et al., 2016; Zhang et al., 2019). In such systems, as the voltage is applied to the wire or needle electrodes, the air surrounding them is ionized, and the ions go toward the plate electrode is linked with the other pole. During the migration of ions to the plate electrode, the momentum of ions is transported to air molecules. When the intensity of the electric field near the corona electrode is elevated, the free electrons of the gas present in this region are accelerated due to electric forces. Once the electrons have gained a sufficient amount of energy, they will interact with the molecules in a neutral gas to form electron-positive ion pairs and photons. The freshly created electrons are accelerated and generate more electron-ion pairs. Hence, the number of electrons is exponentially elevated in the air gap. This process is called the electron avalanche (Dalvi-Isfahan et al., 2016; Rahbari et al., 2018; Zhang and Ding, 2020). If air is the medium for an electrohydrodynamic system, a space charge will be formed for the singly charged ions of $O_2^+$, $N_2^+$, $N^+$, $O_2^-$, and $O^+$. The ions can bind with water molecules and then create small air ions with a specific amount of mobility. Two separate areas exist in the corona discharge. An ionization layer is present near the corona electrode, where avalanche ionization occurs because of the collision between the highly accelerated electrons and the neutral molecules (Cai et al., 2019a). A larger region, also known as the drift region, is mainly full of ions whose polarities are the same as the corona electrode ions. Beyond the ionization layer, the intensity of the electric field is slowly reduced, and electrons lack enough energy for the ionization of the molecules of the gas. As a result, the avalanche ionization is

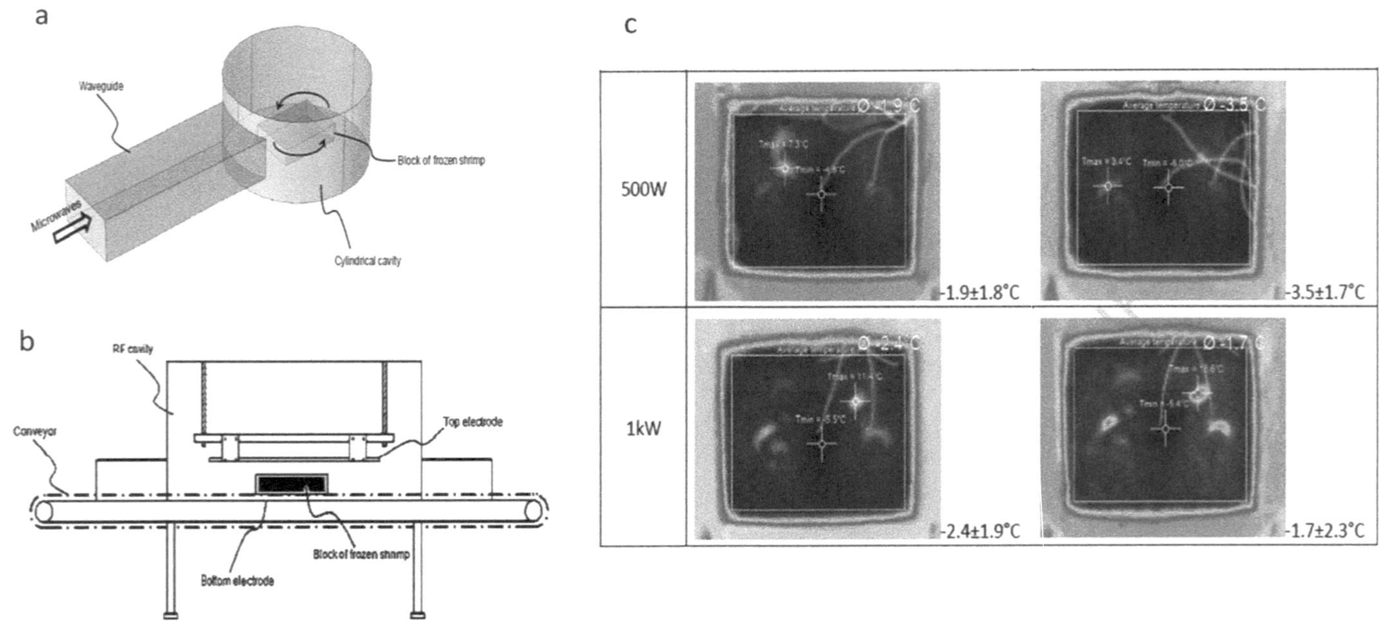


**Fig. 8.33** Schematic of (A) the microwave tempering experimental set-up, (B) the radio frequency tempering experimental set-up, and (C) thermal images (surface temperature distribution) of shrimp block.
From Koray Palazoğlu, T., Miran, W., 2017. Experimental comparison of microwave and radio frequency tempering of frozen block of shrimp. Innov. Food Sci. Emerg. Technol. Elsevier Ltd, 41, 292–300. https://doi.org/10.1016/j.ifset.2017.04.005.

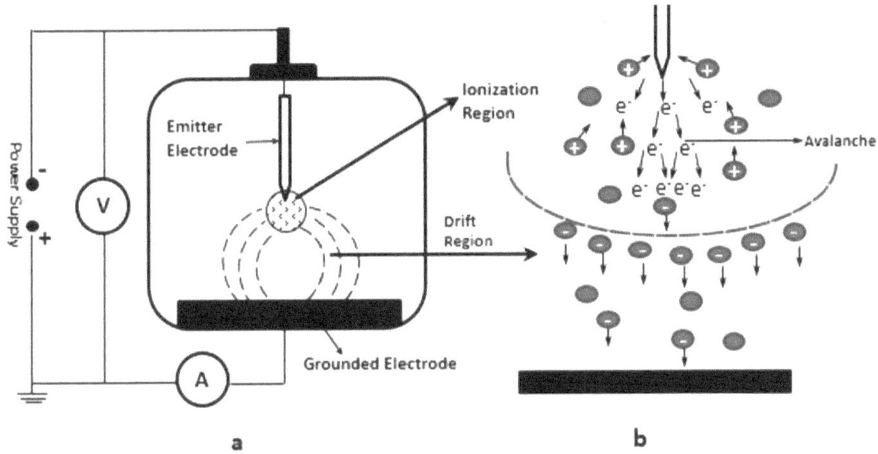


**Fig. 8.34** (A) Schematic illustration of an electrohydrodynamic system with the configuration of needle-to-plate and (B) ionic reactions in corona discharge.
From Dalvi-Isfahan, M., et al., 2016. The principles of high voltage electric field and its application in food processing: a review. Food Res. Int. Elsevier B.V. https://doi.org/10.1016/j.foodres.2016.09.002.

ceased, and the number of electrons slowly lowers in the drift region. Nevertheless, in this region, momentum is transferred due to the collisions of the drifting ions with the electrically neutral molecules of air. As a result, the electrohydrodynamic flow, also called ionic wind or corona wind, is created (Fig. 8.34) (Xie et al., 2021). Selectin of the corona electrodes type is principally dependent on the chamber geometry and the system operation. The systems of HVEF thawing, are equipped with some plate electrodes and needles, and work by AC or DC voltages. Such systems are very efficient in thawing frozen foodstuffs (Dalvi-Isfahan et al., 2016; Cai et al., 2019a; Zhang and Ding, 2020).

## 8.8.1 High-voltage electrostatic fields (HVEF) thawing

Rahbari et al. (2018, 2020) evaluated the effect of HVEF thawing on quality properties and microstructure of chicken breast. They applied HVEF with multiple needle-plate electrode system (Fig. 8.35). The results presented that the HVEF significantly decreased the thawing time and quality loss at the optimum voltage and electrode gaps (2.25 kV/cm). According to the results, the fibers of chicken meat remained relatively intact after thawing under 2.25 kV/cm. Control and samples thawed under 3 kV/cm showed the most deformation and loss of collagen fibrillation due to the great degree of protein denaturation. Thawing under a suitable voltage and electrode gap can effectively minimize the protein denaturation-induced structural damage and preserve the quality of chicken breast protein (Rahbari et al., 2018, 2020).

Zhang and Ding (2020) explored the effect of high-voltage electric field (HVEF) on the thawing properties of frozen beef. They used the HVEF thawing device that

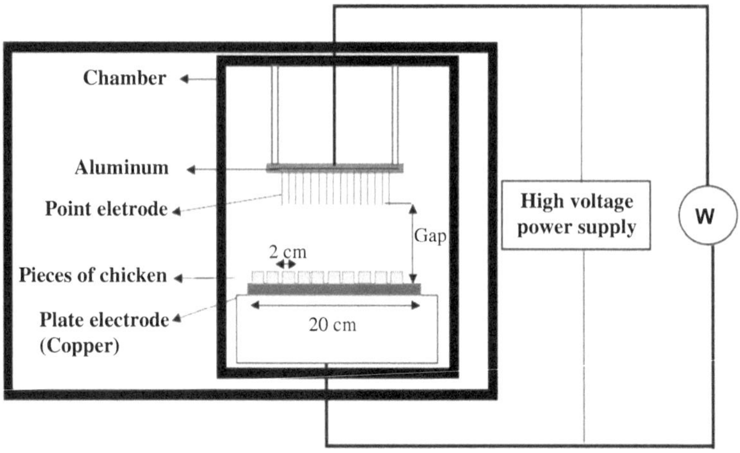


**Fig. 8.35** Diagram of the HVEF treatment apparatus.
From Rahbari, M., et al., 2018. Effects of high voltage electric field thawing on the characteristics of chicken breast protein. J. Food Eng. Elsevier B.V., 216, 98–106. https://doi.org/10.1016/j.jfoodeng.2017.08.006.

consisted of high-voltage power source, voltage controller, ground electrode, needle electrode, thermometer, hygrometer, and temperature sensor (Fig. 8.36). Electrodes mounted vertically with needles (20 mm long, 1 mm diameter, and 60 mm their distance). This electrode was installed onto a grounded plate electrode (rectangular stainless steel), where the beef samples were placed to be thawed. High-voltage power source can provide AC high voltage ranging from 0 to 50 kV with frequency of 50 Hz. This study indicated that compared with the control, HVEF thawing could considerably reduce the thawing time and improve the color parameters of beef. During the HVEF thawing, quality parameters of frozen products are highly influenced by electrode distance and voltage (Zhang and Ding, 2020).

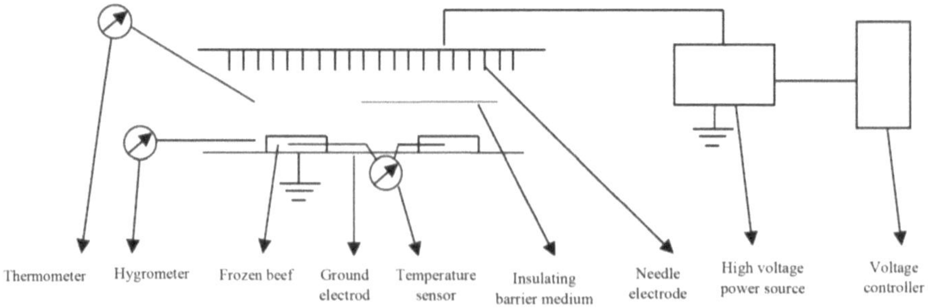


**Fig. 8.36** Diagram of the high-voltage electric field thawing equipment.
From Zhang, Y., Ding, C., 2020. The study of thawing characteristics and mechanism of frozen beef in high voltage electric field. IEEE Access 8, 134630–134639. https://doi.org/10.1109/ACCESS.2020.3010948.

Movahedi et al. (2021) designed and constructed a laboratory continuous high-voltage electrostatic field thawing in wire-plate electrode setups. The frozen food moves inside the tunnel on a 2.5 m-long conveyor in order to thaw continuously. The conveyor surface was utilized as the ground electrode that moved along the ladder-shaped electrodes. The belt was driven by a 375 W three-phase AC electromotor and gearbox (ratio of 1:100). Twenty-eight wired electrodes (0.5 mm-diameter stainless steel wire) were considered as corona electrodes and installed at 7 cm intervals. The voltage was supplied with a DC power supply that could be adjusted between 0 and 70 kV and 2000 μA. The electric field between the two poles was produced by connecting the poles of a power supply to the electrode and rails. The whole conveyor and its components were placed in an insulated chamber (a tunnel of plastic sheet) to avoid the airflow from entering the corona wind zone. The materials could go into and out of the tunnel through the two open ends, covered with thick pads at both ends (Fig. 8.39). According to results, the HVEF continuous system was an effective, rapid and feasible method for thawing food materials like meat. The optimum processing conditions with minimum specific energy consumption (10.33 $kJ\,kg^{-1}$) were achieved with 15 kV and a 10 cm electrode gap (Movahedi et al., 2021).

He et al. (2013, 2014, 2016) evaluated the effect of HVEF thawing on quality and energy consumption of frozen pork meat and compared the results with other thawing methods. Microwave, hot-water and cold-water thawing showed significantly higher specific energy consumption compared to HVEF thawing. The temperature difference between center and surface s was about 0.3–1.2 °C. They stated that, the HVEF method could produce uniform thawing in significantly less time, reduce the total microbial counts by 0.5–1 log CFU/g, save energy, and preserve the quality of food products (He et al., 2013, 2014, 2016).

## 8.8.2 Low voltage electrostatic field (LVEF) thawing

Qian et al. (2019) developed a new thawing method that combined refrigerator thawing with a low voltage electrostatic field (LVEF) device and its effect was investigated on physicochemical properties of frozen beef. As shown in Fig. 8.37, the components of the LVEF device were: voltage power supply, voltage controller, plate electrode, copper electrode, voltmeter, and amperemeter. The applied voltage, discharge gap, and temperature for all treatments were 2500 V, 10 cm, and 4 °C. In order to perform thawing under LVEF and refrigerator, frozen beef samples were placed on a copper plate electrode. Electrodes were fixed in a chamber that was located in a cold incubator at 4 °C. The results presented that LVEF could effectively shorten the thawing process by 42%, decrease quality loss of the beef and promote the renaturation of protein. According to the findings of this study, LVEF-assisted refrigerator is a potential technology for thawing meat that successfully maintains the freshness of meat (Qian et al., 2019).

Hu et al. (2021) evaluated the effects of low voltage electrostatic field thawing (LVEF), high humidity thawing (HT) and their combination (LHT) on the qualitative properties of pork steaks. LVEF used in this study was an experimental device designed by Qian et al. (2019). HT thawing included a polypropylene box with a

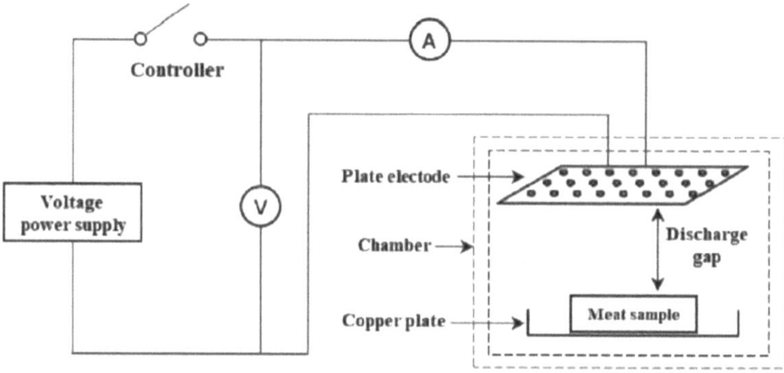


**Fig. 8.37** Schematic diagram of low voltage electrostatic field (LVEF) experimental device. From Qian, S., et al., 2019. Effects of low voltage electrostatic field thawing on the changes in physicochemical properties of myofibrillar proteins of bovine Longissimus dorsi muscle. J. Food Eng. 261 (January), 140–149. https://doi.org/10.1016/j.jfoodeng.2019.06.013.

PVA sponge that was attached to the lid of the box and the relative humidity in the box was 98%. Experimental device for LVEF-assisted high humidity is shown in Fig. 8.38. Findings displayed that the LHT significantly reduced the thawing time and thawing loss, and improved the color stability of pork by preventing the formation of metmyoglobin. Moreover, the LHT process resulted in lower hardness and higher springiness by keeping the muscle microstructures (Hu et al., 2021).

## 8.9 Ohmic thawing

### 8.9.1 Theory

Ohmic heating has drawn growing attention from different food industries because it is regarded as an alternative method to the indirect heating techniques (Cai et al., 2019a). It is also known as electrical resistance heating, Joule heating, and electro-conductive heating (Li and Sun, 2002). Ohmic heating is a thermal process; thus, the residence time, the strength of the electric field, and the product electric conductivity are considered as the main variables of the process, which dramatically affect the heat production and the process performance (Goullieux and Pain, 2005; Varghese et al., 2014). Frozen products are less conductive than the thawed foods; therefore, when the frozen food is thawed, a current goes through its thawed part faster and increases the rate of thawing (Liu et al., 2017). As the current passes through the block-thawed part, the food product in that part might be cooked, whereas the remaining parts are still frozen. This is known as runaway heating or hot spots formation. If this issue is resolved, a novel technique of food thawing will be available (Varghese et al., 2014). Shorter process time, higher quality of the final product, higher efficiency of energy, lower capital cost, and being environmentally friendly are the significant advantages of ohmic heating (Li and Sun, 2002; Cai et al., 2019a).

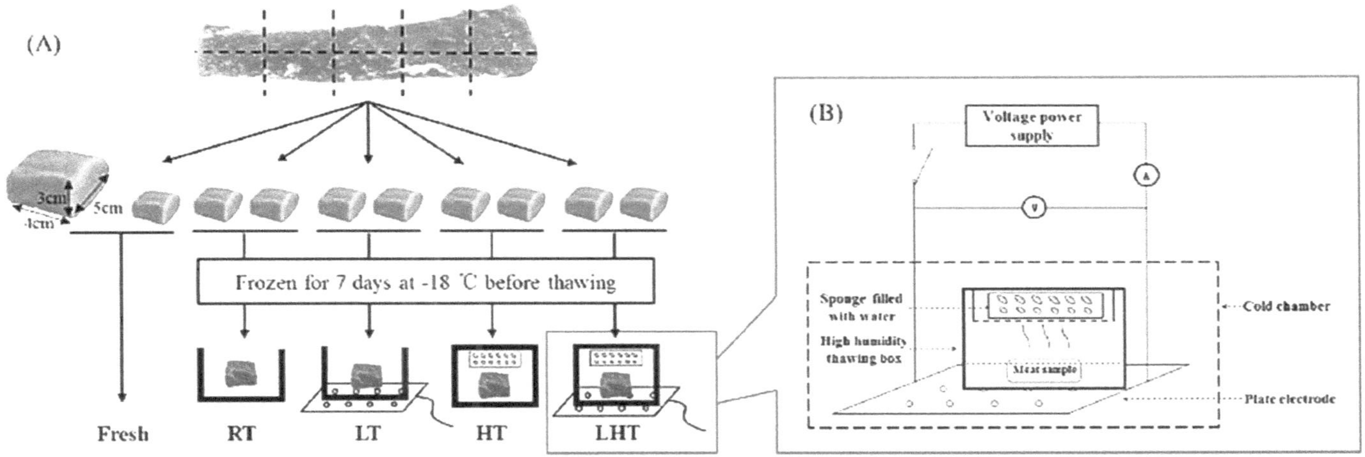


**Fig. 8.38** Schematic diagram of experimental design (A) LVEF-assisted high humidity experimental apparatus (B). Fresh, fresh meat; RT, refrigerator thawing; LT, low voltage electrostatic field thawing; HT, high humidity thawing; LHT, low voltage electrostatic field combined with high humidity thawing.
From Hu, F., et al., 2021. Combined impacts of low voltage electrostatic field and high humidity assisted-thawing on quality of pork steaks. LWT. Elsevier Ltd, 150 (January), 111987. https://doi.org/10.1016/j.lwt.2021.111987.

### 8.9.2 Equipment

The application of ohmic heating in thawing frozen foodstuffs is an innovative technique. Electrical resistance thawing is based on an electric current going through a resistor, leading to heat generation (Cai et al., 2019a). As a result, Ohmic thawing is principally according to the passage of an oscillating current at a proper voltage through frozen food with an electrical resistance that converts into heat. The setup of ohmic thawing is simple from the technical point of view (Fig. 8.39; Kaur and Singh, 2016). However, the effectiveness of this technique depends on the appropriate contact between the electrodes and the product to be thawed (Varghese et al., 2014). Rectangular products are appropriately heated in ohmic systems; while, spherical or oddly-shaped foods might have trouble fulfilling a proper contact. Shape restriction is the most crucial disadvantage of applying ohmic thawing in the food industry. In order to solve this problem, an electrode solution could be employed (Ito et al., 2014; Duygu and Ümit, 2015).

An ohmic heating system can be designed in various ways, but each has several vital elements. Electricity is produced by an electrical source (generator). Electrodes attached to the power supply need to be physically in contact with frozen product to transfer the electric current. The two possible geometries for field modes; the transverse and collinear are illustrated in Fig. 8.40 (Varghese et al., 2014). Transverse configuration: the products flow parallel to electrodes and perpendicular to the electric field and the electrodes are usually plane or coaxial. Although, it appears to have advantages of a simple mechanical design, the transverse mode has a number of problems including localized overheating, boiling, and electrode corrosion. Thus, the development of the transverse mode heater has been limited to fluids with no particles. Collinear configuration: the product flows parallel to the electric field from one electrode to the other. A collinear heater consists of two essential components: electrode housing and spacer tube. The voltages between electrodes and the earthed heater casing can reach 3kV. Therefore, all interior surfaces that touch food be made from insulating material, which can last a long time and provide reliable service. A continuous system, such as a conveyor or tube, can be classified to in field and cross field system depending on the position of electrodes to the flow path (Goullieux and Pain, 2005).

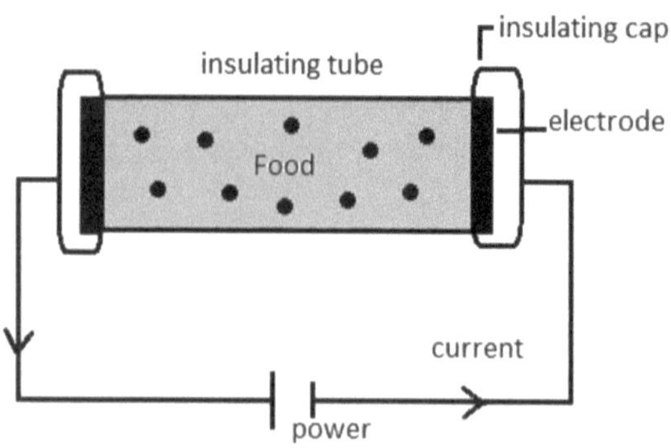


**Fig. 8.39** Schematic view of Ohmic thawing of food.
From Alkanan, Z., Altemimi, A., Al-Hilphy, A.R.S., Watson, D.W., Pratap-Singh, A., 2021. Ohmic Heating in the food industry: developments in concepts and applications during 2013–2020. Appl. Sci. 11 (6), 2507. https://doi.org/10.3390/app11062507.

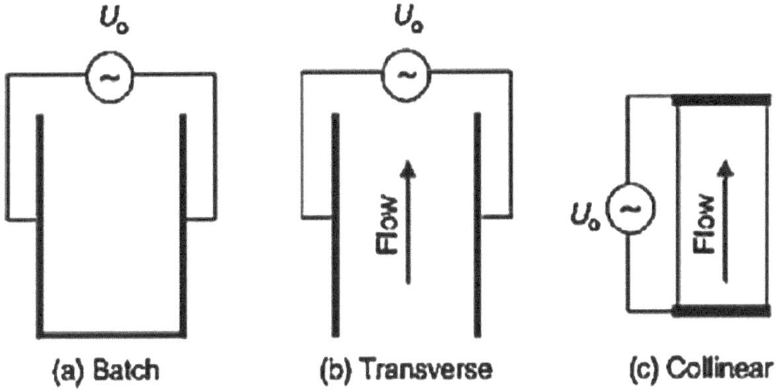


**Fig. 8.40** Types of geometries for field modes. (A) Batch, (B) transverse, and (C) collinear. From Alkanan, Z., Altemimi, A., Al-Hilphy, A.R.S., Watson, D.W., Pratap-Singh, A., 2021. Ohmic Heating in the food industry: developments in concepts and applications during 2013–2020. Appl. Sci. 11 (6), 2507. https://doi.org/10.3390/app11062507.

Min et al. (2016) compared the effects of conventional, ohmic, and a combination of high-pressure and ohmic thawing on melting rate of frozen foods. The pressure ohmic thawing (POT) system consisted of a flexible ohmic cell, an electric field supply, and a pressure chamber (Fig. 8.41). The shortest thawing time was achieved with POT (0.8 min), followed by Ohmic (5.5 min), and conventional (43.3 min). Based on this study, POT technology has the potential to thaw frozen foods rapidly by utilizing a synergistic effect of high pressure (lowering the phase transition point of water) and Ohmic effect (volumetric thawing) (Min et al., 2016).

Alkanan et al. (2021a, b) designed a batch ohmic-vacuum combination heater for thawing food. This device (Fig. 8.42) consisted of (1) pressure vacuum gage; (2) pressure vacuum calibration valve; (3) heating cylinder; (4) electricity source; (5) variac; (6) stainless steel electrodes; (7) thermocouple; (8) temperature controller; (9) cold water circulation unit; (10) pipes; (11) glass heat exchanger; (12) moisture trap; (13) piping; (14) moisture trap; (15) vacuum pump; and (16) valve. Heat-resistant plastic was used to construct the cylindrical ohmic heating chamber. The cylindrical ohmic chamber was lined with stainless steel rectangular electrodes. The system generated heat with alternating current (AC) and voltage ranging from (0–220 V) and frequency (50 Hz) (Alkanan et al., 2021a).

## 8.10 Infrared (IR) radiation thawing

### 8.10.1 Theory

In various food industries, convection and conduction have widely been utilized for heat transfer in thermal processing (Yadav et al., 2020). IR radiation with short wavelengths has been used as a radiant heat transfer. IR is a fraction of the electromagnetic

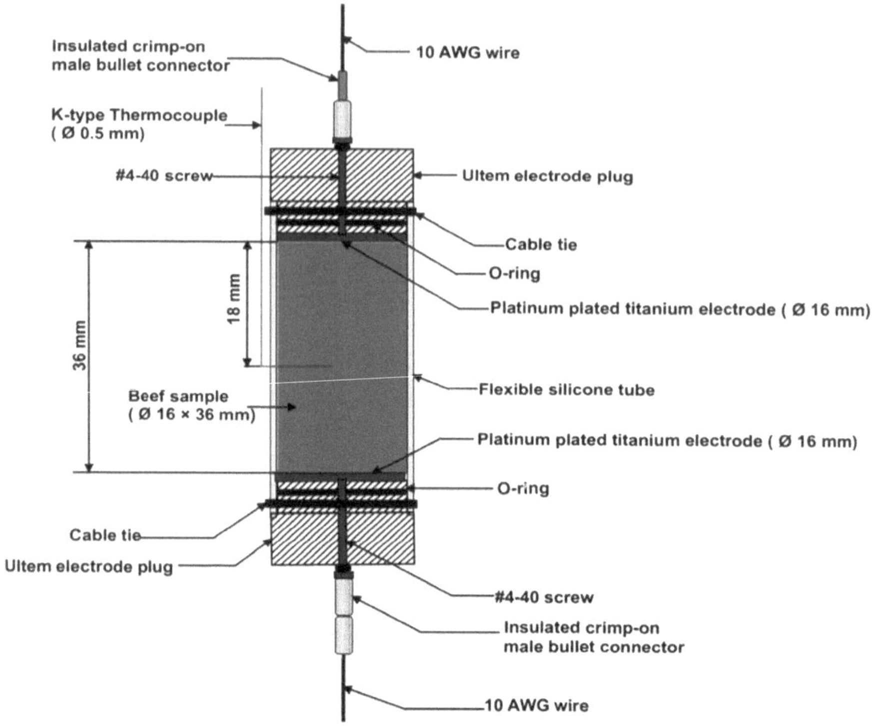


**Fig. 8.41** Cross-sectional view of pressure Ohmic thawing (POT) cell.
From Park, H.S., et al., 2014. Quality of shelf-stable low-acid vegetables processed using pressure–ohmic–thermal sterilization. LWT Food Sci. Technol. 57 (1), 243–252. https://doi.org/10.1016/j.lwt.2013.12.036.

spectrum lying between the visible region and MW, and its wavelength is in the range of 0.5–100 μm (Fig. 8.43) (Aboud et al., 2019). IR rays are divided into three groups: near-IR (NIR), with wavelength from 0.75 to 1.4 μm at temperatures lower than 400 °C, mid-IR (MIR), with wavelength from 1.4 to 3 μm at 400–1000 °C, and far-IR (FIR) with a wavelength of 3–1000 μm at temperatures more than 1000 °C (Krishnamurthy et al., 2008; Yadav et al., 2020). According to a research performed by the National Research Council, there was no correlation between IR cooking and cancer, which also revealed that the energy of the IR radiation was not sufficient to halve or damage DNA (Krishnamurthy et al., 2008; Cai et al., 2019b).

Recently, there has been an increasing interest in heating by FIR radiation due to the development of commercial FIR heaters with high emissivity. As the heater temperature is elevated, the radiant energy level delivered to the food is raised, and the peak wavelength is reduced. The maximal operating temperature of FIR heaters, typically manufactured from fine ceramics, lies in the range of 600–950 K and the highest temperature of a short-wave radiator produced by a tungsten filament lamp in the range of 2400–2500 K (Rastogi, 2012; Yadav et al., 2020). Radiant electromagnetic energy can alter the vibrational, electronic, and rotational states of atoms and molecules when colliding with a food surface. The mechanism for the absorption of energy

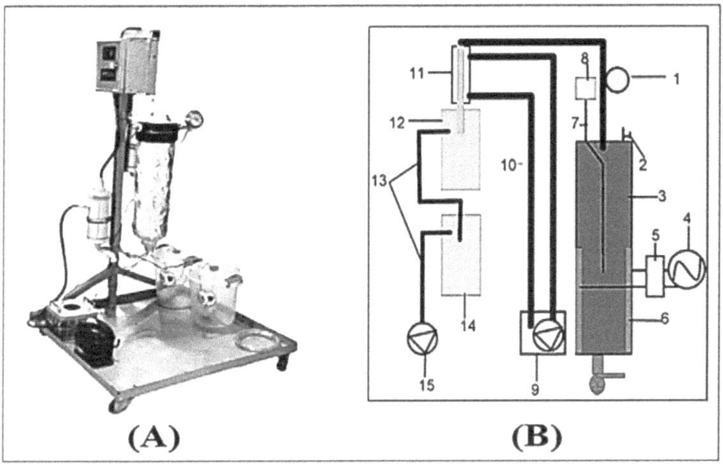


**Fig. 8.42** (A) Photograph of the actual Ohmic-vacuum (OH-VC) combination heating device and (B) schematic diagram of (OH-VC).
From Alkanan, Z.T., Altemimi, A.B., Al-Hilphy, A.R.S., et al., 2021. Application and effects of ohmic-vacuum combination heating on the quality factors of tomato paste. Foods 10 (12), 1–19. https://doi.org/10.3390/foods10122920.

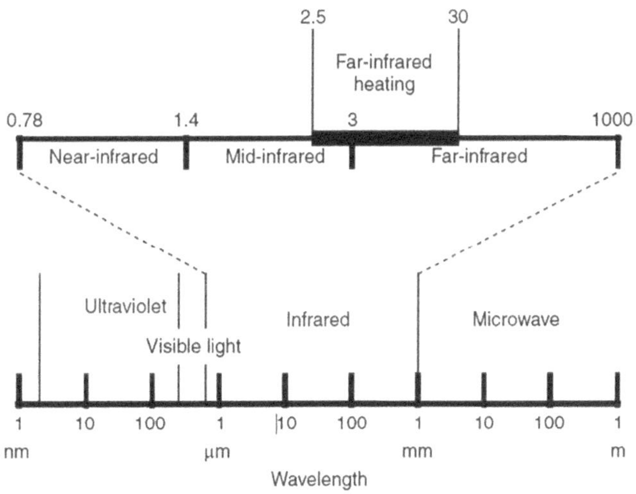


**Fig. 8.43** Classification of electromagnetic wave spectrum.
From Rastogi, N.K., 2012. Recent trends and developments in infrared heating in food processing. Crit. Rev. Food Sci. Nutr. 52 (9), 737–760. https://doi.org/10.1080/10408398.2010.508138.

is governed by the range of the wavelength of the incident beam energy (Rastogi, 2015): alterations in the electronic state correspond to wavelengths ranging from 0.2 to 0.7 μm (ultraviolet [UV] and visible lights); alterations in the vibrational state correspond to wavelengths varying from 2.5 to 100 μm (FIR), and alterations in the rotational state correspond to wavelengths higher than 100 μm (MW). IR radiation has some advantages over conventional heating, including high coefficients of heat transfer, short process duration, and low energy cost. Its equipment is compact and automated, and the process variables can be easily controlled (Krishnamurthy et al., 2008; Rastogi, 2015; Aboud et al., 2019). Recent progress in FIR equipment presents fast and economical techniques for developing food products with high sensory and nutritional values.

### 8.10.2 Equipment

IR emitters are classified into electrical and gaseous type. Gas emitters produce infrared radiation between 343 and 1100 °C for and electric emitters between 1100 and 2200 °C (Yadav et al., 2020). Infrared electric heaters are more popular than gaseous heaters due to their simplicity, ease of use, and fast heating rate. An electrical infrared emitter includes a metal filament enclosed inside a sealed container containing inert gas or an empty container. Electric heaters produce infrared radiation by heating wires such as chrome wires, iron-chromium wires and tungsten filaments to a specific temperature using an electric current (Yadav et al., 2020). When the metal wire reaches the glow temperature, it will emit infrared radiation NIR type (0.7 and 1.4 μm). It is common for industrial applications to use gaseous IR emitters powered by propane or natural gas, which are cheaper to operate and more durable than electrical emitters. Gaseous IR emitters are divided into two general categories: direct flame emitters and catalytic flameless emitters. Industrial examples for electrical IR emitter (Fig. 8.44A) and gaseous IR emitter (Fig. 8.48B) are as follows: The dehydration of vegetables using a combined mid-IR and hot air dryer consists of three insulated chambers with quartz IR heaters installed on either side of a wire mesh conveyor (Hebbar et al., 2004; Rastogi, 2015).

Seyhun et al. (2009) evaluated the effect of microwave tempering assisted infrared on properties of frozen food products. Microwave ovens equipped with a halogen lamp were used to temper frozen potato puree (Fig. 8.45). Results showed that time required for tempering was significantly reduced with higher power levels. Thus, microwave-assisted infrared heating offers significant benefits such as increasing the product quality and energy efficiency compared to conventional methods. MW-IR heating can become a highly profitable commercial application (Seyhun et al., 2009; Aboud et al., 2019).

Hong et al. (2009) examined the effects of infrared radiation (IR) combined with air blast thawing on the physical properties of pork. IR-assisted air thawing apparatus is shown in Fig. 8.46. Aluminum foil was applied to the inner wall of the thawing chamber to increase its IR efficiency. At the top of the thawing chamber, 17 cm from the sample top surface, an IR lamp was hung. In order to achieve air convection, a fan was placed in front of the sample holder. Thawing time and loss were significantly lower with IR treatments. In addition, increasing IR power and air velocity decreased the cooking loss of pork (Hong et al., 2009). According to the results, IR combined with air blast can enhance thawed meat quality and control of humidity can also prevent surface drying of the meat. In agreement, Hong (2009, 2014) and Hebbar et al. (2004)

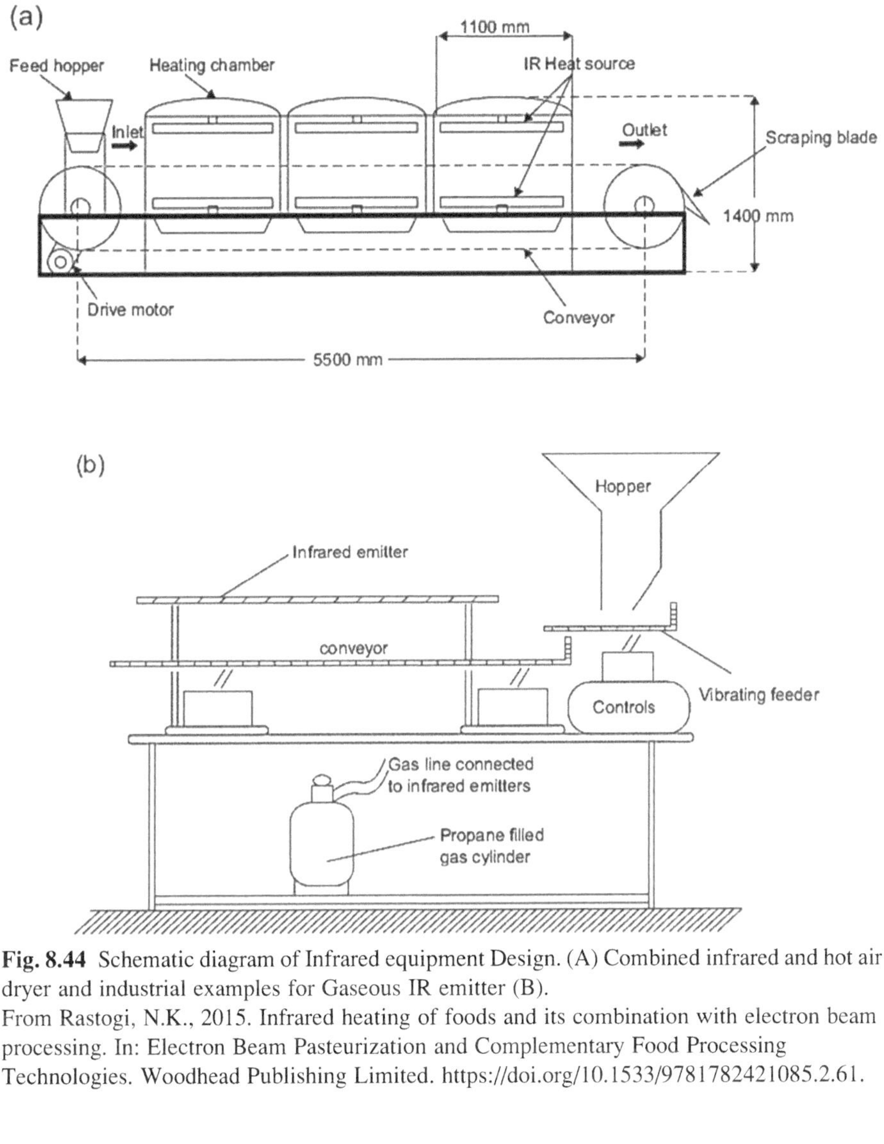


**Fig. 8.44** Schematic diagram of Infrared equipment Design. (A) Combined infrared and hot air dryer and industrial examples for Gaseous IR emitter (B).
From Rastogi, N.K., 2015. Infrared heating of foods and its combination with electron beam processing. In: Electron Beam Pasteurization and Complementary Food Processing Technologies. Woodhead Publishing Limited. https://doi.org/10.1533/9781782421085.2.61.

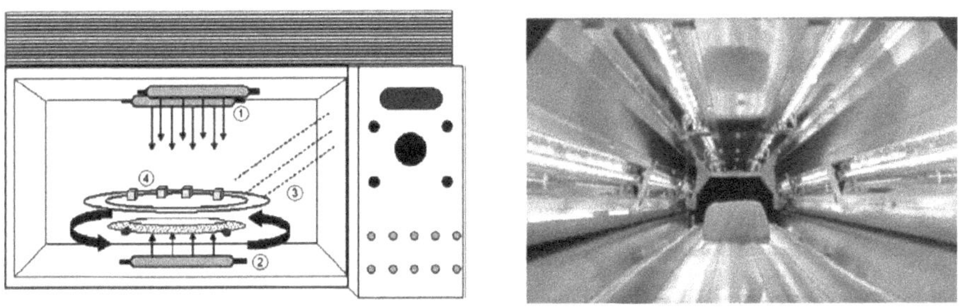

**Fig. 8.45** Schematic diagram of IR-microwave combination oven.
From Aboud, S.A., et al., 2019. A comprehensive review on infrared heating applications in food processing. Molecules 24 (22), 1–20. https://doi.org/10.3390/molecules24224125.

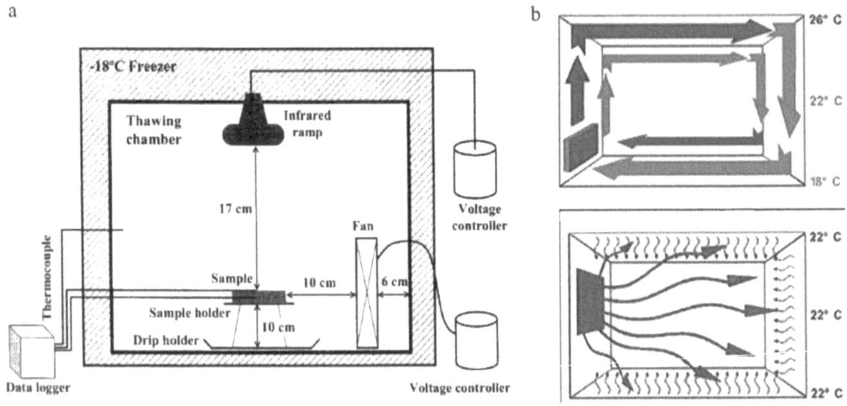


**Fig. 8.46** Schematic diagram of infrared radiation treatment combined with an air blast thawing apparatus (A) heat distribution with convective heating and infrared heating (B). From Aboud, S.A., et al., 2019. A comprehensive review on infrared heating applications in food processing. Molecules 24 (22), 1–20. https://doi.org/10.3390/molecules24224125.

demonstrated that infrared radiation combined with air blast could results in the rapid thawing of pork and prevents excessive overheating (Hebbar et al., 2004; Hong et al., 2009, 2014).

## 8.11 Plasma-activated water thawing

### 8.11.1 Theory

Plasma-activated water (PAW) as an innovative method can be an alternative to other types of food sanitizers. It has been shown that PAW can inactivate some foodborne pathogens, spoilage microbes, and harmful chemicals. Plasma-activated water (PAW) is a delivery medium for plasma-generated reactive species in which the targeted product is not exposed directly to the plasma discharge but instead comes into contact with water that has been preexposed to the plasma discharge. Inactivation of PAW can result from reactive oxygen and nitrogen species, which damage intracellular components and the integrity of the cell. PAW and hurdle technologies improve the inactivation rate and expand its applications. PAW thawing scaling up is still in its early stages, and more research is needed before the technology is applied to industry (Wang and Salvi, 2021; Qian et al., 2022).

There are two methods to generate PAW from plasma and water interactions: a discharge of plasma over the water surface and directly in the water (Fig. 8.47). In food processing, batch plasma systems are widely used to a fixed quantity of water (Liao et al., 2020). PAW can be generated using several plasma sources including dielectric barrier discharge (DBD), corona discharge, gliding arc discharge, plasma streamer bubble, and plasma jet. The reactive oxygen and nitrogen species (RONS) generated by plasma discharge that are transferred from plasma to liquid at the

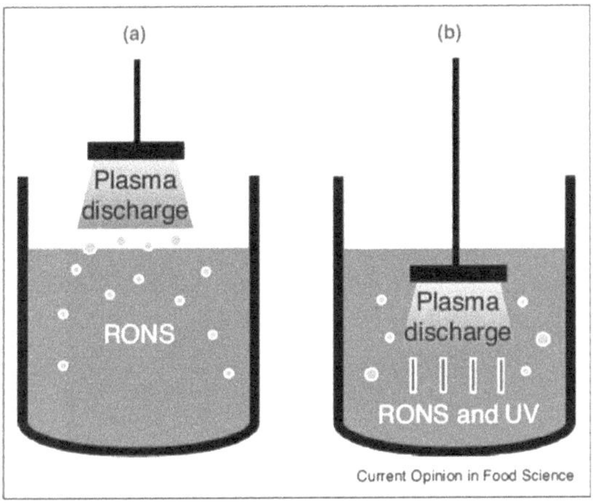


**Fig. 8.47** Illustration of PAW generation by (A) plasma discharge over water surface and (B) plasma discharge beneath water.
From Wang, Q., Salvi, D., 2021. Recent progress in the application of plasma-activated water (PAW) for food decontamination. Curr. Opin. Food Sci. Elsevier Ltd, 42, 51–60. https://doi.org/10.1016/j.cofs.2021.04.012.

gas–liquid interface contribute significantly to the decontamination ability of PAW (Wang and Salvi, 2021). RONS mainly consists of ozone ($O_3$), nitrate ($NO_3^-$), nitrite ($NO_2^-$), hydrogen peroxide ($H_2O_2$) (long-term species), peroxy nitric ($OONO_2^-$), nitric oxide (NO·), superoxide ($O_2^-$), singlet oxygen ($^1O_2$), and hydroxyl radicals (·OH) (short-lived species). The type and concentration of RONS in PAW depend on the plasma generation system and operating conditions. Liquid properties, such as pH, oxidation–reduction potential (ORP), and electrical conductivity in combination with reactive species (RONS), produce synergistic inactivation effects (Liao et al., 2020; Wang and Salvi, 2021).

## 8.11.2 Equipment

Qian et al. (2022) applied plasma-activated water (PAW) as a novel method for chicken thawing to improve the protein quality and reduce the microbial contamination. The instrument used for PAW thawing is presented in Fig. 8.48A. As it is observed, the instrument consisted of four components: water inlet, treatment channel of four plasma jets, water outlet, and water circulation system. Every plasma jet was connected to an electric field generator (voltage: 19kV and frequency: 20kHz) and PAW thawing was performed according to the following parameters: the total water volume: 20L, the treatment time: 25min, and the water flow rate: 2L/min. Results showed that PAW thawing and ultrasound combined with PAW thawing reduced the protein loss compared to conventional thawing. Furthermore, PAW thawing had no noticeable impact on the apparent quality or composition of the chicken meat, and even enhanced the protein digestion (Fig. 8.48B). Based on this study, this thawing method can be developed to sterilize poultry meat while minimizing protein loss (Qian et al., 2022).

Liao et al. (2020) compared the effects of thawing with plasma-activated water (PAW), slightly acidic electrolyzed water (SAEW) and conventional thawing on frozen beef quality. This study used atmospheric pressure plasma jets (APPJs) to produce

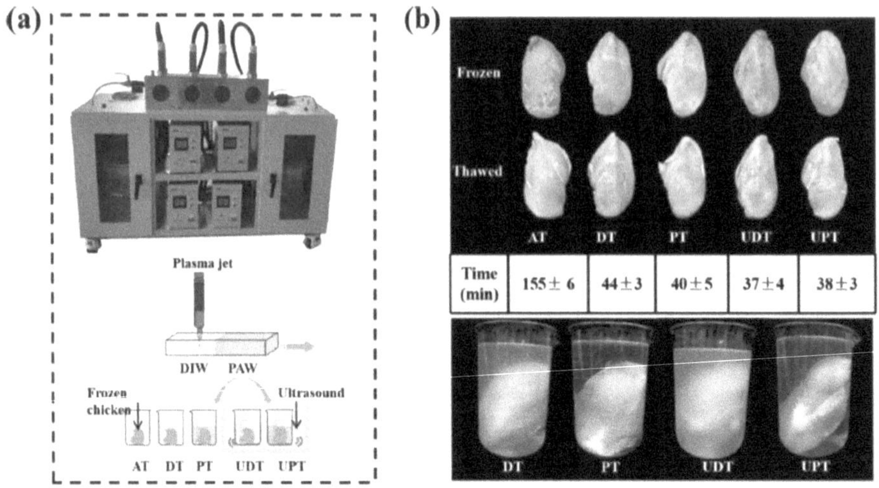


**Fig. 8.48** (A) Diagram of thawing chicken with various treatments. (B) The photograph of frozen and thawed chicken breasts.
From Qian, J., et al., 2022. Plasma-activated water: a novel frozen meat thawing media for reducing microbial contamination on chicken and improving the characteristics of protein. Food Chem. Elsevier Ltd, 375 (November 2021), 131661. https://doi.org/10.1016/j.foodchem.2021.131661.

PAW, and distilled water was treated with APPJs at 600 W for 1 min. The SAEW solution was prepared using 0.01% hydrochloric acid and 0.05% sodium chloride electrolyzed using a redox potential water generator at 220 V for 6 min. Fig. 8.49 depicted thawing beef using various treatments. PAW and SEAW both effectively

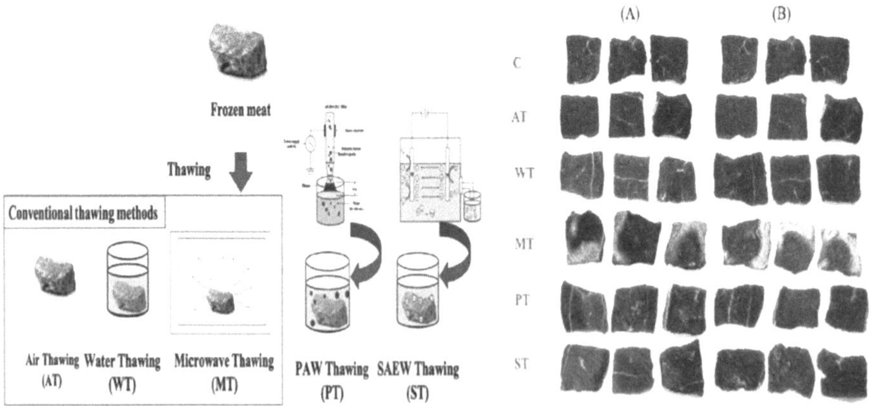


**Fig. 8.49** (*Left*) Schematic diagram of thawing beef with various treatments and (*Right*) photographs of the external (A) and internal (B) surface of beef treated by various thawing method.
From Liao, X., et al., 2020. Plasma-activated water (PAW) and slightly acidic electrolyzed water (SAEW) as beef thawing media for enhancing microbiological safety. LWT 117 (September 2019). https://doi.org/10.1016/j.lwt.2019.108649.

controlled microbial load on beef during thawing. The PT and ST treatments retarded protein oxidation and degradation in beef and preserved the texture, pH, and color of beef better than other treatments (Liao et al., 2020).

## 8.12 Conclusion

The thawing process involves holding frozen material at ambient temperature, i.e., above 0 °C, to become free of ice for the next mechanical processing. Thawing is an important process in the food industry and which is performed with different equipment and methods. The thawing methods can be classified into two basic functional categories, methods that absorb heat mainly via the product's boundary and methods that generate heat within the product. The combination of boundary methods with other thawing-accelerating technologies, such as HPP, HVEF, MW, etc., are promising and may be applied for large-scale thawing in the future. It is expected that the current thawing methods become more energy efficient and fast. On the other hand, the improved thawing technologies, would become more integrated and sophisticated within the next few years.

## References


Aboud, S.A., et al., 2019. A comprehensive review on infrared heating applications in food processing. Molecules 24 (22), 1–20. https://doi.org/10.3390/molecules24224125.

Aidani, E., et al., 2014. Effect of chilling, freezing and thawing on meat quality: a review. Int. J. Biosci. 5 (4), 159–169. https://doi.org/10.12692/ijb/5.4.159-169.

Alkanan, Z.T., Altemimi, A.B., Al-Hilphy, A.R.S., et al., 2021a. Ohmic heating in the food industry: developments in concepts and applications during 2013–2020. Appl. Sci. 11 (6). https://doi.org/10.3390/app11062507.

Alkanan, Z.T., Altemimi, A.B., Al-Hilphy, A.R.S., et al., 2021b. Application and effects of ohmic-vacuum combination heating on the quality factors of tomato paste. Foods 10 (12), 1–19. https://doi.org/10.3390/foods10122920.

Anderson, B.A., Singh, R.P., 2006a. Effective heat transfer coefficient measurement during air impingement thawing using an inverse method. Int. J. Refrig. 29 (2), 281–293. https://doi.org/10.1016/j.ijrefrig.2005.05.016.

Anderson, B.A., Singh, R.P., 2006b. Modeling the thawing of frozen foods using air impingement technology. Int. J. Refrig. 29 (2), 294–304. https://doi.org/10.1016/j.ijrefrig.2005.05.003.

Backi, C.J., 2018. Methods for (industrial) thawing of fish blocks: a review. J. Food Process Eng. 41 (1). https://doi.org/10.1111/jfpe.12598.

Backi, C.J., 2022. Industrial Thawing of Fish Blocks. Sustainable Fish Production and Processing Inc, https://doi.org/10.1016/b978-0-12-824296-4.00001-3.

Bedane, T., Marra, F., Wang, S., 2017. Performance comparison between batch and continuous thawing of food products assisted by radio frequency heating. Chem. Eng. Trans. 57, 2017–2022. https://doi.org/10.3303/CET1757337.

Bhargava, N., et al., 2021. Advances in application of ultrasound in food processing: a review. Ultrason. Sonochem. 70 (June 2020), 105293. https://doi.org/10.1016/j.ultsonch.2020.105293.

Boonsumrej, S., et al., 2007. Effects of freezing and thawing on the quality changes of tiger shrimp (Penaeus monodon) frozen by air-blast and cryogenic freezing. J. Food Eng. 80 (1), 292–299. https://doi.org/10.1016/j.jfoodeng.2006.04.059.

Bozzato, A., et al., 2021. Air impingement to reduce thawing time of chicken fingers for food service. J. Food Process. Preserv. 45 (11), 1–10. https://doi.org/10.1111/jfpp.15962.

Cai, L., et al., 2018. Ultrasound or microwave vacuum thawing of red seabream (Pagrus major) fillets. Ultrason. Sonochem. 47 (March), 122–132. https://doi.org/10.1016/j.ultsonch.2018.05.001.

Cai, L., Zhang, W., et al., 2019b. Effects of ultrasonics combined with far infrared or microwave thawing on protein denaturation and moisture migration of Sciaenops ocellatus (red drum). Ultrason. Sonochem. 55 (March), 96–104. https://doi.org/10.1016/j.ultsonch.2019.03.017.

Cai, L., Cao, M., et al., 2019a. Recent advances in food thawing technologies. Compr. Rev. Food Sci. Food Saf. 18, 953–970. https://doi.org/10.1111/1541-4337.12458.

Cai, L., et al., 2020. Effects of different thawing methods on the quality of largemouth bass (*Micropterus salmonides*). LWT. https://doi.org/10.1016/j.lwt.2019.108908.

Cheftel, J.C., Lévy, J., Dumay, E., 2000. Pressure-assisted freezing and thawing: principles and potential applications. Food Rev. Int. 16 (4), 453–483. https://doi.org/10.1081/FRI-100102319.

Chemat, F., Zill-E-Huma, Khan, M.K., 2011. Applications of ultrasound in food technology: processing, preservation and extraction. Ultrason. Sonochem. 18 (4), 813–835. https://doi.org/10.1016/j.ultsonch.2010.11.023.

Chen, Y., et al., 2021. Model food development for tuna (Thunnus Obesus) in radio frequency and microwave tempering using grass carp mince. J. Food Eng. 292 (July 2020), 110267. https://doi.org/10.1016/j.jfoodeng.2020.110267.

Choi, E.J., et al., 2017. Effect of tempering methods on quality changes of pork loin frozen by cryogenic immersion. Meat Sci. 124, 69–76. https://doi.org/10.1016/j.meatsci.2016.11.003.

Dalvi-Isfahan, M., et al., 2016. The principles of high voltage electric field and its application in food processing: a review. Food Res. Int. https://doi.org/10.1016/j.foodres.2016.09.002.

Duygu, B., Ümit, G., 2015. Application of ohmic heating system in meat thawing. Procedia Soc. Behav. Sci. 195, 2822–2828. https://doi.org/10.1016/j.sbspro.2015.06.400.

Elamin, W.M., et al., 2015. High pressure processing technology and equipment evolution: a review. J. Eng. Sci. Technol. Rev. 8 (5), 75–83. https://doi.org/10.25103/jestr.085.11.

Ersoy, B., Aksan, E., Özeren, A., 2008. The effect of thawing methods on the quality of eels (*Anguilla anguilla*). Food Chem. 111 (2), 377–380. https://doi.org/10.1016/j.foodchem.2008.03.081.

Gambuteanu, C., Alexe, P., 2015. Comparison of thawing assisted by low-intensity ultrasound on technological properties of pork longissimus dorsi muscle. J. Food Sci. Technol. 52 (4), 2130–2138. https://doi.org/10.1007/s13197-013-1204-7.

Ganessingh, V., Sahibdeen, R., Maharaj, R., 2018. An evaluation of the impact of novel processing technologies on the phytochemical composition of fruits and vegetables. In: Phytochemicals—Source of Antioxidants and Role in Disease Prevention. IntechOpen, https://doi.org/10.5772/intechopen.77730 (November 2018).

Goullieux, A., Pain, J.P., 2005. Ohmic heating. In: Emerging Technologies for Food Processing., https://doi.org/10.1016/B978-012676757-5/50020-7.

Guo, Z., et al., 2021. Ultrasound-assisted thawing of frozen white yak meat: effects on thawing rate, meat quality, nutrients, and microstructure. Ultrason. Sonochem. 70 (1), 105345. https://doi.org/10.1016/j.ultsonch.2020.105345.

Guzik, P., et al., 2021. Microwave applications in the food industry: an overview of recent developments. In: Critical Reviews in Food Science and Nutrition. Taylor & Francis, pp. 1–20, https://doi.org/10.1080/10408398.2021.1922871.

He, X., et al., 2013. Effect of high voltage electrostatic field treatment on thawing characteristics and post-thawing quality of frozen pork tenderloin meat. J. Food Eng. 115 (2), 245–250. https://doi.org/10.1016/j.jfoodeng.2012.10.023.

He, X., et al., 2014. Factors affecting the thawing characteristics and energy consumption of frozen pork tenderloin meat using high-voltage electrostatic field. Innov. Food Sci. Emerg. Technol. 22, 110–115. https://doi.org/10.1016/j.ifset.2013.12.019.

He, X., et al., 2016. Effect of corona wind, current, electric field and energy consumption on the reduction of the thawing time during the high-voltage electrostatic-field (HVEF) treatment process. Innov. Food Sci. Emerg. Technol. 34, 135–140. https://doi.org/10.1016/j.ifset.2016.01.006.

Hebbar, H.U., Vishwanathan, K.H., Ramesh, M.N., 2004. Development of combined infrared and hot air dryer for vegetables. J. Food Eng. 65 (4), 557–563. https://doi.org/10.1016/j.jfoodeng.2004.02.020.

Hong, G.P., et al., 2009. Effects of air blast thawing combined with infrared radiation on physical properties of pork. Korean J. Food Sci. Anim. Resour. 29 (3), 302–309. https://doi.org/10.5851/kosfa.2009.29.3.302.

Hong, G.P., et al., 2014. Effects of water or brine immersion thawing combined with ultrasound on quality attributes of frozen pork loin. Korean J. Food Sci. Anim. Resour. 34 (1), 115–121. https://doi.org/10.5851/kosfa.2014.34.1.115.

Hu, F., et al., 2021. Combined impacts of low voltage electrostatic field and high humidity assisted-thawing on quality of pork steaks. LWT 150 (January), 111987. https://doi.org/10.1016/j.lwt.2021.111987.

Huang, Z., Marra, F., Wang, S., 2016. A novel strategy for improving radio frequency heating uniformity of dry food products using computational modeling. Innov. Food Sci. Emerg. Technol. 34, 100–111. https://doi.org/10.1016/j.ifset.2016.01.005.

Ito, R., Fukuoka, M., Hamada-Sato, N., 2014. Innovative food processing technology using ohmic heating and aseptic packaging for meat. Meat Sci. 96 (2), 675–681. https://doi.org/10.1016/j.meatsci.2013.10.012.

Jafari, F., Saien, J., 2021. Utilizing low-voltage pulsed electric fields for mass transfer intensification of drops in liquid-liquid extraction. Ind. Eng. Chem. Res. 60 (4), 1532–1541. https://doi.org/10.1021/acs.iecr.0c05952.

James, S.J., James, C., 2002. Meat Refrigeration. Elsevier, https://doi.org/10.1533/9781855736535.

James, S.J., James, C., Purnell, G., 2017. Microwave-assisted thawing and tempering. In: The Microwave Processing of Foods, second ed. Elsevier Ltd., https://doi.org/10.1016/B978-0-08-100528-6.00012-7.

Kaur, N., Singh, A.K., 2016. Ohmic heating: concept and applications—a review. Crit. Rev. Food Sci. Nutr. 56 (14), 2338–2351. https://doi.org/10.1080/10408398.2013.835303.

Kerry, J.F., 2011. Effects of novel thermal processing technologies on the sensory quality of meat and meat products. In: Processed Meats: Improving Safety, Nutrition and Quality. Woodhead Publishing Limited, https://doi.org/10.1533/9780857092946.3.617.

Khadhraoui, B., et al., 2021. Review of ultrasound combinations with hybrid and innovative techniques for extraction and processing of food and natural products. Ultrason. Sonochem. 76, 105625. https://doi.org/10.1016/j.ultsonch.2021.105625.

Kissam, A.D., et al., 1982. Water-thawing of fish using low frequency acoustics. J. Food Sci. 47 (1), 71–75. https://doi.org/10.1111/j.1365-2621.1982.tb11029.x.

Koray Palazoğlu, T., Miran, W., 2017. Experimental comparison of microwave and radio frequency tempering of frozen block of shrimp. Innov. Food Sci. Emerg. Technol. 41, 292–300. https://doi.org/10.1016/j.ifset.2017.04.005.

Krishnamurthy, K., et al., 2008. Infrared heating in food processing: an overview. Compr. Rev. Food Sci. Food Saf. 7 (1), 2–13. https://doi.org/10.1111/j.1541-4337.2007.00024.x.

Lee, S.H., et al., 2015. Development of a dual cylindrical microwave and ohmic combination heater for minimization of thermal lags in the processing of particulate foods. LWT 63 (2), 1220–1228. https://doi.org/10.1016/j.lwt.2015.04.042.

Lee, S.Y., et al., 2021. Two-stage air thawing as an effective method for controlling thawing temperature and improving the freshness of frozen pork loin. LWT 140 (November 2020), 110668. https://doi.org/10.1016/j.lwt.2020.110668.

Lelieveld, H.L.M., Hoogland, H., 2016. Continuous high-pressure processing to extend product shelf life. In: Food Engineering Series. Elsevier, pp. 67–72, https://doi.org/10.1007/978-1-4939-3234-4_4.

Li, B., Sun, D.W., 2002. Novel methods for rapid freezing and thawing of foods—a review. J. Food Eng. 54 (3), 175–182. https://doi.org/10.1016/S0260-8774(01)00209-6.

Liao, X., et al., 2020. Plasma-activated water (PAW) and slightly acidic electrolyzed water (SAEW) as beef thawing media for enhancing microbiological safety. LWT 117 (September 2019). https://doi.org/10.1016/j.lwt.2019.108649.

Liu, L., et al., 2017. Electrical conductivity and ohmic thawing of frozen tuna at high frequencies. J. Food Eng. 197, 68–77. https://doi.org/10.1016/j.jfoodeng.2016.11.002.

Liu, Y., et al., 2019. Ultrasound-assisted thawing of mango pulp: effect on thawing rate, sensory, and nutritional properties. Food Chem. 286 (October 2018), 576–583. https://doi.org/10.1016/j.foodchem.2019.02.059.

Llave, Y., Erdogdu, F., 2022. Radio frequency processing and recent advances on thawing and tempering of frozen food products. Crit. Rev. Food Sci. Nutr. 62 (3), 598–618. https://doi.org/10.1080/10408398.2020.1823815.

Llave, Y., Sakai, N., 2018. Dielectric Defrosting of Frozen Foods, Food Processing for Increased Quality and Consumption. Elsevier Inc., https://doi.org/10.1016/B978-0-12-811447-6.00013-8.

Min, S.G., et al., 2016. Pressure Ohmic thawing: a feasible approach for the rapid thawing of frozen meat and its effects on quality attributes. Food Bioprocess Technol. 9 (4), 564–575. https://doi.org/10.1007/s11947-015-1652-3.

Mousakhani-Ganjeh, A., Hamdami, N., Soltanizadeh, N., 2016. Thawing of frozen tuna fish (Thunnus albacares) using still air method combined with a high voltage electrostatic field. J. Food Eng. 169, 149–154. https://doi.org/10.1016/j.jfoodeng.2015.08.036.

Movahedi, E., et al., 2021. A continuous high voltage electrostatic field system for thawing food materials. J. Agric. Eng. 52 (2). https://doi.org/10.4081/jae.2021.1163.

Natarajan, S., Ponnusamy, V., 2020. A review on the applications of ultrasound in food processing. Mater. Today Proc. https://doi.org/10.1016/j.matpr.2020.09.516.

Nguyen, L.T., et al., 2013. Exploring the heating patterns of multiphase foods in a continuous flow, simultaneous microwave and ohmic combination heater. J. Food Eng. 116 (1), 65–71. https://doi.org/10.1016/j.jfoodeng.2012.11.011.

Okamoto, A., Suzuki, A., 2002. Effects of high hydrostatic pressure-thawing on pork meat. Prog. Biotechnol. 19 (C), 571–576. https://doi.org/10.1016/S0921-0423(02)80155-1.

Orsat, V., Raghavan, G.S.V., Krishnaswamy, K., 2017. Microwave technology for food processing: an overview of current and future applications. In: The Microwave Processing of Foods, second ed. Elsevier Ltd., https://doi.org/10.1016/B978-0-08-100528-6.00005-X.

Peng, Z., et al., 2021. Physicochemical and structural changes in myofibrillar proteins from porcine longissimus dorsi subjected to microwave combined with air convection thawing treatment. Food Chem. 343, 128412. https://doi.org/10.1016/j.foodchem.2020.128412.

Qian, S., et al., 2019. Effects of low voltage electrostatic field thawing on the changes in physicochemical properties of myofibrillar proteins of bovine Longissimus dorsi muscle. J. Food Eng. 261 (January), 140–149. https://doi.org/10.1016/j.jfoodeng.2019.06.013.

Qian, J., et al., 2022. Plasma-activated water: a novel frozen meat thawing media for reducing microbial contamination on chicken and improving the characteristics of protein. Food Chem. 375 (November 2021), 131661. https://doi.org/10.1016/j.foodchem.2021.131661.

Qiu, L., et al., 2020. Application of power ultrasound in freezing and thawing processes: effect on process efficiency and product quality. Ultrason. Sonochem. https://doi.org/10.1016/j.ultsonch.2020.105230.

Rahbari, M., et al., 2018. Effects of high voltage electric field thawing on the characteristics of chicken breast protein. J. Food Eng. 216, 98–106. https://doi.org/10.1016/j.jfoodeng.2017.08.006.

Rahbari, M., et al., 2020. Investigation of the histological and textural properties of chicken breast thawed by high voltage electric field. J. Food Process Eng. 43 (11). https://doi.org/10.1111/jfpe.13543.

Rastogi, N.K., 2012. Recent trends and developments in infrared heating in food processing. Crit. Rev. Food Sci. Nutr. 52 (9), 737–760. https://doi.org/10.1080/10408398.2010.508138.

Rastogi, N.K., 2015. Infrared heating of foods and its combination with electron beam processing. In: Electron Beam Pasteurization and Complementary Food Processing Technologies. Woodhead Publishing Limited, https://doi.org/10.1533/9781782421085.2.61.

Rouillé, J., et al., 2002. High pressure thawing of fish and shellfish. J. Food Eng. 53 (1), 83–88. https://doi.org/10.1016/S0260-8774(01)00143-1.

Seyhun, N., et al., 2009. Comparison and modeling of microwave tempering and infrared assisted microwave tempering of frozen potato puree. J. Food Eng. 92 (3), 339–344. https://doi.org/10.1016/j.jfoodeng.2008.12.003.

Shen, G., et al., 2020. Preparation of potato flour by freeze-thaw pretreatment: effect of different thawing methods on hot-air drying process and physicochemical properties. LWT 133 (June), 110157. https://doi.org/10.1016/j.lwt.2020.110157.

Singla, M., Sit, N., 2021. Application of ultrasound in combination with other technologies in food processing: a review. Ultrason. Sonochem. 73, 105506. https://doi.org/10.1016/j.ultsonch.2021.105506.

Sumnu, G.D.D., Sahin, S., 2005. Recent developments in microwave heating. In: Emerging Technologies for Food Processing. Elsevier, pp. 419–444, https://doi.org/10.1016/B978-012676757-5/50018-9.

Svendsen, E.S., et al., 2022. Industrial methods of freezing, thawing and subsequent chilled storage of whitefish. J. Food Eng. 315. https://doi.org/10.1016/j.jfoodeng.2021.110803.

Swain, M., James, S., 2005. Thawing and Tempering Using Microwave Processing, The Microwave Processing of Foods. Woodhead Publishing Limited, https://doi.org/10.1533/9781845690212.2.174.

Takai, R., Kozhima, T., Suzuki, T., 1991. Low temperature thawing by using high pressure. In: 17th Congress of the International Institute of Refrigeration, Montréal, Québec, Canada. International Institute of Refrigeration.

Tang, T., et al., 2021. Radio frequency thawing chamber for rapidly thawing applications. Microw. Opt. Technol. Lett. 63 (1), 175–180. https://doi.org/10.1002/mop.32596.

Uyar, R., et al., 2015. Radio-frequency thawing of food products—a computational study. J. Food Eng. 146, 163–171. https://doi.org/10.1016/j.jfoodeng.2014.08.018.
Varghese, K.S., et al., 2014. Technology, applications and modelling of ohmic heating: a review. J. Food Sci. Technol. 51 (10), 2304–2317. https://doi.org/10.1007/s13197-012-0710-3.
Verma, D.K., et al., 2020. High pressure processing (HPP): fundamental concepts. In: Emerging Scope, and Food Application, Emerging Thermal and Nonthermal Technologies in Food Processing., https://doi.org/10.1201/9780429297335-9.
Wang, Q., Salvi, D., 2021. Recent progress in the application of plasma-activated water (PAW) for food decontamination. Curr. Opin. Food Sci. 42, 51–60. https://doi.org/10.1016/j.cofs.2021.04.012.
Wang, Y.Y., et al., 2021. Effect of multi-frequency ultrasound thawing on the structure and rheological properties of myofibrillar proteins from small yellow croaker. Ultrason. Sonochem. 70 (August 2020), 105352. https://doi.org/10.1016/j.ultsonch.2020.105352.
Watanabe, T., Ando, Y., 2021. Evaluation of heating uniformity and quality attributes during vacuum microwave thawing of frozen apples. LWT 150 (June), 111997. https://doi.org/10.1016/j.lwt.2021.111997.
Wray, D., Ramaswamy, H.S., 2015. Novel concepts in microwave drying of foods. Dry. Technol 33 (7), 769–783. https://doi.org/10.1080/07373937.2014.985793.
Xie, Y., et al., 2021. Applying low voltage electrostatic field in the freezing process of beef steak reduced the loss of juiciness and textural properties. Innov. Food Sci. Emerg. Technol. 68 (October 2020), 102600. https://doi.org/10.1016/j.ifset.2021.102600.
Xu, B., et al., 2021. Effect of different thawing methods on the efficiency and quality attributes of frozen red radish. J. Sci. Food Agric. 101 (8), 3237–3245. https://doi.org/10.1002/jsfa.10953.
Yadav, G., et al., 2020. Infrared heating and its application in food processing. Pharma Innov. 9 (2), 142–151.
Yagi, S., Shibata, K., 2002. Microwave defrosting under reduced pressure. US Patent Application 20020195447.
Zhang, Y., Ding, C., 2020. The study of thawing characteristics and mechanism of frozen beef in high voltage electric field. IEEE Access 8, 134630–134639. https://doi.org/10.1109/ACCESS.2020.3010948.
Zhang, Y., et al., 2019. Effects of high-voltage electric field process parameters on the water-holding capacity of frozen beef during thawing process. J. Food Qual. 2019. https://doi.org/10.1155/2019/9140179.
Zhu, S., Ramaswamy, H.S., Simpson, B.K., 2004. Effect of high-pressure versus conventional thawing on color, drip loss and texture of Atlantic salmon frozen by different methods. LWT Food Sci. Technol. 37 (3), 291–299. https://doi.org/10.1016/j.lwt.2003.09.004.
Zhu, S.M., et al., 2014. Water phase transition under pressure and its application in high pressure thawing of agar gel and fish. J. Food Eng. 125 (1), 1–6. https://doi.org/10.1016/j.jfoodeng.2013.10.016.
Zhu, M.M., et al., 2020. Physicochemical properties and protein denaturation of pork longissimus dorsi muscle subjected to six microwave-based thawing methods. Foods 9 (1). https://doi.org/10.3390/foods9010026.

# Section D

# Frying operations

# Fundamentals of food frying processes

9

*Georgia Frakolaki, Tryfon Kekes, Angelos-Panagiotis Bizymis, Virginia Giannou, and Constantina Tzia*
Laboratory of Food Chemistry and Technology, School of Chemical Engineering, National Technical University of Athens, Athens, Greece

## 9.1 Introduction

Frying is a food preparation process used at both industrial and household scale. The frying process is applied to various foods such as roots and tubers, fruits, meat, fish, as well as food preparations such as dough products and snacks. Fried foods can be consumed directly after frying or they are preserved for longer times under packaging. Fried food products are liked by consumers of all ages mainly due to their appealing taste, and oily and crunchy texture. For example, fried potatoes, known as "French fries," is a product that is widely consumed throughout the world. However, fried foods are often associated with safety hazards that induce an adverse effect on human health. Despite the hype about healthy and low-fat diets, fried foods are also considered as comfort foods and are included in the menu of almost all fast-food restaurants with growing consumption during the last decades (i.e., fried chicken). They are also served worldwide by food services and prepared at home or are produced and packaged in a shelf-stable form by the food industry under growing demand.

Frying is one of the oldest food preparation processes. Deep-fat frying describes the process in which a food of vegetable or animal origin or a food preparation is immersed into a fryer, filled with a hot (~180°C) oil or fat medium, where it remains for a certain time and is then pulled away, is allowed to cool in the environment, and is consumed directly or distributed as a packaged product with a longer shelf life. The mechanisms affecting the phenomena involved with the frying process and result in changes in fried foods have been investigated and are of great interest to the food industry.

The raw materials are basically the foods selected to be fried and the oil to be used as the frying medium. As far as the food is concerned, it can be a vegetable, fruit, meat, or fish in a raw state which may be subjected to pretreatments depending on its type (cutting/slicing for size reducing, blanching, pre-drying, coating, etc.) or it can be prepared by mixing the ingredients, according to product's formulation, and properly shaped. The oil or fat selected must be suitable for frying; as it is absorbed in the food, it should be stable at high temperatures and have desirable sensorial properties.

During frying, heat and mass transfer phenomena take place evoking changes in both the food and frying oil medium. The food is heated by the hot oil while absorbing

**High-Temperature Processing of Food Products. https://doi.org/10.1016/B978-0-12-818618-3.00001-X**

an amount of oil and is then allowed to cool. The heat transfer is affected by the food unit's geometry (size and dimensions, density), composition (water content), and thermal properties (which vary during frying). Therefore, the food water is evaporated, and other physical (i.e., color, texture crunchiness, etc.) and chemical changes (Maillard reactions, oxidation, etc.) occur, and its microbiological load is decreased. The food composition is also altered, mainly its moisture content is decreased while its oil content (in dry basis) is increased. Due to the intense heat treatment, the fried products exhibit low water activity and decreased microbial load and therefore have a longer shelf life.

The oil is submitted to various chemical alterations (oxidation, polymerization, isomerization, aromatic cyclic compounds, etc.), and new degradation products (primary and secondary oxidized, polymerized oil substances, polar compounds, trans fatty acids, etc.) are formed with undesirable flavor and taste (volatile fat degradation substances) or with adverse health effect (trans fatty acids). The extent of the aforementioned alterations depends on the oil type (unsaturation level, antioxidants) and the frying processing conditions (temperature, residence time, renewing of oil in the fryer, etc.).

Fried products are popular tasteful products that consumers like due to their sensorial properties mainly their golden yellow to dark brown color, crunchy texture, and pleasant flavor, but at the same time they are involved with nutritional and safety issues. The fried food products are characterized by a high oil content that lowers their nutritional value according to the recommendations for a healthy diet. In addition, the oil absorbed in the food, depending on its oxidative and degradation state during frying, will significantly affect the sensorial quality and safety of the fried food.

Frying is assessed as a critical process according to the HACCP approach because during processing, there is a high possibility of developing hazardous substances (i.e., trans fatty acids, polar compounds, acrylamide), which are considered of high severity in terms of their adverse effects on human health. Thus, monitoring of the frying operation (frying temperature and time, renewal of oil in the fryer) should be performed on an industrial scale to control the existence of potential hazards to acceptable limits and assure the safety of the final product. Apart from the safety, the quality of fried products should be also assured by respective controls. Concluding, industries producing fried foods and snacks should implement both quality and safety management in order to provide safe products with high acceptance by the consumers.

In the case of fried foods intended to be preserved as packaged and standardized products with prolonged shelf life, their size, shape, and composition should be controlled to maintain their quality characteristics during their commercial shelf life. The use of antioxidants and/or the application of modified atmosphere packaging function protectively against oil auto-oxidation in the packaged fried products.

New technologies, such as edible films and coating of foods or pre-drying of foods before their frying, can help decrease the oil absorbed and provide low-fat fried products. The opportunity is also given to incorporate antioxidants in these films and coating (i.e., natural extracts, which at the same time provide fried products with a specific taste and flavor character). Additionally, par-fried food products are produced and distributed to the market to facilitate and minimize the fried food preparation by consumers.

Finally, alternative processes, such as toasting, are investigated to substitute frying and its adverse health effects while still developing the respective frying characteristics in foods that are attractive to consumers.

### 9.1.1 Frying: An overview

Frying is a complex unit operation, used extensively in food processing, that can be categorized as follows: (1) home frying, (2) restaurant or food service frying, and (3) industrial frying. In homes and restaurants, mostly pan or griddle frying is applied, while in large-scale frying processes, deep fryers are used, commonly of batch operation in food services and continuous operation in the frying industry, respectively. The fried foods are ready to be consumed almost immediately, whereas industrial products with prolonged shelf life are available in the market in packaged form (Rossell, 2001a; Gupta et al., 2004; Sahin and Sumnu, 2009).

Although frying characteristics may vary, depending on the operation scale, the process is basically almost the same. Frying, in general, describes the process where the food is brought in direct contact with hot oil. Moreover, in deep frying or immersion frying processes, the food is placed in a bed of hot oil under atmospheric conditions. The oil is heated at a high temperature of about 160–190°C and the food, due to the thermal energy of the oil, is dehydrated. At such temperatures, various changes in foods' components occur which modify the appearance of the fried products. Different phenomena take place simultaneously during frying, such as heat, moisture, and fat transfer between the food and the heating medium (frying oil) and the formation of a crust layer on the outer surface of the product; also, the composition of the oil and the properties of the products steadily change throughout the process. Deep-fat frying is the most commonly used frying technique in the food industry.

In pan-frying or griddle frying that is the traditional frying procedure, a thin layer of oil is heated on a skillet or a griddle and the food is fried inside this layer of oil until completion. During batch frying in food services, the food is placed in a wire basket, immersed in a bed of hot oil, and removed from the hot oil when fully fried. In both cases, the frying temperature and time vary depending on the products fried, and they are usually determined through observing the appearance of fried foods; the color of the food surface becomes golden yellow to dark brown, a desired moisture content and texture is developed, and a pleasant fried food flavor and mouthfeel is produced by the oil absorbed in the fried food (Rossell, 2001a; Gupta et al., 2004; Sahin and Sumnu, 2009).

In industrial-scale batch frying, the food is added into a large pan of hot oil, which is continuously recirculated and stirred, and the fried product is removed and spun to remove the excess oil from its surface. In continuous industrial frying, the food to be fried enters the fryer at one end, submerged in a bed of hot oil for a specific time interval, until it attains the desired color and texture, and removed from the other end. In both cases, the residence time in the fryer depends on the type of food being fried. The fried products are commonly packaged and distributed for sale. Nowadays, the frying industry can use advanced frying techniques and equipment, ingredients, and packaging materials, and methods to provide fried products which preserve their freshness for extended shelf life.

### 9.1.2 Raw materials: Food fried and frying oil—Fried food products

Foods being most commonly fried are vegetables as roots and tubers (i.e., sliced potatoes, vegetables mixed with seasonings or binders, batter-coated vegetables, etc.), fruits, meat, fish and seafood, and others. Raw food materials are cut or sliced in pieces and subjected to necessary pretreatments depending on the type of food (blanching, predrying, coating, etc.). Food preparations, such as dough products and snacks, are also included in fried food products. For their production, ingredients are mixed in proper amounts, according to the recipe, and pieces suitably shaped through extrusion are fried. Coated products, and par-fried products with or without seasoning, have also become popular across the world (Rossell, 2001a; Gupta et al., 2004; Sahin and Sumnu, 2009).

The oil or fat used for frying is of high importance for both the frying process and the fried foods. During frying, a complicated interaction between the frying oil and the food is developed, and as a result, the oil becomes part of the food being fried thus increasing its caloric content and contributing to its nutritional value (fat-soluble vitamins, essential fatty acids). The absorbed oil also affects the fried food's taste, flavor, and mouthfeel while it accelerates the salivation process, which enables the release of oil-soluble flavors from the fried food (O'Brien, 2004; Asokapandian et al., 2020). Fat solids affect the acceptability of fried foods. For this reason, the frying medium is selected based on its sensorial properties in order to provide a pleasant result and make the fried food even more appealing to the taster. At the same time, during deep frying, the oil is exposed to highly adverse conditions; it is subjected continuously or repeatedly to high temperatures (160–190°C) in the presence of air and moisture. Under such conditions, a variety of oil degradation reactions may occur, such as autoxidation, thermal polymerization, thermal oxidation, and hydrolysis leading to changes in viscosity and color, and the development of off-tastes and off-flavors or even more hazardous substances (trans fatty acids, acrylamide). The extent of these reactions depends on various factors related to food, oil, and processing (Rossell, 2001a; Gupta et al., 2004; Sahin and Sumnu, 2009). Common vegetable oils exhibit different thermal stabilities influenced by their unsaturation level as well as the presence of minor components with antioxidative activity (sterols, tocopherols, polyphenols, etc.) (Kochhar, 2001; Angelo et al., 1996).

Foods, when fried, can introduce various components in the oil which may be deleterious (i.e., trace metals, etc.) catalyzing oil oxidations or reacting with the oil or its breakdown products (carbohydrates, phosphates, sulfur compounds). Moreover, the temperature and time of frying, the technique (oil mixing, reusing, renewing, etc.) applied, and/or the use of antioxidants may influence oil oxidation (Zhang et al., 2012; Asokapandian et al., 2020). Finally, the quality of the frying oil influences oil absorption and the types of by-products and residues absorbed by the food. Furthermore, the frying oil absorbed in packaged fried products may undergo autoxidation during their storage (depending on the packaging material and method used, i.e., MAP), thus degrading the quality of fried products and reducing their storage life. Consequently, for all the above reasons, the quality of the frying oil plays important role in determining the quality of the fried foods and their shelf life.

The selection of a frying oil for industrial use (i.e., snack foods) is also based on the local availability. Several vegetable oils are distributed across the world sufficiently supplying the frying industry (Gupta, 2004; Asokapandian et al., 2020). They are recovered from oilseeds by mechanical extraction or by solvent extraction and become suitable for use after refining. Olive oil, in special, is separated from the olive fruit by a physical method (centrifugation of olive paste) and it is almost the only vegetable oil that can be directly consumed without requiring refining. Regional preferences are also observed for particular oils in fried foods. Thus, cottonseed oil is considered as the "gold standard" for potato chips in many countries, while sesame seed oil, safflower oil or peanut (groundnut) oil are occasionally preferred for snack foods. In certain regions, palm oil or palmolein oil is favored for fried foods and snacks, while olive oil is popular among the consumers of the Mediterranean region. Other oils used, due to availability or cost, are sunflower, canola, soybean, coconut, peanut, and corn oil, as well as shortenings (produced by the fractionation of palm oil and/or the interesterification process) (Gupta, 2004; O'Brien, 2004).

### 9.1.3 Fried foods—Frying industry

Fried foods prepared and served in restaurants and food services are mainly potatoes known as "French fries," and other vegetables, as well as breaded chicken and fish. "French fries" is the most popular fried product among the consumers in many countries; they are served in waxed paper packages by food services. Restaurants and food services, apart from fresh foods, also use a variety of frozen par-fried products, such as French fries, potato wedges, vegetable/cheese stuffing, batter-coated vegetables, fish sticks, breaded shrimps, and chicken. Other freshly fried products, being very popular in many countries, are donuts, used as breakfast servings, and popped corn (Rossell, 2001b; Gupta et al., 2004; Sahin and Sumnu, 2009).

A variety of shelf-stable packaged products are produced by the frying industry with salty snack foods becoming common in every country and considered a source of culinary delight throughout the world. Potato chips are the most popular and predating savory snacks sold in various countries, while other salty snack products include tortilla chips or corn chips, extruded products, nuts, and a variety of other products. Traditional snack foods are produced in many countries (popped corn kernels, nuts, grains, leguminous products, extruded grains, and fried wafers) (Rossell, 2001b; Gupta et al., 2004; Sahin and Sumnu, 2009).

The frying industry is a dynamic food sector, which has significantly evolved during the last decades. It has gradually been introduced with larger and more sophisticated equipment and continuous fryers and systems with high production rates and temperature control. By reducing the oil turnover times, better oil quality is attained along with shorter frying times. Modified atmosphere packaging (MAP) has also been applied using metallized film packaging with nitrogen flush that achieves longer shelf life. Salty snack foods have respectively gone through significant changes during the last few decades. These changes were enforced by consumers' trends toward hygiene diet and health protection. Thus, the food industry was directed to the development of "low-fat" and "fat-free" snack foods. Moreover, current trends suggest the use of more "healthful oils" (instead of oils high in saturated fats, such as palm oil, palmolein, and coconut oil), oils with low or zero linolenic acid content, or interesterified (instead of hydrogenated) oils for snack products. The snack food industry is now entering the areas of "natural," to some degree "organic" (fried in nonhydrogenated vegetable oils), or "no-salt" products as well as to products with novel flavors.

## 9.2 The frying process: Heat and mass transfer

Deep-fat frying or immersion frying are the most popular unit operations to process foods. It is defined as the process of cooking and dehydration of foods by immersing them in hot oil, typically, at 160–190°C. During this time, various chemical and physical changes occur in the components of the "frying system," that is the food material and the frying oil or fat. Chemical and structural changes take place, such as starch gelatinization, protein denaturation, and flavor development. Physical changes are manifested as a decrease in moisture content, increase in products' temperature and oil content, and crust formation. Fried products absorb oil during frying and as they cool.

Deep-fat frying is a complex process that involves heat and mass transfer between the product and the frying oil at the same time. During frying, heat is initially transferred by convection between the oil and the food surface and then by conduction within the food. When the food is immersed in the hot oil, its water is evaporated due to the high temperatures, and the vapor formed is transferred through the surface of the product due to pressure and concentration gradients. The evaporation is more intense at the beginning and continues as the frying proceeds until the characteristic hydrophilicity of the food is lost resulting in crust formation and pores development; both crust and pores contribute to a higher rate of oil absorption. Oil enters into the pores formed by moisture loss mainly during cooling of the fried foods while shrinkage may additionally occur during frying. If a thick crust is formed, the evaporation is restricted with the excess of vapor accumulating and increasing the pressure inside the food; thus cracks are formed in the crust allowing oil to enter into the product (Sahin and Sumnu, 2009).

Moisture loss, oil uptake, and crust and cracks formation on the surface of the fried products are important quality parameters. Therefore, these changes are examined at the microlevel in order to explain the mechanism of deep frying. The understanding of these mechanisms is important for food industries to optimize and control the process (i.e., increase frying oil life, decrease oil absorption by the product, control process specifications, and reduce operating and energy costs) as well as formulate sustainable strategies toward safe and healthy fried products.

During deep frying, heat and mass transfer in the product is governed by the heat transferred from the oil to the product through its external surface (Fig. 9.1). Since the understanding of the mechanisms of deep frying or immersion frying is of special interest, heat and mass transfer are described in the following paragraphs (Achir et al., 2009).

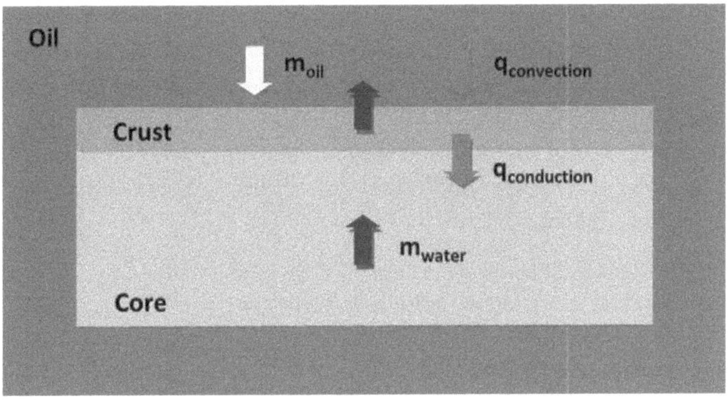


**Fig. 9.1** Heat transfer ($q$) and mass transfer ($m$) during deep-fat frying ($q_{cond}$: heat transfer by conduction, $q_{cond}$: heat transfer by convection).

## 9.2.1 Heat transfer

During deep-fat frying, heat transfer takes place through convection between the food and the frying oil and conduction within the food. When the food at a certain initial temperature (ambient, refrigeration, or freezing) is immersed in the hot oil, heat is transferred by convection from the hot oil to the product through its external surface and then via conduction from the surface to the thermal center of the food.

The temperature of the oil or fat is well above the boiling point of the water within the food material. Therefore, heat transferred from the oil to the food results in water evaporation and changes in the food product. Since most foods have high moisture contents water evaporation plays a significant role in defining heat transfer during deep-fat frying. From the moment the food is immersed in the hot oil it is rapidly heated (heat-up) and in a few seconds, its wet surface reaches the boiling point of the surface water. The evaporation of water occurs rapidly at this initial stage of frying. As steam is the only gas phase inside the product, vaporization of free water occurs at the saturation temperature of the water. Then, the water escaping from the inner part of the product reaches the surface, comes in contact with the frying oil, and forms bubbles (surface boiling) which transfer vigorously throughout the oil thus creating turbulence in the oil. The bubbling water at the surface of the product results in a rapid increase of the heat transfer coefficient by free convective (from the oil to the food surface) and conductive (within the food) heating. The product's surface remains at that temperature until all surface liquid has flashed off. Meanwhile, the rapid evaporation (explosion) of the bubbles continues and a crust starts to build up in the form of a porous matrix that increases in thickness over time. Therefore, as the frying time increases, the amount of water vapor bubbles that escape from the food decreases (falling rate) due to the decreased amount of remaining moisture within the product. This relatively long stage includes food core heating, thickening of the crust layer, moisture loss, and other reactions (starch gelatinization, protein denaturation) as well. Maintaining the temperature at the saturation point of water, prevents charring and burning and excessive dehydration while evaporation of the food moisture also has implications on oil absorption. Finally, the bubble end-point is observed when water vaporization nearly ends due to the complete removal of all liquid water in the food, or the reduction of the heat transfer to the crust/core interface (Sahin and Sumnu, 2009).

Therefore, frying could include four stages: heat-up, surface boiling, falling rate, and bubble end-point (Farkas and Hubbard, 2000; Mallikarjunan et al., 2010; Asokapandian et al., 2020).

1. *Initial heating (heat-up)*: It is the shortest of all stages and lasts for only a few seconds. During this stage the food temperature increases to the boiling point of water just before vaporization initiates; a very negligible amount of water is lost from the product.
2. *Surface boiling*: In this stage, when the surface moisture evaporates a crust begins to form at the surface of the product. Also, during this stage, an increase in the surface heat transfer coefficient occurs.
3. *Falling rate*: This stage is considered the longest of all. When the internal core temperature raises, more moisture moves from the core toward the outer part of the food, and evaporation of the bulk moisture takes place. During this period, the temperature of the center core of the product approaches the boiling temperature of the water. Most physicothermal changes, like protein denaturation, starch gelatinization, and internal cooking of the food take place during this step. Additionally, the crust is formed and becomes thicker with processing time.
4. *Bubble end-point*: It is the final stage after a considerably long period of frying. The rate of moisture removal and water vapor bubble formation rapidly decreases.

As far as the temperature of the frying food is concerned, most of its interior remains at approximately 100°C (phase change temperature of water) for most of the frying time. Thus, from a heat transfer point of view, deep-fat frying can be divided into two major regimes: non-boiling, when the temperature of the product is below the boiling temperature of the water and boiling. For a given food product, these regimes can be characterized in order to determine the required frying time. Stages 2 to 4 (as described above) can be combined to represent the boiling regime while most foods are sufficiently cooked by stage 3 (Achir et al., 2009).

### 9.2.1.1 Convective heat transfer

The heat is transferred almost uniformly to the product at a very high rate due to the larger volume (density and heat capacity) of oil compared to that of the food. Throughout the frying process, the core region is heated by conduction to a fixed (maximum) temperature at the crust/core interface. Due to the isotherm at the crust/core interface, the increase in the oil temperature results in a small increase in the heating rate of the core region.

The convective heat transfer during deep-fat frying may include two stages: (i) free convection during the initial heating of the food and (ii) forced convection during the boiling phase (Farkas and Hubbard, 2000). Free convective heat transfer occurs as the product's surface is heated and currents form around the product due to the density differences in the frying oil as it is cooled by contact with the food material. The convective heat transfer coefficient for this stage may have values in the range of 250 to 300 W/($m^2$ °C), while it increases at higher oil temperatures and pressure. It is affected

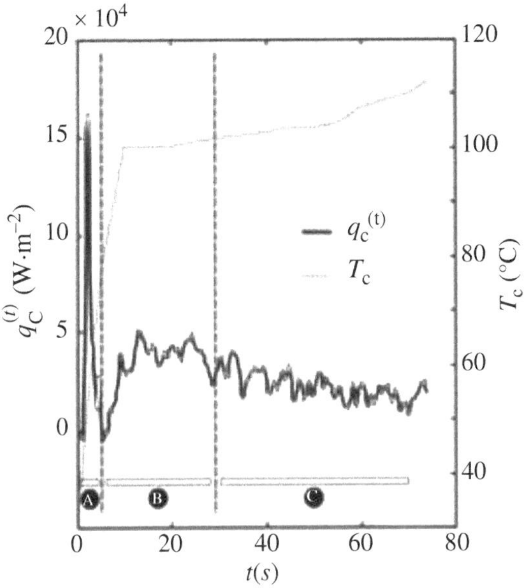


**Fig. 9.2** Kinetic convective heat flux ($q_C$) and temperature at the geometric center of a fried food ($T_C$) (A, B, C regimes are denoted) (Achir et al., 2009).

by heat capacity and viscosity but not affected by the oil type and the level of degradation. The convective heat transfer coefficient follows a trend similar to the moisture rate; the maximum value is attained at the early stages of frying, when the moisture rate is high, while it subsequently decreases to its minimum values when moisture loss becomes constant (Fig. 9.2). The boiling phase of frying that constitutes a substantial portion of the processing time is related to high temperatures and heat flux, and it is associated with the characteristic properties of the fried products.

High convective heat fluxes are obtained during frying by large temperature differences and high conductances between the oil and the fried product. Convective heat transfer coefficients ($h$) can be measured during frying at different distances from the product's surface to obtain their distribution in time and space. For example, during frying of a model food, in the form of a slice of 5 mm thickness with a high moisture content of 13.2%, a very high initial $h$ ($>1500\,W/(m^2\,K)$) is initially presented which decreases rapidly ($<450\,W/(m^2\,K)$) and then progressively to the values of the bulk flow ($300–350\,W/(m^2\,K)$) (Achir et al., 2009; Vitrac et al., 2002).

### 9.2.1.2 Heat and mass transfer equations

Both conduction and convection modes of heat transfer are possible during deep-fat frying. The magnitude of heat transfer greatly depends upon specific heat, density, thermal diffusivity, and thermal conductivity. Heat is transferred from the oil to the surface of a product by convection, whereas conductive heat transfer takes place within the food material (Mallikarjunan et al., 2010).

Heat conduction in a product can be expressed as follows:

$$\rho c_p = \frac{\partial T}{\partial t} = \nabla \cdot (k \nabla T) \tag{9.1}$$

where $T$ is the temperature (°C), $\rho$ is the product's density (kg/m$^3$), $c_p$ is the heat capacity (J/kg K), $k$ is the thermal conductivity (W/(m K), and $\nabla$ is the nabla gradient (that can be expanded for different coordinates).

Convective heat transfer occurs as the circulating oil contacts the frying food with the temperature gradient between the food product and the frying medium being the driving force as described by

$$q'' = h(T_\infty - T_s) \tag{9.2}$$

where $q''$ is the convective heat flux (W/m$^2$), $h$ is the convective heat transfer coefficient (W/(m$^2$ K)), $T_\infty$ (K) is the oil temperature, and $T_s$ (K) is the surface temperature.

Conduction heat transfer is influenced by the thermal properties of the food such as thermal conductivity, thermal diffusivity, specific heat, and density. In convective heat transfer, either in forced or free mode, the convective heat transfer coefficient is the major factor that affects the heat transfer rate to the frying product. It is mainly related to the transport boundary layer developed between the frying medium and the product. Thermal properties of foods, such as specific heat, thermal conductivity, and density, change depending on the moisture level of the food and the temperature variation. Data on the thermal properties of most foods can be obtained from the literature (Saravacos and Maroulis, 2001; Rao et al., 2005; Mittal, 2009; Safari et al., 2018).

As far as the convective heat transfer coefficient between the product and the oil is concerned, it is a very important parameter for the modeling and design of frying systems for foods. The heat transfer coefficient is defined as the rate of heat that will be convected during the food surface-oil (fluid) interaction through a unit surface area of the material if a unit temperature gradient exists between the product's surface and the surrounding oil. Data on convective surface heat transfer coefficients ($h$-value) have be obtained during food frying through the use of model metal transducers or food, and model foods have been reported in the literature (Alvis et al., 2009; Erdogdu and Dejmek, 2010; Safari et al., 2018). These methods assume that an appropriate heat transfer model is used and that the experimental data are reliable and consistent. In practice, heat transfer coefficients can be based on products' shape and size in addition to fluid characteristics.

Heat transfer coefficients can be estimated from experimental moisture evaporation rates by assuming that heat transferred to the frying product is used in evaporating moisture (Mallikarjunan et al., 2010).

$$q = \Delta H_{\text{vap}} \frac{dm}{dt} \tag{9.3}$$

where $q$ is the heat transferred (W), $\Delta H_{\text{vap}}$ is the heat of vaporization (J/kg), $m$ is the moisture, and $t$ is the frying time ($s$). Thus:

$$h = \frac{dm}{dt} \frac{\Delta H_{\text{vap}}}{A(T_o - T_s)} \tag{9.4}$$

where $T_o$ is the oil temperature (K), $T_s$ is the surface temperature of the product (K), and $A$ is the area of the product ($m^2$).

### 9.2.2 Mass transfer

During frying, mass transfer takes place as moisture and oil transfer. Although some reference has already been made to the moisture or fat transfer, as the mass transfer phenomena occur simultaneously with the heat transfer ones, these issues are also examined separately in the following paragraphs.

#### 9.2.2.1 Moisture transfer

The moisture transfer occurs initially from the product's surface, in the form of surface water evaporation, and then at the interface between the dry surface layer and the wet core of the food. As frying proceeds, the evaporated moisture and steam escape as bubbles through the food structure, with their amount decreasing over time due to the decreasing moisture content of the food. Respectively, the moisture is removed at a rapid rate within the first few seconds of processing and then at a much reduced rate reaching a constant value toward the end of frying.

The moisture is transferred in the liquid form when the frying process begins and also when the temperature of the frying oil is low; moisture transfer in the liquid form is considerably slower compared to that in the vapor form. The driving force for the moisture transfer is created due to the diffusion gradient between the dry surface and the wet core of the food as well as the pressure gradient resulting from the evaporation of the inner moisture. As the frying process proceeds, the variation of the moisture loss rate (initially rapid, later much reduced, and finally constant) can be justified by the gradual decrease in the moisture content of the food or moisture loss (Achir et al., 2009; Asokapandian et al., 2020). The moisture loss during deep-fat frying can include three different periods as follows:

1. Heating the food to the boiling point of water and evaporation of the superficial moisture content.
2. Evaporation of free or capillary water at a temperature near the boiling point of water.
3. Increase of the food's temperature to the frying oil temperature along with reduced drying rates. At the frying oil temperature, no further evaporation occurs or may be solely initiated in case that a high temperature is maintained.

Moisture transfer from the product is considered a process controlled by diffusion, and therefore, it is governed by Fick's second law of diffusion (Mallikarjunan et al., 2010).

$$\frac{\delta(\rho_s m)}{\delta t} = \nabla(D\rho_s \nabla m) \tag{9.5}$$

where $m$ is the moisture content on a dry basis (kg/kg), $\rho_s$ is the density of the dry solids (kg/m$^3$), $t$ is the time ($s$), and $D$ is the effective moisture diffusivity (m$^2$/s). If moisture diffusivity and density are constant, the previous equation can be reduced to

$$\frac{\delta m}{\delta t} = D\nabla^2 m \tag{9.6}$$

The solution of the moisture transfer equation can be achieved by defining appropriate boundary conditions for certain food products (i.e., potato chips, French fries, etc.), or composite food products (i.e., breaded or coated foods).

It has been established by several researchers that the instantaneous rate of moisture loss at any time during frying is proportional to the moisture content at that instant and that it increases by increasing the frying oil temperature, thus allowing the use of a first-order reaction kinetics to describe moisture loss during frying. Several studies correlating moisture loss with frying time suggest that the moisture loss is proportional to the square root of the frying time. Drying of fried foods significantly affects the moisture loss and consequently their texture; rapid drying can ensure the desired texture of the final product; however, excessive moisture loss may result in greater oil absorption by the product or in poor texture. The use of a coating (batter/breading) in foods can minimize moisture loss. For example, bread coatings serve to maintain higher moisture content in the final products while also resulting in lower final fat content (Mallikarjunan et al., 2010).

### *9.2.2.2 Fat/oil transfer*

The oil transfer occurs during frying from the oil medium to the food. Due to the moisture loss, cavities, capillary pores, voids, and channels are created in the cellular structure of the food through which the oil can penetrate the fried food. The oil uptake can be controlled by water loss and adsorption during the cooling of the fried food. Thus, correlation with water loss is substantial for modeling and controlling deep-fat frying operations and therefore the properties of the fried foods.

The mechanism of moisture loss and oil adsorption during deep-fat frying can include the following stages (Achir et al., 2009; Mallikarjunan et al., 2010; Asokapandian et al., 2020):

1. High temperatures during frying generate "explosive" boiling of the water contained in the food which bursts cell walls and forms capillary pores and voids.
2. Oil is absorbed into the food through these pores and voids.
3. The amount of oil absorbed (oil uptake) increases as the internal pressure is reduced due to the moisture loss and the subsequent cooling which create a "vacuum effect." The voids formed, allow the oil to enter and hold the capillaries open so that the food structure is kept from shrinking or collapsing. The frying oil entry through the capillary pores occurs during the first few seconds of frying.

Based on the proposed mechanisms (moisture loss and oil absorption), it is obvious that capillary displacement plays a central role in defining fat absorption during deep-fat frying. Oil absorption occurs mainly during products' cooling after frying as a result of adhesion and capillary action and can be explained by the increase in the internal pressure. During frying, a pressure gradient is caused due to steam generation by moisture vaporization inside the food, thus preventing the oil from filling the voids with only the large ones being filled by the oil. During cooling, the internal pressure decreases due to condensation, thus creating a vacuum effect that precedes the oil into the product (Dana and Saguy, 2006; Mousavi Kalajahi and Asefi, 2014). Several studies on various fried products have shown that during frying most of the oil content remains at the surface of the fried food (i.e., 80%) and only a minor part is absorbed (i.e., 20%), while during cooling the largest part is absorbed (i.e., 64%) compared to that remaining at the surface of the fried food (i.e., 36%).

Oil absorption into a product is influenced by oil temperature, frying time, product's surface moisture and area, and pressure (Asokapandian et al., 2020). Moreover, the oil uptake generally reduces at higher frying temperatures due to the reduction of frying time (residence in the frying oil), while it is proportional to the amount of moisture lost from the food. The distribution and the absorption of the oil depend on several factors including the pre-drying treatment, frying time, surface area, and the thickness of the products. More specifically, based on researches there is a linear relationship between product's thickness and the amount of oil absorbed (i.e., in potato chips), and the oil uptake is also related to the surface area (i.e., in potato slices). The final oil content decreases linearly with increased thickness, the distribution of the oil is not uniform and most of the oil may be concentrated around the edges of the product (i.e., in tortilla chips).

Most of the studies on deep-fat frying that examine oil transfer are engaged with non-fatty foods (i.e., potato products and tortilla chips) while only a few studies deal with fatty products (i.e., meat and chicken products). The mechanism of fat transport in fat containing products has been investigated through simultaneous modeling of the fat, heat, and moisture transfer. The fat transfer could include two periods: fat adsorption and fat desorption; fat absorption is characterized by oil diffusion into the food, whereas fat desorption by the migration of the fat from the food to its surroundings. In foods lacking an initial fat content, a fat desorption period is not observed. In fatty foods, the fat is absorbed to the product's surface into the space vacated by evaporated moisture and fills these void spaces, while the fat desorption may occur due to the pressure difference. The driving force is considered to be the tension gradient in the capillaries and is caused by the expansion of fat and the shrinkage of the capillaries in the product probably due to collagen shrinkage and denaturation of meat proteins (Mallikarjunan et al., 2010).

### 9.2.3 Coupled heat and mass transfer inside the food and coupling reactions

During frying, the convective heat transfer from the oil to the product is responsible for the temperature increase in the product and the internal vaporization of the water. The temperature profile and the profile of the residual water are interconnected for each product and can explain its drying kinetics. These phenomena occur independently of the geometry (i.e., chips or French fries) or the composition of the product; however, the intensity of the transfers and the final residual water content are affected by both the geometry and the composition. The microstructure and the mechanical properties of the fried food, as well as the coupling between heat and mass transfer reactions during drying and cooling, influence the oil uptake and its properties (Achir et al., 2009).

### 9.2.3.1 *Internal liquid water vapor transition*

#### Temperature and water profiles

The initial superficial vaporization (regime A) and the subsequent in-depth vaporization (regime B) create a dried porous region, known as crust, with characteristic sensorial properties (texture, color, flavor). The core, on the other hand, containing free or capillary water, according to the product's geometry, remains unchanged and soft. Respectively, the crust is overheated (above the saturation temperature) while the core temperature remains below or close to the saturation temperature. The main transformations occurring in this region are starch gelatinization and protein denaturation resulting in increased digestibility of the fried food. Typical fried foods, such as French fries, are removed from the oil during regime C, whereas potato chips are removed after regime C (Achir et al., 2009).

The position of the vaporization front is controlled and interpreted by the relationship between local temperature and local water content in the hygroscopic domain and the internal liquid water transport inside the core region due to a total pressure gradient (regime B).

#### Thermodynamical control of drying in the hygroscopic domain

During frying, the food is dried at or above the boiling point of water. The boiling point of water inside the product is in practice higher due to the osmotic pressure generated by solutes and the interactions with the solid matrix, which both decrease the vapor pressure in equilibrium. Thus, the temperature inside the product follows the isobar of water desorption. However, residual water remains in the fried food even when it is at thermal equilibrium with the oil; it increases with the external pressure and decreases with oil temperature.

In thin products (i.e., chips), the core temperature versus the residual water content may follow the theoretical boiling curve governing the thermodynamical equilibrium between adsorbed water and steam. In thin or highly porous foods, if the total pressure is equal to the external pressure, the liquid water-steam equilibrium depends either solely on the local temperature or the residual water content, or the solid mass respectively.

The boiling curve or the desorption isobar describes the change in water content versus temperature at a given pressure; it characterizes the specific physicochemical interactions between water and the solid matrix and is not affected by the oil content or the thermo-mechanical history of the food. The effect of pressure on the boiling point of water is higher for free and capillary water than for dispersed water in strong interactions with the solid matrix. Below the boiling point of free water, the local temperature does not depend on the water content while just above this point the water

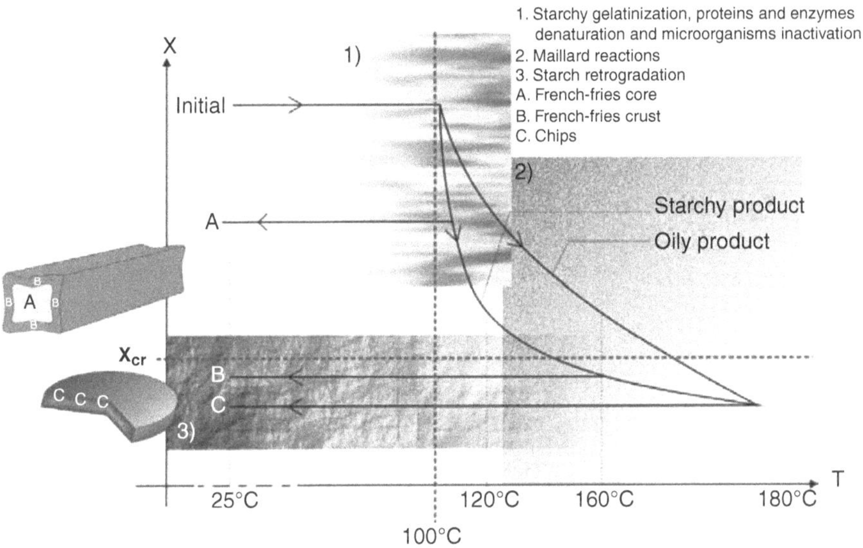


**Fig. 9.3** State diagram of the hydrothermal history and the related food transformations for different geometries (A, B, C) and food compositions (starchy and oily products) ($X$, water content, $X_{cr}$, crust water content). (A) French fries core and (B) French fries crust, (C) Chips, 1, 2, 3: transformations (Achir et al., 2009).

content decreases sharply due to evaporation of capillary water. Its temperature is close to the temperature of free water. At higher temperatures, the variation of the residual water content with temperature is lower, and asymptotic behavior is observed.

A state diagram of the hydrothermal history and the related food transformations for different geometries and food compositions is presented in Fig. 9.3. The isobar shape varies significantly with the composition and the cohesive energy of the solid matrix; the drying of materials such as gels is much faster than in oily/fatty products while it is intermediate for starchy products (Achir et al., 2009; Vitrac et al., 2002).

## Internal liquid water vapor transport

The liquid water transport is responsible for the cooling of the food surface (regime B) thus preserving it from rapid overheating and darkening. The driving force is the total pressure gradient, and the amount of liquid water, assuming a microchannel crossing in a model food, is proportional to

$$q_m^1 = \frac{S}{C}\frac{\phi_1 - \phi_2}{\Delta H_v} \tag{9.7}$$

where $\{\phi_i\}_{i=1,\,2}$ are the measured flux densities in W/m$^2$, $S$ is the surface area of the flux meter, $C$ is cross-sectional surface area, and $\Delta H_v$ is the latent heat of vaporization. At this stage, the measured liquid flux is maximal, with a liquid mass flow rate up to 0.01 kg/(m$^2$ s), and is progressively decreasing. An almost insignificant liquid water flux is observed, while the vaporization front does not cross the channel. In foods with a weak structure, due to their high water content and/or due to the absence of cell structure, liquid water transport can be so intense that liquid water escapes the product surface without vaporization (Achir et al., 2009).

When the food material is transformed into a porous medium, tiny void spaces (or small pores) are obtained which are interconnected and filled with fluid (liquid or vapor). The flow of water vapor through the porous structure of fried foods can be described by Darcy's law, taking into account the significant pressure build up inside them due to the evaporation of the internal water during frying and the resistance of the porous structure depending on their thickness (Singh and Verma, 2016; Datta, 2007). Thus, the water loss due to vapor flow through the crust can be estimated, as follows:

$$Q_w = \frac{Ak\Delta P}{\mu L} \tag{9.8}$$

where $Q_w$ is the flow rate of water vapor, $A$ is the cross-sectional area of the fried food, $k$ is the permeability of the crust layer (related to porosity), $\Delta P$ is the pressure difference/drop over a given distance, $\mu$ is the viscosity of the water vapor, and $L$ is the thickness of the fried food.

### 9.2.3.2 Fat/oil uptake

Oil uptake occurs mainly during cooling immediately after frying. The oil distribution inside the product occurring after frying can be observed macroscopically in the crust. By increasing the frying time, the oil is transmitted inside the food to the core region filling the voids created by the vaporized water. Oil uptake during and after frying (drying and cooling) is controlled by the moisture loss and condensation, and the oil absorption phenomena and mechanisms. More specifically, an overpressure generated during frying prevents the migration of oil inside the food material. However, after frying, the crust in the hygroscopic domain of the food rapidly cools generating the immediate condensation of the steam inside the solid matrix. This results in a pressure drop that depends on the apparent density and porosity of the food (Dana and Saguy, 2006; Achir et al., 2009).

During and after frying the pores of the food can be either filled with water, vapor, or air thus affecting oil/fat penetration. The wetting of food pores with oil and the oil/fat uptake depend on the pore radius and pore depth, the oil viscosity, the oil/food surface tension, and the food type. In general, wider or deeper pores, depending on the food type, and particularly long continuous channels will lead to increased oil uptake. Additionally, contact angles greater than 90° result in lower capillary pressures; narrow pores exhibit higher fat uptake than wide pores but this is also affected by oil viscosity; for pore radii up to approximately 1 mm, pressures comparable to the atmospheric pressure are obtained, while for smaller radii, the capillary pressure can lead to fat uptake even if the pores are filled with vapor (Dana and Saguy, 2006; Achir et al., 2009).

The estimation of the oil uptake is based on mechanisms such as water replacement, capillary penetration, surface-activity, and cooling-phase effect, which are associated with water transfer and/or crust formation (Baumann and Escher, 1995; Mousavi Kalajahi and Asefi, 2014; Arslan et al., 2018). The capillary pathways and voids created by water evaporation increase the surface porosity of the food material, and oil enters through them, thus water loss is considered the basis of oil uptake. Moreover, when the food is removed from the frier, oil sticks to its surface and fills the accessible voids in competition with air. Depending on the drying conditions, the food can be wetted and impregnated with the oil due to the capillary pressure and the pressure drop caused by steam condensation. During cooling of the fried food in the air, the maximum capillary rise is obtained through the combined effect of the mechanical equilibrium and the capillary force.

Capillarity penetration describes the oil that can move upward through narrow pores in fried foods when the adhesive intermolecular forces between the oil and the food materials are stronger than the cohesive intermolecular forces in the oil. The pressure difference (DP*) between both ends of the capillary pathway mainly drives the capillarity penetration phenomena (Bouchon and Pyle, 2005):

$$\Delta P^* = P_2 - P_1 = P_{\text{atm}} - \left(P_v - \frac{2\sigma\cos\theta}{r} \pm \rho g h \cos\alpha\right) \tag{9.9}$$

where $P_i$ is the pressure at the point $i$ ($P_2$: pressure at the pore surface, $P_1$: pressure at the deepest pore point inside the food material, $P_{\text{atm}}$: atmospheric pressure, $P_v$: water vapor pressure), $\theta$ is the contact angle between the oil and the food material, $r$ is the pore radius, $\sigma$ is the surface tension of the oil, $\rho$ is the oil density, $g$ is the acceleration gravity, $h$ is the height of the capillary motion, and $\alpha$ is the angle between the capillary pathway and the vertical direction.

Capillary penetration can also be described by the Washburn equation:

$$Q_{\text{oil}} = \frac{\pi r^4 \Delta P^*}{8\mu h} \tag{9.10}$$

where $Q_{\text{oil}}$ is the volumetric flow of laminar oil, $r$ is the pore radius, and $\mu$ is the oil viscosity.

The penetration of oil over time can be calculated by the above Eqs. (9.8), (9.9) as (Bouchon and Pyle, 2005):

$$\frac{dh}{dt} = \frac{r^2}{8\mu h}\left(P_{\text{atm}} - P_v + \frac{2\sigma\cos\theta}{r}\right) \pm \rho g h \cos\alpha \tag{9.11}$$

However, the voids or capillary pathways filled with water may resist oil penetration, while the oil may be absorbed during cooling due to water vapor condensation and internal pressure reduction. Oil uptake upon removal of a plant based food from the oil is a balance between oil drainage and oil adhesion.

Oil adhered to the surface of a fried material is absorbed due to the "vacuum effect" caused by the condensation of steam during the cooling period. When the fried food is removed from the oil, an oil film is formed on the surface and its thickness (H) can be calculated by the Landau-Levich-Derjaguin equation:

$$H = 0.944\frac{(\mu U)^{\frac{2}{3}}}{\gamma^{\frac{1}{6}}(\rho g)^{\frac{1}{2}}} \tag{9.12}$$

where $\mu$ is the oil viscosity, $\gamma$ is the surface tension, $U$ is the speed of oil removal after frying; $\rho$ is the oil density, and $g$ is the acceleration gravity.

Finally, the momentum transfer can be estimated based on the equilibrium of the forces acting due to all the above mechanisms. The oil uptake at equilibrium will be higher in the case of large pores, but oil impregnation will be deeper in the case of small pores while decreasing the pore radius will increase the final oil content of the food.

#### *9.2.3.3 Crust formation*

Crust formation is important for fried foods in providing their color, flavor, and crispy texture. The crust begins to form as a porous matrix once the surface moisture has been removed, and over the frying process, it increases in thickness and extends from the surface to the core.

The porous structure of the crust allows the energy flow from the surface to the core region as well as the vapor loss from the core region to the surface. However, crust formation makes the coupled heat and mass transfer mechanism more complex due to the changes in the material's properties. The crust serves as an insulating layer between the surface and the core region as it has low thermal conductivity ($k = 0.1$ W/m °C). When the free water from the core region is vaporized at the crust/core interface, the vapor reaching the surface crosses the surface/oil interface while the oil enters the food in the form of discrete bubbles. These phenomena must be taken into account in the transport equation that describe the frying system. The continuous increase in crust thickness corresponds to a moving boundary or a Stefan problem, similar to the one-dimensional freezing/melting of water/ice where heat transfer with phase change (under constant thermophysical properties in each phase) occur. When the crust layer thickens enough at increased frying time, it no longer permits oil to pass through (Achir et al., 2009; Mallikarjunan et al., 2010).

The golden brown, crispy layer which is formed on the outer surface of the food is generally characteristic of fried products. The crust is formed within seconds after the product comes into contact with the hot oil followed by both chemical and structural changes in the product. The controlling reactions have been recognized as well as the determining parameters either in terms of the frying conditions or the food properties. The golden color of the crust can be attributed to the Maillard reactions; these chemical changes are attributed to the sugar compounds found on the food's surface and are favored under certain conditions such as low water content in combination with high temperatures. The crust limits water diffusion from the center of the food and, due to the pressure buildup inside the food, leads to pore expansion and provides the fried product with a crunchy texture. The dry crust developed may limit heat and mass transfer acting as a barrier between the inner part of the food material and the surrounding oil, but it also influences the oil uptake. The oil uptake during deep-fat frying tends to be mostly localized in the products' crust and concentrated near the edges, corners, and broken "slots" (Mallikarjunan et al., 2010).

Crust thickness in fried foods, as well as the porosity, and the gel strength of the crust section are related to the frying parameters and the initial properties of the food material. Based on the literature, the crust thickness increases during frying, while the porosity of the crust region decreases linearly with frying time. An increase in the porosity of the fried product is normally expected to increase oil uptake. Crust porosity and fat uptake are also related to the initial structure of the unfried product; crust porosity decreases whereas fat uptake increases in products with lower deformability modulus in their unfried state, while the use of a batter coating may contribute to the control of the porosity and fat absorption. In addition, the accelerated formation of the crust either by applying higher frying temperatures or by frying foods with low moisture content may reduce the amount of the oil absorbed (Moreira, 2007).

Concluding, the characteristics of the crust may depend on the frying conditions and the type and nature of the product being fried. Crust formation in fried foods has a vital influence on fat absorption and texture, thus in order to model the changes that take place during frying the interrelated factors have to be taken into account.

### 9.2.4 Frying model

Frying is a complex process that simultaneously involves heat and mass transfer, moisture loss, and oil uptake. Heating and drying of foods during frying are governed by heat and mass transfer phenomena, while oil adhesion and oil capillary migration during cooling by physical phenomena, respectively. Steam bubbling, initially at the oil-food interface, intensifies the convective heat transfer and eventually heat and mass transfer inside the food. For the above reasons, both food composition and structure play an important role in the oil uptake. The combination of the transport engineering models with the reaction kinetics (i.e., microbial destruction, and color, flavor, or texture development) can contribute to product and process optimization that is of great interest to the food industry (Blumenthal and Stier, 1991; Mousavi Kalajahi and Asefi, 2014).

Several mathematical models of various complexities have occasionally been developed (Yamsaengsung et al., 2008). Some models deal with the frying of certain food products assuming constant physical properties. Other models are based on simple energy and mass transfer with various approximations (i.e., accounting for evaporation or ignoring it altogether, and not including the oil phase transport). The frying models can be divided into heat and mass balance (semi-empirical) and differential equation (DE) models. Model of coupled heat and mass transfer are also proposed for thin (chips) or thick (French fries) geometries (Fig. 9.4).

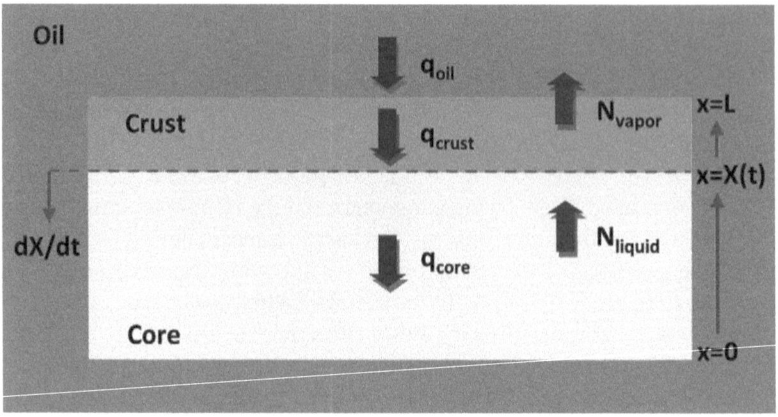


**Fig. 9.4** Cross-section of potato during immersion frying.
Based on Farkas, B.E., Singh, R.P., Rurnsey, T.R., 1996. Modeling heat and mass transfer in immersion frying. I, model development. J. Food Eng. 29, 211–226.

### 9.2.4.1 Semi-empirical models

These models make the following assumptions: (1) both physical and thermal properties are constant; (2) oil absorption is neglected; (3) sensible heating is neglected in the core and crust; and (4) transient temperature and moisture profiles in the core are not addressed. For crustless food products (chips), moisture content during frying is proportional to the square root of the frying time and the difference between the oil temperature and the boiling temperature of the water. For crust-forming products (French fries), an ordinary DE similar to Plank's equation for freezing is proposed to predict the crust thickness of the product during frying.

### 9.2.4.2 DE models—Deterministic models

The models' assumptions include: (1) the product is initially isotropic and isothermal; (2) the initial moisture and temperature distributions in the product are uniform; (3) the heat required for chemical reactions (Maillard reactions, starch gelatinization, protein denaturation) is limited compared to that required for water evaporation; (4) the thermal and physical properties are dependent on local temperature and moisture content during the frying process; (5) a "microscopically uniform" porous medium is formed after frying; (6) the food surface is covered with a uniform oil layer after frying, and (7) most of the oil diffuses into the product after frying during the cooling period.

DE frying models, based on the laws of heat and mass transfer, lead to complicated systems of equations (Fig. 9.4), mainly concerning heat transfer by convection from the oil to the food surface, and by conduction into the food center, coupled heat and mass transfer by water vapor removal and oil uptake, and shift of the crust/core interface toward the food center with the interface temperature remaining at the boiling point allowing water evaporation.

Characteristic DE models, one for chips (all crust) and one for French fries (crust and core), presenting satisfactory accuracy and applicability over a wide range of frying problems, are presented below in as much detail as possible (Farkas et al., 1996; Farid and Chen, 1998; Moreira, 2007; Mallikarjunan et al., 2010).

(i) *DE model for all crust products (chips)* (Moreira, 2007)

The governing equation for temperature change in the product taking also into account the heat transfer caused by the diffusion of water vapor is

$$\nabla \cdot \left(k \nabla T_p\right) - \nabla \cdot \left(N_w c_{pv} T_p\right) = \frac{\partial\left(\rho_p c_p T_p\right)}{\partial t} \tag{9.13}$$

$$N_w = \nabla \cdot \left(\nabla \rho_p D_w M'\right) \tag{9.14}$$

where $N_w$ is the water vapor flux, $\rho_p$ is the product density, $M'$ is the product moisture content, $D_w$ is the mass diffusivity of water vapor, $k$ is the thermal conductivity of the product, $c_p$ is the specific heat of the chip, $T_p$ is the product temperature, and $c_{pv}$ is the specific heat of the water vapor.

The governing equations for mass change in the product using Fick's diffusion law for the mass transfer rate of moisture (water vapor) from the chip to the oil and of oil from the surface to the center of the chip are

$$\nabla \cdot \left(D_v \rho_p \nabla M'\right) = \frac{\partial\left(\rho_p M'\right)}{\partial t} \tag{9.15}$$

$$\nabla \cdot \left(D_f \rho_p \nabla F\right) = \frac{\partial\left(\rho_p F\right)}{\partial t} \tag{9.16}$$

where $D_f$ is the mass diffusivity of oil, and $F$ is the oil content.

The evaporation zone moves toward the center of the product for a constant period (depending on the moisture content) under constant temperature until the water content reduces to the equilibrium moisture content ($M_e$), and then the temperature increases rapidly and the crust is formed. The different thermal properties for the two different zones in the product are used, while at temperatures higher than the boiling temperature of water the properties of vapor instead of those of the liquid water are used to predict the temperature and moisture changes in the product.

Changes in the enthalpy of the oil, over time, as the sum of energy required for heating the product, evaporating water from the chips, heating the water vapor evaporated from the chips, and exchanging energy to the surrounding oil, are described by the following equation:

$$\begin{aligned} m_{\text{fat}} c_{p\text{fat}} \frac{\partial T_{\text{fat}}}{\partial t} &= \left[-h \nabla T_p + h_{fg} D_w \nabla \rho_b M' + N_w c_{pv}\left(T_{psur} - T_{fat}\right)\right] \frac{A_p}{(1-\Phi)} \\ &\quad + h' A_o\left(T_{fo} - T_{\text{fat}}\right) \end{aligned} \tag{9.17}$$

where $h_{fg}$ is the convective heat transfer coefficient between the frying oil (in the bed of products) and the oil surrounding the food, $A_p$ is the specific surface area of the chips (batch of chips), $A_o$ is the specific surface area of the oil (surface area of the frying oil per unit volume of oil), and $\Phi$ is the porosity of the batch of products.

At the onset of the frying process, the temperature and moisture content of the product are uniform. The initial conditions in equation form are thus,

$$T_P\big|_{n,t=0} = T_{P_o} \quad \text{and} \quad M'\big|_{n,t=0} = M'_o \tag{9.18}$$

The boundary conditions are defined by the product's surface and the center of the stack of products. Therefore, due to the symmetry, there is no temperature or moisture concentration gradient at the product's center ($n=0$), and the following conditions exist:

$$\left.\frac{\partial M'}{\partial n}\right|_{n=0,t} = 0\left.\frac{\partial F}{\partial n}\right|_{n=0,t} = 0 \quad \text{and} \quad \left.\frac{\partial T}{\partial n}\right|_{n=0,t} = 0 \tag{9.19}$$

At the surface ($n=L$) for any time, the energy transferred by convection from the oil to the chip's surface is equal to the sum of energy required for transferring heat to the center of the product by conduction, for evaporating water from the chips, and for heating the water vapor evaporated from the chips at temperature $T_p$ to the oil temperature $T_{\text{fat}}$:

$$h\left(T_{p\,\text{sur}} - T_{\text{fat}}\right) = -k\frac{\partial \theta}{\partial n} + N_w\left[h_{fg} + c_{pv}\left(T_{p\,\text{sur}} - T_{\text{fat}}\right)\right] \tag{9.20}$$

where $n$ represents the normal to the surface, $h$ is the convective heat transfer coefficient, and $h_{fg}$ is the latent heat of vaporization. The second term on the right side of the Eq. (9.20) can be eliminated at chips' temperatures higher than the boiling point of water,

$$h\left(T_{p\,\text{sur}} - T_{\text{fat}}\right) = -k\frac{\partial \theta}{\partial n} + N_w c_{pv}\left(T_{p\,\text{sur}} - T_{\text{fat}}\right) \tag{9.21}$$

with two mass transfer boundary conditions,

$$k_d \rho_p\left(M'_{\text{sur}} - M'_{\infty}\right) = -N_w \tag{9.22}$$

where $M_\infty = M_e \approx 0$ (i.e., the moisture content of the surrounding oil), and

$$k_d \rho_p\left(F_{\text{sur}} - F_{\infty}\right) = -D_f\frac{\partial F}{\partial n} \tag{9.23}$$

where $F_\infty = M_e \approx 1$ (i.e., the oil content of the surrounding oil).

(ii) *DE model for crust and core products (French fries)* (Moreira, 2007)

In this case, the process is viewed as a moving boundary problem similar to the freeze-drying situations. The mass and energy equations are developed for each of the two regions of the product: (I) crust (free of liquid water and above the boiling point) and (II) core (vapor free and below the boiling temperature, but not saturated) (Fig. 9.4). The vapor moves in the crust region due to pressure differences while the mass fraction of oil in the fried product is assumed negligible with negligible effect on other mass and energy fluxes.

The governing equations for temperature change in the product are

Core region $0 < x < X(t)$

$$k^{\mathrm{II}}\left(\nabla^2 T_p\right) - N_{\beta x} c_{p\beta}\left(\nabla T_p\right) = \left(\varepsilon_\beta \rho_\beta + \varepsilon_\sigma \rho_\sigma c_{p\sigma}\right)\frac{\partial T_p}{\partial t} \tag{9.24}$$

Crust region $X(t) < x < L$

$$\left(\varepsilon_\gamma \rho_\beta c_{p\gamma} + \varepsilon_\sigma \rho_\sigma c_{p\sigma}\right)\frac{\partial T_p}{\partial t} = k^{\mathrm{I}}\left(\nabla^2 T_p\right) + N_{\gamma x} c_{p\gamma}\left(\nabla T_p\right) \tag{9.25}$$

where subscripts $\beta$, $\gamma$, and $\sigma$ refer to liquid water, water vapor, and solid, respectively; superscripts I and II are crust and core regions, $X(t)$ is the position of the crust/core interface, and $\varepsilon$ is the volume fraction.

The governing equations for mass change in the product are

Core region $0 < x < X(t)$

$$\frac{\partial C_\beta}{\partial t} = D_{\beta\sigma}\nabla^2 C_\beta \tag{9.26}$$

Crust region $X(t) < x < L$

$$\nabla \cdot \left(\rho_\gamma \nabla P_\gamma\right) = 0 \tag{9.27}$$

where $C$ is the concentration.

These equations are solved by specifying sets of eight boundary and three initial conditions. The boundary condition at the crust-core interface is defined as:

$$\begin{aligned} -k^{\mathrm{I}}\left(\nabla T_p\right) + k^{\mathrm{II}}\left(\nabla T_p\right) - \frac{\rho_\gamma K_\gamma}{\mu_\gamma}\left(\nabla P_\gamma\right)\left(h_\beta - H_\gamma\right) \\ = \left[\left(\varepsilon_\sigma \rho_\sigma\left(h_\sigma^{\mathrm{I}} - h_\sigma^{\mathrm{II}}\right) + \varepsilon_\gamma \rho_\gamma\left(H_\gamma - h_\gamma\right)\right]\frac{dX}{dt} \end{aligned} \tag{9.28}$$

where the interfacial mass balance is

$$\frac{dX}{dt} = \frac{N_{\beta x} - N_{\gamma x}}{\varepsilon_\gamma \rho_\gamma - \varepsilon_\beta \rho_\beta} \tag{9.29}$$

$$N_{\beta x} = -D_{\beta\sigma}\nabla^2 c_\beta \quad \text{and} \quad N_{\gamma x} = \frac{\rho_\gamma K_\gamma}{\mu_\gamma}\left(\nabla P_\gamma\right) \tag{9.30}$$

where $K$ is the permeability, $\mu$ is the viscosity, and $P$ is the pressure.

The final mass transfer boundary condition relates the pressure of the vapor, $P_\gamma$, at the crust/core interface to the temperature through the Clausius-Clapeyron equation:

$$\frac{\partial P_\gamma}{\partial T} = \frac{h_{fg}}{T\left(V\gamma - V_\beta\right)} \tag{9.31}$$

at $x = X(t)$, $t > 0$.

$$P_\gamma = \exp\left[\frac{-h_{fg}}{RT} + B\right] \tag{9.32}$$

It should be noted that the mathematical models require physical and thermal data on the food material to be fried and the process as well, such as the heat transfer coefficient, the mass diffusion coefficient, the crust and core thermal properties, and densities. This is foreseeable since the rate of heat and mass transfer during frying of foods is designated by their physical properties. The thermal properties of foods depend on their temperature and composition which change (moisture and oil content) during frying. Their geometric properties, such as shape, size, surface area, volume, and density also change during frying.

Models for coupled heat and mass transfer have been developed for complex systems in which the heterogeneity in the porous structure is taken into account (i.e., doughnuts) (Ghaitaranpour et al., 2020) and for composite layer systems (i.e., coated, or breaded foods) (Mallikarjunan et al., 2010).

## 9.3 Frying oil—Deterioration reactions during frying

Frying, though the physical and chemical reactions it involves, results in the development of desirable sensorial attributes in the fried foods. However, the high temperatures developed during the frying process can negatively affect the frying oil leading to the production of degradation compounds which in turn reduce the sensorial quality and more importantly the safety of the fried products. The progress of the physical and chemical reactions that take place during frying has been extensively studied and is considered significant for the control of the frying process and the assurance of both quality and safety of the fried products.

During the frying process, the oil undergoes degradation as it is exposed to high temperatures in the presence of air and moisture. Although a complex series of reactions also occur during contact of the oil with the food, in order to understand oil degradation, it is necessary to study the changes that occur in the main oil/fat components in the absence of food.

The edible fat or oils are mainly composed of triglycerides (triacylglycerols—TAGs) which are esters of glycerin with fatty acids (FA). The edible oils and fats consist of fatty acids either saturated or unsaturated with different degrees of unsaturation. The common fatty acids are the saturated stearic acid (C18:0), and the unsaturated:

oleic acid (C18:1, $\Delta_{9-10}$), linoleic (C18:2, $\Delta_{9-10}$, $\Delta_{12-13}$), and linolenic (C18:3, $\Delta_{9-10}$, $\Delta_{12-13}$, $\Delta_{15-16}$) with 1, 2, or 3 double bonds, respectively. These influence the oxidative stability and thermal resistance during the frying of oil. Various fats and oils also contain minor components some of which possess antioxidative activity (tocopherols, polyphenols, sterols, etc.) thus protecting the oil from oxidation during heating and storage. Antioxidant additives can also be used to enhance the heat resistance of oils (Kochhar, 2001).

The reactions involved in fat/oil degradation during frying—oxidation, polymerization, and hydrolysis of the main fat constituents—are well-known as also are the resulting degradation products (volatile and nonvolatile). The rate of fat degradation varies depending on factors associated with the frying process (temperature, time of heating), the frying system (continuous or batch, turnover rate), the food (lipid composition, main and minor constituents), and the oil or fat (degree of unsaturation, initial quality, and additives) itself. The degradation products formed (hydroperoxides, aldehydes, ketones, hydrocarbons, fatty acids, polymers, etc.) are generally polar compounds that are used to indicate fat deterioration. The volatile or nonvolatile compounds absorbed in fried foods affect flavor stability and quality, and the texture of fried foods during storage.

### *9.3.1 Oxidative reactions—Thermal oxidation*

The oxidation of fats and oils at frying temperatures is highly complex and includes oxidative and thermal reactions resulting in the formation of various products. The chemical mechanism of thermal oxidation is principally the same as the autoxidation mechanism, for this reason, the basic autoxidation issues are firstly presented (Angelo et al., 1996; Warner, 2004; Velasco et al., 2009).

#### *9.3.1.1 Autoxidation*

Autoxidation is the spontaneous reaction of the unsaturated fatty acids of oils with atmospheric oxygen. It is the most common process leading to oxidative deterioration. The reactant fatty acids could be contained in a triglyceride, diglyceride, monoglyceride, or phospholipid molecule, or can be present in a free form in the fresh oil, or produced in the fried oil by hydrolysis. It is widely accepted that oxidation proceeds via a free radical mechanism of chain reactions and involves the formation of a hydroperoxide on a methylene group adjacent to a double bond following the three distinct steps below (Frankel, 1980; Angelo et al., 1996; Gordon, 2001; Shahidi and Zhong, 2010):

$$\text{Initiation}: \quad RH + X^{\bullet}(\text{initiator}) \rightarrow R^{\bullet} + XH$$

$$\begin{aligned}\text{Propagation}: \quad & R^{\bullet} + O_2 && \rightarrow ROO^{\bullet} \\ & ROO^{\bullet} + RH && \rightarrow ROOH + R^{\bullet}\end{aligned}$$

$$\begin{aligned}\text{Termination}:\quad & R^{\bullet} + R^{\bullet} && \rightarrow R - R \\ & ROO^{\bullet} + R^{\bullet} && \rightarrow ROOR \\ & ROO^{\bullet} + ROO^{\bullet} && \rightarrow ROOR + O_2 (R : \text{lipid alkyl})\end{aligned}$$

where RH represents the triglyceride (TG) molecule undergoing oxidation in one of its unsaturated fatty acyl groups.

In the first step free radicals are formed from lipid molecules. This occurs when an unsaturated fatty acid is exposed to oxygen in the presence of an initiator (such as heat, or metal ions). Thus, during initiation, from an unsaturated lipid molecule, RH, the free radical R· is formed by abstracting a hydrogen atom at an ally lie methylene group (α to the double bond). In the propagation step, the free radical R· can then react with oxygen to form a peroxy radical ROO·, which further reacts with another lipid molecule generating a hydroperoxide, ROOH, and another lipid radical, R·. In these reactions, one lipid radical is converted into a different lipid radical commonly involving the abstraction of a hydrogen atom from a lipid molecule or addition of oxygen to an alkyl radical. The propagation reactions, due to their relatively low enthalpy, occur more rapidly than the initiation reaction. In the presence of oxygen, the reaction of alkyl radicals is very rapid, as a result, the peroxy radicals are present at higher concentrations than the alkyl radicals. Abstraction of hydrogen is favored at carbon atoms where the bond dissociation energy is low. For example, in a polyunsaturated fatty acid (PUFA) it appears at the methylene group between two alkene groups; therefore, the radical formed initially is delocalized across five carbon atoms of the 1,4-pentadienyl moiety, while the addition of oxygen occurs preferentially at one of the end carbons of this structure.

In the termination phase, two radicals react providing products that do not sustain the propagation phase. These reactions are of low energy but are limited by the low concentration of radicals and by the requirement for radicals with the correct orientation for the reaction to collide. When the radical concentration has reached a certain limit, the chain reaction is gradually stopped by a mutual combination of radicals. Termination also occurs when antioxidants or free radical scavengers react with free radicals generated during propagation. Antioxidants are used to prolong the induction period by reacting with the intermediate products of the chain reaction, forming inactive radicals, while minor, naturally occurring substances of oils, such as tocopherols and polyphenols, work the same way (Shahidi and Zhong, 2010).

This reaction may be repeated during propagation until no hydrogen source is available, or the chain is interrupted by antioxidants. Therefore, lipid oxidation is a self-propagating and self+accelerating process. Oxidation normally proceeds very slowly at the initial stage (induction period—IP), until it reaches a certain limit followed by a sudden increase. At the end of the IP, the off-flavors become most noticeable as a consequence of the sharp rise in their concentration. As temperature increases, the initiation, and therefore, all the oxidation reactions are accelerated. Oil hydrolysis products, such as monoacylglycerol, or diacylglycerol, promote autoxidation as the contact of oil with oxygen increases due to the reduction of the interfacial tension between oil and air. Pro-oxidants (i.e., metal ions) in small concentrations shorten the IP, while antioxidants make it slower (by radical scavenging, metal chelation, or other mechanisms).

Lipid hydroperoxides produced during propagation are primary products of oxidation. The intermediate hydroperoxides are very unstable and break down to a wide range of secondary oxidation products (aldehydes, ketones, esters, alcohols, hydrocarbons, volatile organic acids, and epoxy compounds, isomerization, and polymerization products). Some of these compounds (i.e., volatile carbonyl compounds) have undesirable odors with very low threshold values and therefore are considered responsible for the off-taste and odor or rancidity of the oxidized oils. The off-flavors developed during lipid oxidation, in addition to making the oil no longer edible, may lead to adverse health effects (hydroperoxides absorbed and free radicals from their decomposition may cause damaging effects in vivo and also generate carcinogens). Autoxidation is one of the major reactions taking place during frying, as well as in the oil present in the packaged product during storage.

Meanwhile, new radicals generated from the decomposition of intermediate hydroperoxides help to further continue the chain reactions (secondary initiation); homolytic cleavage of hydroperoxides is a relatively low energy reaction commonly catalyzed by metal ions (Gordon, 2001).

$$\text{Secondary initiation}: \quad \begin{aligned} &\text{ROOH} &&\rightarrow \text{RO}^{\bullet} + {}^{\bullet}\text{OH} \\ &2\text{ROOH} &&\rightarrow \text{RO}^{\bullet} + \text{ROO}^{\bullet} + \text{H}_2\text{O} \end{aligned}$$

$$\text{Metal}-\text{catalyzed initiation}: \quad \begin{aligned} &\text{M}^{n+} + \text{ROOH} &&\rightarrow \text{RO}^{\bullet} + {}^{-}\text{OH} + \text{M}^{(n+1)+} \\ &\text{M}^{(n+1)+} + \text{ROOH} &&\rightarrow \text{ROO}^{\bullet} + \text{H}^{+} + \text{M}^{(n)+} \end{aligned}$$

Moreover, the lipid-free radicals generated during the initiation and propagation are stabilized by radical resonance, leading to the shift of double bonds and cis-trans isomerization. As a result, conjugated dienes and trienes are produced by the rearrangement of the methylene-interrupted double bonds. Hydroperoxides and conjugated dienes and trienes are used as indicators for oxidation as well as for secondary oxidation by hydroperoxides decomposition.

Following, more details on the oxidation reactions and mechanisms, as well as the formation of primary and secondary oxidation products from the known unsaturated fatty acids are presented. In Fig. 9.5, the initiation, propagation, and termination reactions involved in the thermal oxidation of unsaturated fatty acids are presented using unsaturated linoleic acid as an example.

The oxidation rate during thermal oxidation or autoxidation varies among the common fatty acids. Thus, linoleic acid oxidizes 10 to 20 times faster than oleic acid and linolenic acid about three times faster than linoleic acid. The reactivity of a methylene group in forming a hydroperoxide depends on its position in the lipid molecule in relation to the position of the double bond indicating that it is enhanced by a second adjacent double bond. Strength variation in the hydrogen carbon bond of stearic, oleic, linoleic, and linoleic acids explain these differences.

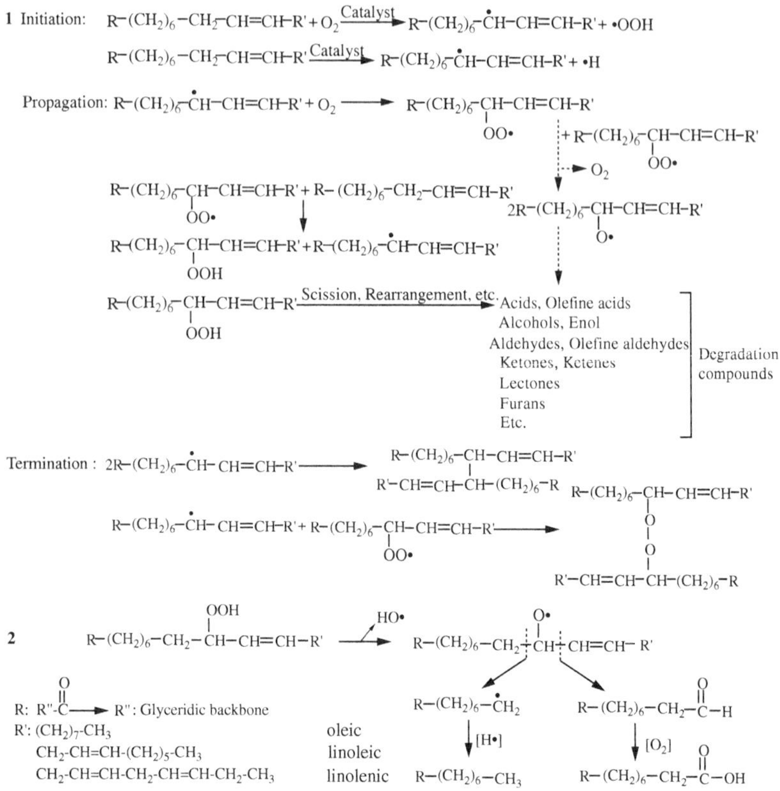

**Fig. 9.5** The initiation, propagation, and termination of the thermal oxidation of unsaturated fatty acids (Zhang et al., 2012).

The energy required to remove hydrogen from fatty acids is dependent on its position in the molecules; the hydrogen adjacent to a double bond and especially the hydrogen attached to a carbon between 2 double bonds, is removed easily. The energy required to break the carbon-hydrogen bond on the C11 of linoleic acid is 50 kcal/mol; this bond is the weakest and will be removed first to form a radical at C11. The double bonds at C9 and C12 decrease the strength of the carbon-hydrogen bond at C11 by withdrawing electrons. The energy requirement for the carbon-hydrogen bond on C8 or C14, which is $\alpha$ to the double bond of oleic acid, is about 75 kcal/mol, while for the carbon-hydrogen bond on a saturated carbon without a double bond next to it is approximately 100 kcal/mol (Choe and Min, 2006; Fig. 9.5). The radical at C11 (facilitated by heat, or metals) rearranges forming a conjugated pentadienyl radical at C9 or C13 with a *trans* double bond. The site of radical formation in saturated fatty acids is different from that in unsaturated oleic or linoleic acids and different positional isomers of hydroperoxides (conjugated or nonconjugated) are formed at certain carbon atoms in the autoxidation of oleic, linoleic, and linolenic acid as shown

**Table 9.1** Hydroperoxides of fatty acids by autoxidation.

| Fatty acid | Hydroperoxide at | Amount range (%) |
|---|---|---|
| Oleic acid | C8 | 26–28 |
| | C9 | 22–25 |
| | C10 | 22–24 |
| | C11 | 26–28 |
| Linoleic acid | C9 | 48–53 |
| | C13 | 48–53 |
| Linolenic acid | C9 | 28–35 |
| | C12 | 8–13 |
| | C13 | 10–13 |
| | C16 | 28–35 |

Based on Akoh, C.C., 2017. Food Lipids: Chemistry, Nutrition, and Biotechnology. Taylor & Francis Group, LLC, Boca Raton, FL.

in Table 9.1. Linoleic acid forms the *trans* 9- and 13-hydroperoxides as shown in Fig. 9.6.

The decomposition pathway of hydroperoxides is more likely through the homolytic cleavage of the O—O bond to form alkoxyl and hydroxyl radicals, as it is thermodynamically favored since it requires lower activation energy (DE 44 kcal/mol) than the O—H bond (DE 90 kcal/mol). Although hydroperoxides are relatively stable at room temperature and in the absence of metals, they are readily decomposed to alkoxy radicals in the presence of metals or at high temperatures forming secondary oxidation products (aldehydes, ketones, acids, esters, alcohols, and short-chain hydrocarbons) (Frankel, 1980; Gordon, 2001; Shahidi and Zhong, 2010; Fig. 9.7).

The type of secondary oxidation products of oils depends on the composition of the hydroperoxides and the pattern of oxidative cleavage based on the composition of their fatty acids. The mechanism of hydroperoxide decomposition with one double bond to form secondary oxidation products is presented in Fig. 9.8. The alkoxy radical then undergoes hemolytic β-scission of the carbon-carbon bond and produces oxo-compounds and saturated or unsaturated alkyl radicals. After electron rearrangement, the addition of hydroxyl radical, or hydrogen transfer, the ultimate secondary lipid oxidation products are mostly low molecular-weight aldehydes, ketones, alcohols, and short-chain hydrocarbons, as shown in Table 9.2. The decomposition of alkoxy radicals releases volatile hydrocarbons, alcohols, or aldehydes (no longer bound to the glyceride backbone), while non-volatile alcohols and ketones may also be formed. The volatile compounds contribute to the off-flavor of oxidized oils as perceived by sensory assessment. Individual carbonyl compounds are used as indicators for assessing the oxidative deterioration of different lipid systems.

Various vegetable oils exhibit different oxidative behavior depending on their concentration in unsaturated fatty acids and endogenous antioxidant substances. Therefore, the time until the formation of secondary products from hydroperoxide decomposition varies with different oils; these may be formed immediately (i.e., in

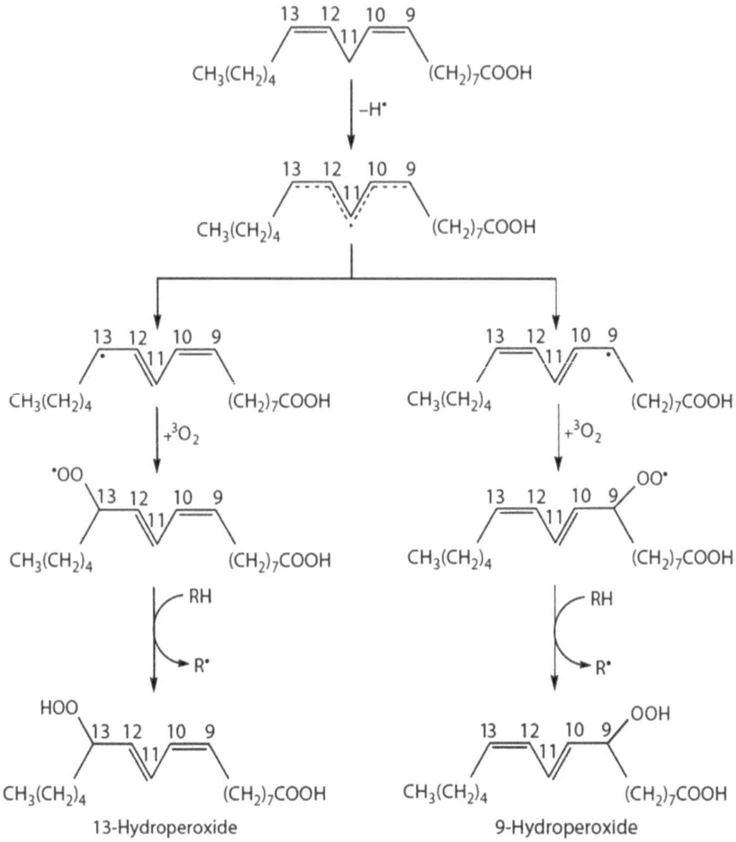


**Fig. 9.6** Hydroperoxide formation in linoleic acid autoxidation (Akoh, 2017).

$$R_1O^{\bullet} + R_2O^{\bullet} \longrightarrow R_1OOR_2$$ dimer peroxides

$$R_1O^{\bullet} + R_2^{\bullet} \longrightarrow R_1OR_2$$ eathers

$$R_1{-}\underset{\displaystyle O^{\bullet}}{\underset{|}{CH}}{-}R_2 + R^{\bullet} \longrightarrow R_1{-}\underset{\displaystyle O}{\underset{\|}{C}}{-}R_2 + RH$$ ketones, alkanes

$$R_1{-}\underset{\displaystyle O^{\bullet}}{\underset{|}{CH}}{-}R_2 + RO^{\bullet} \longrightarrow R_1{-}\underset{\displaystyle O}{\underset{\|}{C}}{-}R_2 + ROH$$ ketones, alcohols

$$R_1^{\bullet} + R_2^{\bullet} \longrightarrow R_1{-}R_2$$ alkane polymers

**Fig. 9.7** Formation of secondary oxidation products by hydroperoxide decomposition (Shahidi, 2005).

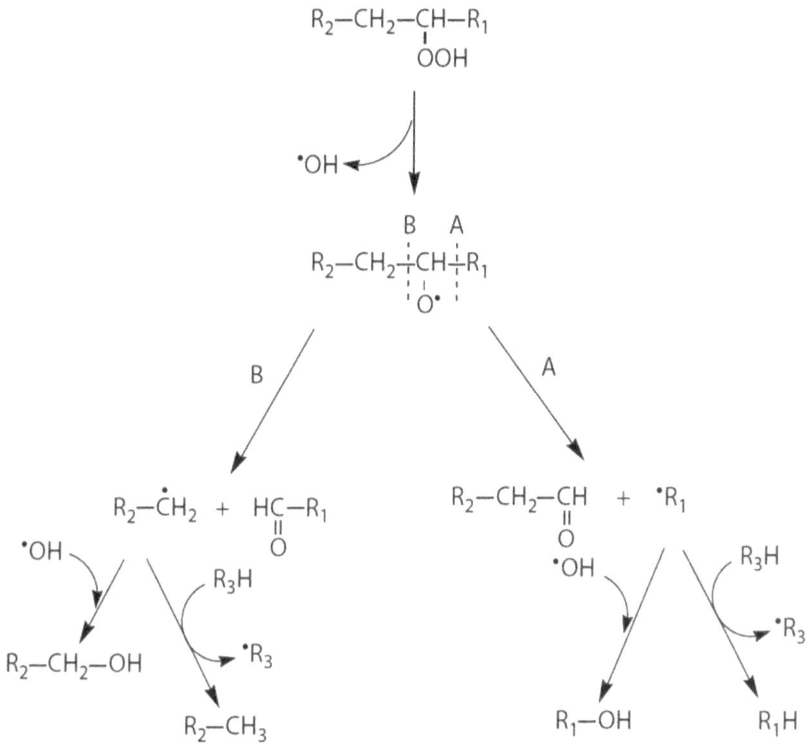


**Fig. 9.8** Mechanism of hydroperoxide decomposition toward secondary oxidation products from oleic acid (Akoh, 2017).

**Table 9.2** Secondary oxidation products of fatty acids methyl esters by autoxidation Choe and Min, 2006(Akoh, 2017).

| Fatty acid | Carbonyl compounds | Hydrocarbons |
|---|---|---|
| *Oleic acid* | 1-Heptanol, octanal, methyl heptanoate, nonanal, methyl octanoate, methyl 8-oxooctanoate, decanal, 2-decenal, methyl 9-oxononanoate, methyl 10-oxodecanoate, methyl 10-oxo-8-decenoate, methyl 11-oxo-9-undecenoate | Heptane, octane |
| *Linoleic acid* | Pentanal, 1-pentanol, hexanal, 2-octenal, 1-octene-3-ol, methyl heptanoate, 2-nonenal, methyl octanoate, methyl 8-oxooctanoate, 2,4-decadienal, methyl 9-oxononanoate, methyl 10-oxodecanoate | Pentane |
| *Linolenic acid* | Propanal, butanal, 2-butenal, 2-pentenal, 2-hexenal, methyl heptanoate, 3,6-nonadienal, methyl octanoate, decatrienal, methyl nonanoate, methyl 9-oxononanoate, methyl 10-oxodecanoate | Ethane, pentane |

**Table 9.3** Composition of the volatile fractions produced by the autoxidation of linoleic and linolenic acid substrates[a].

| Linoleic acid substrate | | Linolenic acid substrate | |
|---|---|---|---|
| **Product** | **%** | **Product** | **%** |
| *n*-Pentanal | 0.7 | Propanal | – |
| *n*-Hexanal | 67 | *n*-But-2-enal | 1 |
| *n*-Hept-2-enal | 6 | *n*-Pent-2-enal | 10 |
| *n*-Oct-2-enal | 18 | *n*-Hex-2-enal | 1 |
| *n*-Nona-2,4-dienal | 0.4 | *n*-Hex-3-enal | 14 |
| *n*-Deca-2,4-dienal | 5 | *n*-Hepta-2,4-dienal | 50 |

[a] The concentrations of *cis* and *trans* isomers are summed.

Based on Gordon, M.H., 2001. The development of oxidative rancidity in foods.
In: Pokorny, J., Yanishlieva, N., Gordon, M. (Eds.), Antioxidants in Food, Practical Applications. Woodhead Publishing Ltd. and CRC Press LLC, Boca Raton, FL, pp. 7–21 (Chapter 2).

olive and rapeseed oil), or when the concentration of hydroperoxides reaches appreciable levels (i.e., in sunflower and safflower oil) (Frankel, 1980). The volatile compounds contribute to the off-flavor of oxidized oils depending on their relative amounts but more on their flavor threshold. The threshold of a volatile component is determined as the concentration required in order to detect its contribution to flavor and depends on the distribution medium. For example, non-polar components have normally a higher flavor threshold in non-polar media such as edible oils than in water. In Table 9.2, the secondary oxidation products by the autoxidation of common unsaturated fatty acids are presented. In Table 9.3, the composition of the volatile fraction produced by the autoxidation of linoleic and linolenic acid substrates is presented and in Table 9.4 the flavor thresholds of potential linoleic acid oxidation products are shown.

As shown in Tables 9.2 and 9.3, although hexanal holds a large percentage of the total volatile substances of the oxidized linoleic acid (via the 13-hydroperoxide), other volatile products may have a greater contribution than hexanal to the off-flavor of the oxidized oils. Therefore, while most decomposition products of hydroperoxides may be responsible for the off-flavor in oxidized edible oils, aliphatic carbonyl compounds have more influence on the oxidized/rancid oil flavor due to their low thresholds.

More specifically, in canola oil, pentane, hexanal, propenal, and 2,4-decadienal are present in high amounts. The major volatile compounds detected in soybean oil are hexanal (23.5%) and 2-decenal (34.3%), while in corn oil, 2-heptenal (29.5%) and *trans*-2-octenal (18.1%), respectively. Also, *trans,cis*-2,4-decadienal has been reported as the most significant compound in determining the oxidized flavor of the oil, followed by *trans,trans*-2,4-decadienal, *trans,cis*-2,4-heptadienal, 1-octen-3-ol, butanal, and hexanal. Moreover, hexanal, pentane, and 2,4-decadienal have been suggested as indicators to determine the extent of oil oxidation. The *trans*-2-hexenal, and *trans,cis,trans*-2,4,7-decatrienal, and 1-octen-3-one are reported to give grass-like

**Table 9.4** Flavor thresholds of linoleic acid oxidation products.

| | Threshold (mg/kg) | |
|---|---|---|
| **Compound** | **Gordon (2001)** | **Choe and Min (2006)** |
| Alkanals | | 0.04–1 |
| Hexanal | 0.08–0.6 | |
| Heptanal | 0.04–0.055 | |
| Octanal | 0.04–0.6 | |
| 2-Alkenals | | 0.04–2.5 |
| *trans*-2-Nonenal | 0.04–0.4 | |
| *cis*-2-Decenal | 0.1 | |
| *trans,trans*-2,4-Alkadienals | | 0.04–0.3 |
| *trans,trans*-2,4-Nonadienal | 0.46 | |
| *trans,cis*-2,4-Decadienal | 0.02 | |
| Hydrocarbons | | 90–2150 |

Based on Gordon, M.H., 2001. The development of oxidative rancidity in foods. In: Pokorny, J., Yanishlieva, N., Gordon, M. (Eds.), Antioxidants in Food, Practical Applications. Woodhead Publishing Ltd. and CRC Press LLC, Boca Raton, FL, pp. 7–21 (Chapter 2); Choe, E., Min, D.B., 2006. Mechanisms and factors for edible oil oxidation. Compr. Rev. Food Sci. Food Saf. 5, 169–186.

and fish-like flavor in oxidized soybean oil, respectively (Choe and Min, 2006; Gordon, 2001).

The flavor generally developed during frying is characterized as fruity, grassy, buttery, burnt, nutty, and fishy and it is affected mainly by the oil and the frying intensity; sensory quality decreases with the number of fryings. The concentration of oxygen plays a significant role in the development of a desirable flavor; low amounts of oxygen contribute to poor and weak flavor, while high levels of oxygen to off-flavor, respectively. The volatile compounds produced mainly by linoleic acid oxidation are responsible for the desirable flavor of frying oils; during frying the fishy odor increases, while the fruity and nutty odor decreases. The main products of linoleic acid oxidation are: dienals, alkenals, lactones, hydrocarbons, and various cyclic compounds. Desirable flavor compounds produced from linolenic or linoleic acid oxidation are: 4-Hydroxy-2-nonenoic acid and its lactone, 4-hydroxy-3-nonenoic acid and its lactone, *trans,trans*-2,4-decadienal, *trans,trans*-2,4-nonadienal, *trans,trans*-2,4-octadienal, *trans*-2-heptenal, *trans*-2-octenal, *trans*-7-octenal, nonalactone, and trienals, while compounds with off-odors are: butanal, pentanal, hexanal, heptane, pentanol, 2-hexenal, heptanal, 1-octen-5-ol, 2-pentylfuran, and 2-decenal. Differences in the flavor quality are observed between vegetable oils. For example, peanut oil was found better than soybean oil, or rapeseed oil in frying potatoes at 160°C, 180°C, and 200°C. The major volatile compounds reported in soybean oil during frying at 200°C are: 2-heptenal, 2-octenal, 1-octen-3-ol, 2,4-heptadienal, and 2,4-decadienal. Finally, some volatile compounds (i.e., carbonyl compounds) formed during frying can react with amino acids, amines, and proteins and produce desirable and nutty pyrazines, whereas other compounds, such as 1,4-dioxane, benzene, toluene, and hexylbenzene, do not contribute to the desired flavor and are considered toxic (Choe and Min, 2007).

#### 9.3.1.2 Thermal oxidation

As previously referred, the mechanism involved in thermal oxidation during frying is the same as in autoxidation; nevertheless, the chain reactions are accelerated and the thermal oxidation rate is faster than in autoxidation. Thus, at low or moderate temperatures the oils are more slowly oxidized presenting a relatively long induction time; hydroperoxides are the major compounds formed, and their concentration increases until the advanced stages of oxidation. Also, polymerization compounds may be formed in the accelerated stage of oxidation after the end of the induction period (Warner, 2004; Gupta, 2005; Velasco et al., 2009). During frying, the oxygen in the air (diradical) reacts with the oil forming radicals which are commonly obtained when the hydrogen with the weakest bond in the carbon chain of the fatty acids is first removed. However, at the elevated temperatures used in frying, the formation of free radicals occurs rapidly and their concentration increases to a much higher level than at moderate temperatures, accelerating the thermal oxidation of oils. Practically above 150°C hydroperoxides are absent, indicating that the rate of their decomposition becomes higher than that of their formation, while the formation of new compounds is very rapid. Respectively, polymeric compounds are formed from the very early stages of heating, and, additionally, significant amounts of non-polar TG dimers (R-R) are formed in the absence of oxygen. Also, hydroperoxides formed from polyunsaturated fatty acids may undergo further oxidation to form dihydroperoxides and molecules which have oxygen-containing rings including hydroperoxy epidioxides and bicycloendoperoxides. The mechanism for the formation of hydroperoxy epidioxides from a-linolenic acid is shown in Fig. 9.9.

Different secondary oxidation products may result from various oils depending on their fatty acids composition. For instance, hexanal is formed (Fig. 9.10) as one of the major secondary oxidation products of linoleic and other ω-6 fatty acids, whereas oxidation of ω-3 fatty acids results in the production of propanal. Mono-oxygenated products, such as 9-keto-10-, 12-, and 13-keto-9-, 11-octadecadienoate, and hydroxyl derivatives, such as 9-hydroxy-10, 12-, and 13-hydroxy-9-, and 11-octadecadienoate have been identified from hydroperoxides of thermally oxidized methyl linoleate at 150°C. The aldehydes formed from hydroperoxide decomposition can be further oxidized to organic acids and other tertiary oxidation products. New substances are produced through radical-radical coupling or radical-radical disproportionation to form stable non-radical products, including a variety of polymer products. Secondary oxidation products may be toxic to tissues and cells of living organisms, while polymers formed during the oxidation of phospholipids in cell membranes reduce their fluidity, affecting their transport and cell signaling, as observed in Alzheimer's disease.

Most volatiles are removed from the oil by steam during deep-fat frying. The amount of volatile compounds present in the oil varies widely depending on the oil, the food, and the frying conditions. A loss of volatile compounds occurs in frying oil due to their evaporation, decomposition, and reaction with other food components.

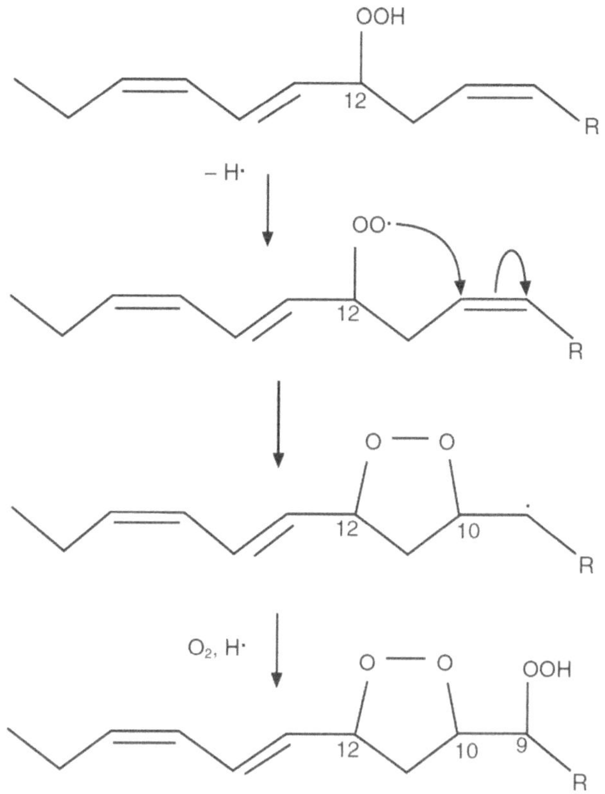


**Fig. 9.9** Reaction of 12-hydroperoxide from α-linolenic acid to form 9-hydroperoxy endoperoxide (Gordon, 2001).

$CH_3CH_2CH_2CH_2CH_2CHCH = CHCH = CH\ (CH_2)_7COOH$ 13- Hydroperoxide

(with O-OH on the CH)

↓ $-OH^{\bullet}$

$CH_3CH_2CH_2CH_2CH_2 - CH \mid CH = CHCH = CH\ (CH_2)_7COOH$ Alkoxy radical

(with $O^{\bullet}$ on the CH)

↓

$CH_3CH_2CH_2CH_2\ CH_2CHO$ + $^{\bullet}CH = CH\ CH = CH\ (CH_2)_7COOH$

Hexanal

**Fig. 9.10** Decomposition of 13-hydroperoxide from linoleic acid to form hexanal (Gordon, 2001).

The volatile compounds undergo further reactions such as oxidation, dimerization, and polymerization. Volatile compounds significantly contribute to the flavor quality of frying oil and fried foods. The amount of mono- and diacylglycerols is small at the beginning of frying. The high interfacial tension in the frying system breaks steam bubbles and forms a steam blanket over the oil surface that reduces the contact between the oil and oxygen, and restrains oil oxidation.

During frying, an increase in the viscosity, foaming, smoking, and free fatty acids content of the oil occurs as well as darkening of the oil. According to the recommendations of the German Society for Fat Research (DGF), a frying fat should not contain more than 24% polar material or 12% polymeric material. Frying temperatures also affect minor components of the oil, and among them the tocopherols that are the major natural antioxidants, and sterols that protect the oil against oxidation. The frying life of the oil also decreases due to the rapid loss of those compounds.

## 9.3.2 Thermal oxidation products and polymers

A simplified scheme explaining the mechanism of oil oxidation and the structure of the products formed during frying is given in Fig. 9.11. The decomposition of hydroperoxides follows different pathways leading to the formation of secondary oxidation products, however at frying temperature, and as the oxygen pressure is reduced, polymeric compounds are predominantly formed (Warner, 2004; Velasco et al., 2009; Zhang et al., 2012; Khor et al., 2019).

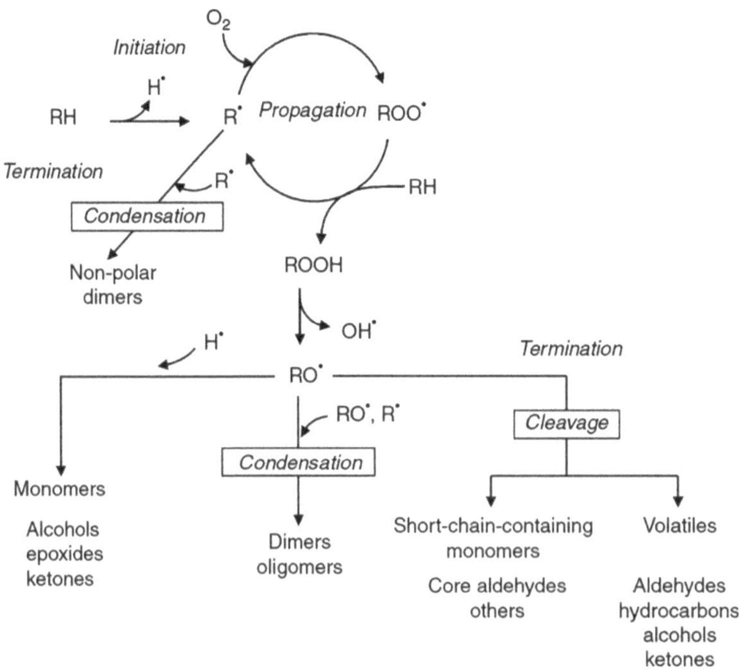


**Fig. 9.11** Mechanism of thermal oxidation (schematic) (Velasco et al., 2009).

In summary, as observed in Fig. 9.11, two TG radicals, R· formed in the initiation reaction and RO· formed by ROOH decomposition, are involved in the set of the termination reactions. This leads to various products of different polarity, stability, and molecular weight. Three main groups of compounds can be distinguished by molecular weight: (i) volatile compounds formed through breakdown of RO· which have molecular weights lower than the RH, (ii) TG dimers and oligomers that are formed through the interaction between TG radicals and have molecular weights higher than the RH, and (iii) oxidized TG monomers, with at least one of their fatty acyl groups oxidized, that are formed through interaction between TG radicals and hydrogen or hydroxyl radicals and have similar molecular weight to the parent non-altered TG (RH).

The compounds produced may differ in volatility that affects the deteriorative sensory assessment of frying oils or in polarity that is used as a criterion for the acceptability of used frying oils (Razali and Badri, 2003). Volatile compounds may be extremely important for the flavor quality of frying oils, but their concentration compared to the sum of all the decomposition products present in the frying oil is at the parts per million level. The major decomposition products of the frying oil are non-volatile polar compounds and triacylglycerol dimers, and polymers while cyclic compounds are produced in relatively small amounts. Dimers and polymers are large molecules with a molecular weight range of 692 to 1600 Da and have hydroperoxy, epoxy, hydroxy, and carbonyl groups, and —C—O—C— and —C—O—O—C— linkages. In used frying fats, TG dimers and oligomers are the most specific compounds while the polymers formed are also of high interest. As previously mentioned, the unsaturated fatty acyl groups attached to the glyceridic backbone react to form stable final products TG monomers, dimers, and oligomers containing modified and non-modified acyl groups as schematically shown in Fig. 9.12 (Velasco et al., 2009).

Oxidative polymers are formed during autoxidation when the free radicals terminate each other. When a triacylglycerol molecule breaks down, the partial triacylglycerol molecules can react with each other, forming dimers, trimers, or polymers which do not always impart an off-flavor to the freshly fried food. Polymerization of oil occurs under heat with or without the presence of oxygen. Heat can cleave the oil molecule or fatty acid and the cleaved compounds can then react with each other, forming large molecules, referred to as thermal polymers. In the frying process, excessive fryer heat and excessive fryer downtime can produce high levels of thermal polymers. Thermal polymers can be easily detected as they generally impart a bitter aftertaste to the fried food.

### *9.3.2.1 Oxidation-polymerization products*

#### Triglyceride dimers and oligomers

The mechanism of polar dimers formation in frying oils has been obtained by FAME (Fatty acid methyl esters) heated under simulated frying conditions (in the absence and presence of air). The products of the polymerization reactions are the most complex developed during frying given that they may contain TGs with more than one unsaturated acyl group in their molecule and each unsaturated fatty acid may follow

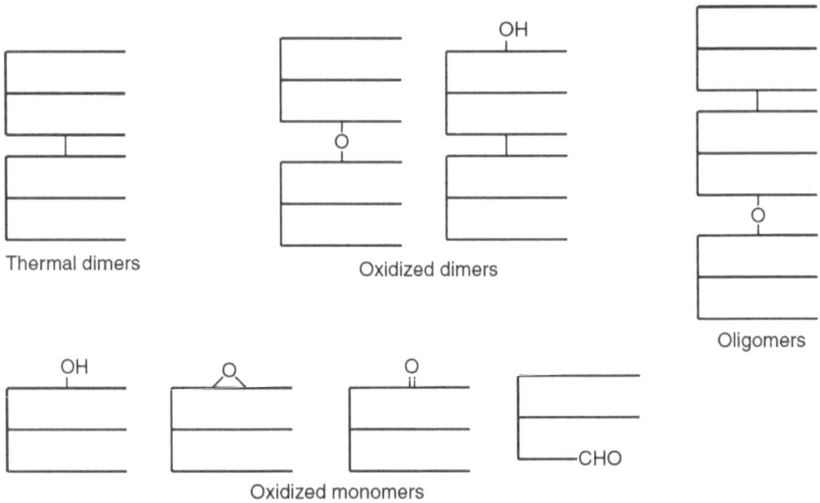


**Fig. 9.12** Representation of the molecular structures of the main groups of compounds formed during frying (schematic) (Velasco et al., 2009).

a different oxidation pathway. Although these are the major decomposition products of the frying oils, dimers formation in the first step of polymerization has also been studied. Compounds with no extra oxygen and further thermal oxidation can be obtained, but compounds containing extra oxygenated functions are still largely unknown. Thus, a large number of products may be obtained by possible combinations due to different oxygenated functions in the dimeric linkage or any other unsaturated fatty acyl group of the molecule, and to the presence of more than one functional group in the same dimeric molecule (Zhang et al., 2012; Khor et al., 2019). Some of the less complex structures in this group are shown in Fig. 9.12.

The main dimeric structures (acyclic or cyclic) found in used frying oils are dehydrodimers, bicyclic, tricyclic, and Diels-Alder non-polar dimers, suggesting the significance of the allyl radicals in the formation of non-polar conjugated dienes, even in the presence of oxygen. Acyclic dimers with C—O—C linkages and cyclic dimers, such as tetrasubstituted tetrahydrofurans have been reported. Structures found for polar dimers were mainly C—C linked dimers containing monohydroxy, dihydroxy, and keto groups. Structures for compounds with molecular weight higher than that of dimers have not been reported in frying fats or model systems (Velasco et al., 2009).

Dimers and trimers are formed from allyl radicals reactions by C—C linkages as shown in the acyclic polymers (dimer or trimer) from oleic acid in Fig. 9.13 and the cyclic dimers and polymers from linoleic acids in Fig. 9.14. Triacylglycerols react with oxygen and produce alkyl hydroperoxides (ROOH) or dialkyl peroxides (ROOR). They are readily decomposed to alkoxy and peroxy radicals by RO-OH and ROO-R scission, respectively. Alkoxy radicals can abstract hydrogen from the oil molecules to produce hydroxy compounds, or combine with other alkyl radicals to produce oxydimers.

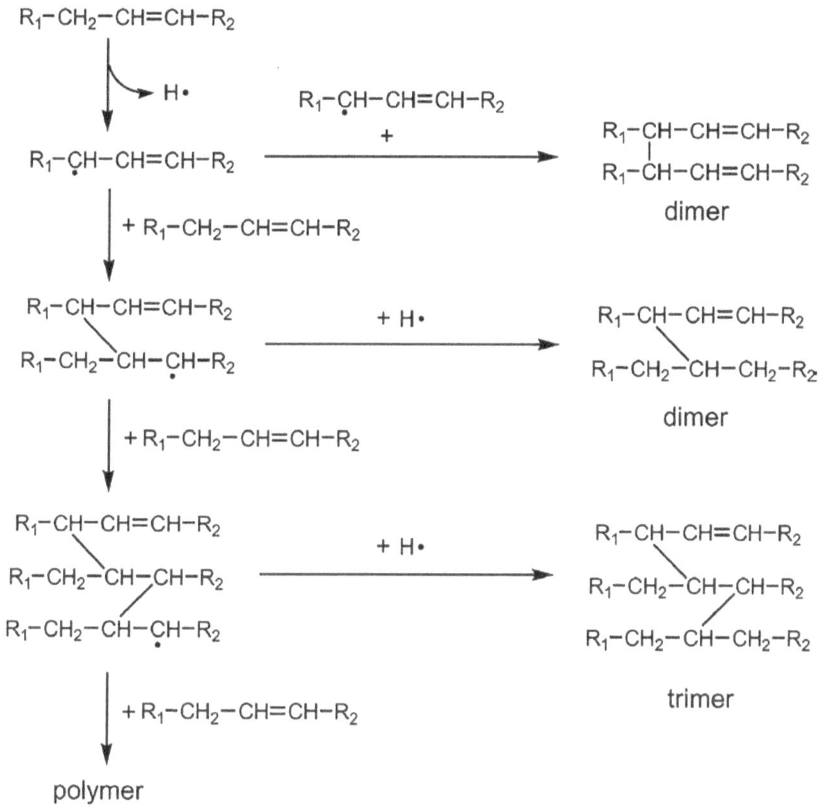


**Fig. 9.13** Formation of acyclic polymers from oleic acid during deep-fat frying (Kochhar, 2016).

The formation of dimers and polymers depends on the oil type, frying temperature, and the number of fryings. As the number of fryings and the frying temperature increase, the amounts of polymers increase as well. Peroxy radicals can combine with alkyl radicals and produce peroxy dimers. Oils rich in linoleic acid are more easily polymerized during deep-fat frying than oils rich in oleic acid. The dimers found during frying of soybean oil at 195°C are dehydroxydimer, ketodehydrodimer, monohydrodimer, dehydrodimer of linoleate, and dehydrodimer of oleate (Choe and Min, 2007).

## Oxidized triglyceride monomers

Oxidized TG monomers are stable final products resulting from the decomposition of the primary oxidation products (ROOH) that contain at least one extra oxygen atom in at least one of their three fatty acyl chains (Zhang et al., 2012; Khor et al., 2019). The main functional groups present in the frying oils are epoxy, keto, and hydroxyl groups, and the schematic representations of the main structures of the oxidized TG monomers are illustrated in Fig. 9.12 (Velasco et al., 2009).

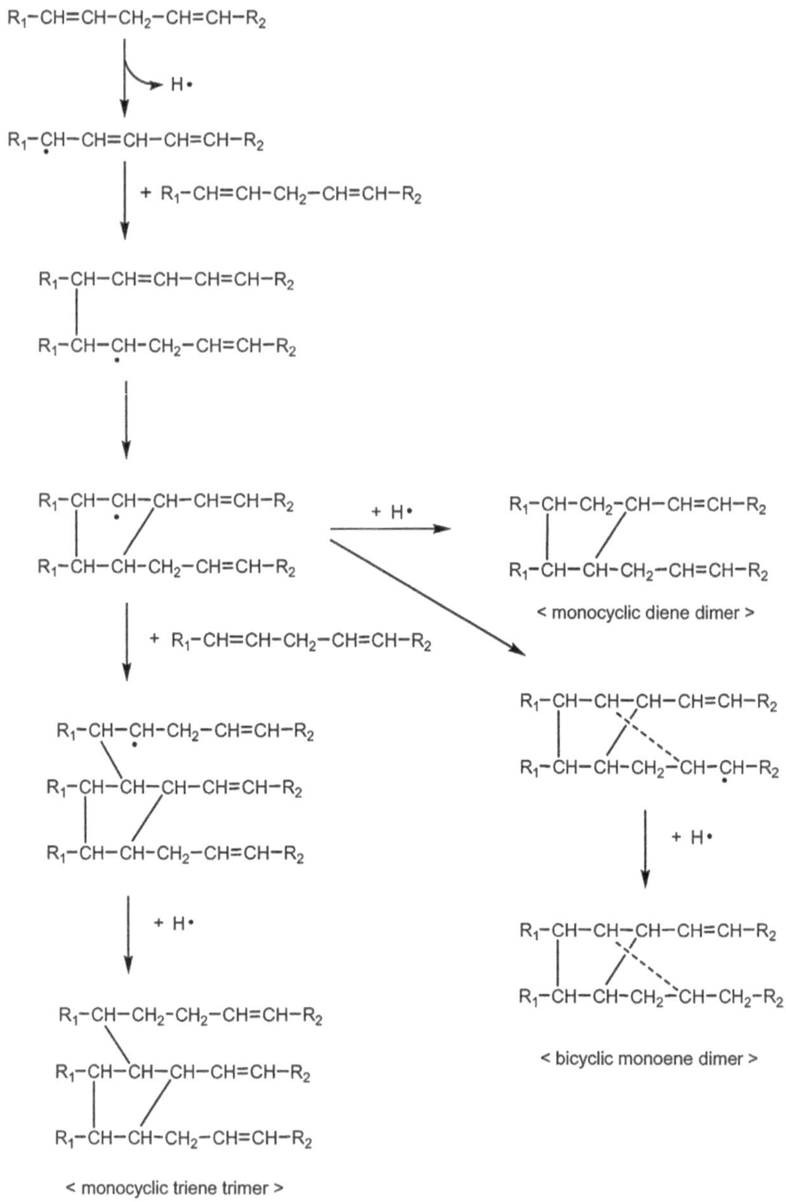


**Fig. 9.14** Formation of cyclic dimers and polymers from linoleic acids during deep-fat frying (Kochhar, 2016).

The formation of the epoxide ring may proceed through the reaction of a double bond with an external ROOH. The main epoxides reported by the thermal oxidation of model systems are: the saturated epoxides, *trans*-9,10- and *cis*-9,10-epoxystearate from methyl oleate and triolein, and the monounsaturated epoxides, *trans*-12,13-, *trans*-9,10-, *cis*-12,13- and *cis*-9,10-epoxyoleate from methyl linoleate and trilinolein.

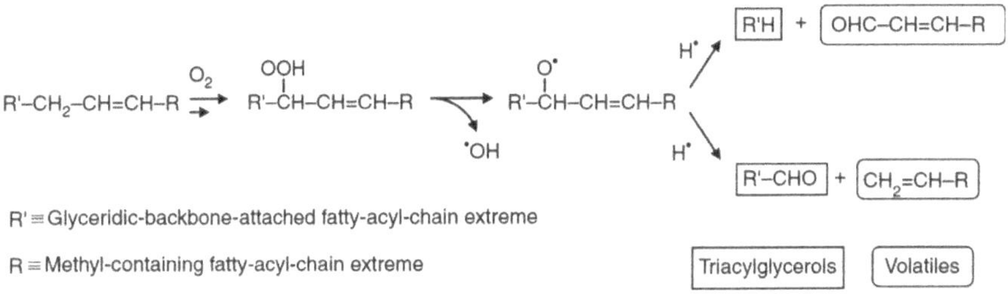


**Fig. 9.15** Formation of volatiles and short-chain-containing triacyloglycerols by scission of hydroperoxides (Velasco et al., 2009).

Also, six more epoxides from the oleyl and linoleyl chains are reported as major oxidized groups in used frying oils. The hydroxy and keto fatty acyl groups present a complex structure containing a higher number of double bonds.

Other important oxidation products are those originating from TG molecules which contain more than one oxidized fatty acyl group by ROOH breakdown. These products are aldehydes that may be formed from lipid hydroperoxides through homolytic scission of the C—C bonds on either side of the RO· according to the main mechanism shown in Fig. 9.15. Thus, two types of aldehydes are formed: volatiles deriving from the methyl side of the fatty acid chain, and aldehydes still bound to the TG. TGs containing linoleic acid contribute to the sensory characteristics of frying oils. As mentioned above, the volatile 2,4-decadienal is formed by the oxidation of linoleic acid in soybean oil (Fig. 9.16). The volatile aldehydes affect the sensory characteristics of the used frying oils while the non-volatile aldehydes exhibit nutritional interest as they remain in the oil and are absorbed by the food.

### 9.3.2.2 Thermal reaction products

High frying temperatures favor thermal reactions which in the absence of oxygen lead to low polar products, while in the presence of oxygen (with low available oxygen) lead to various non-polar degradation products, respectively. The main groups of compounds identified in used frying oils are TG dimers or isomeric TGs, including cyclic or *trans* fatty acyl groups (Velasco et al., 2009; Zhang et al., 2012; Khor et al., 2019; Fig. 9.17). The mechanism behind the formation of these products in the frying oils is generally based on studies in model systems using FAMEs under conditions simulating the frying temperatures (or at higher temperatures >200°C) in the absence of air. Even though thermal conditions in the absence of air are far different from those used in frying, the techniques applied in these studies are helpful for the identification and/or quantification of non-polar products in used frying fats. Oxidized polymer compounds accelerate the oxidation of oil, increase oil viscosity, reduce heat transfer, produce foam during deep-fat frying, and develop undesirable colors in the food.

Cyclic polymers are produced within or between triacylglycerols by radical reactions (Fig. 9.14) and the Diels-Alder reaction (Fig. 9.18). The formation of cyclic compounds in the frying oil depends on its degree of unsaturation and the frying

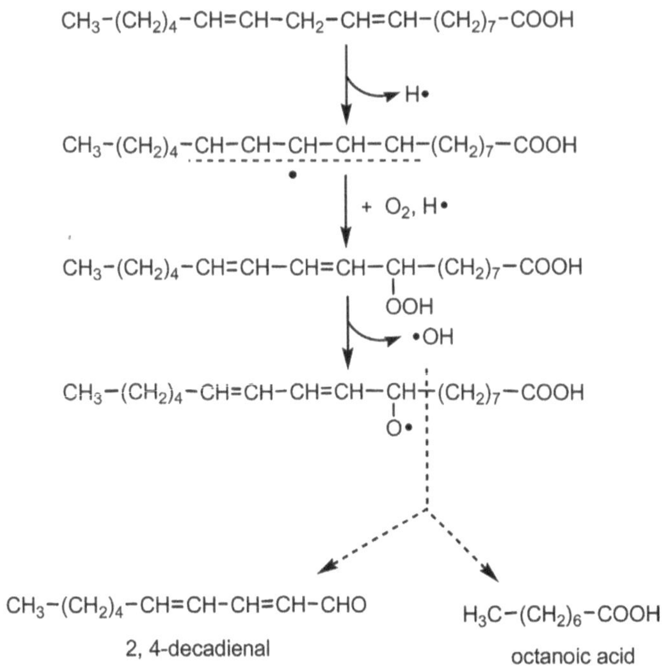


**Fig. 9.16** Formation of 2,4-decadienal from the oxidation of linoleic acid (Kochhar, 2016).

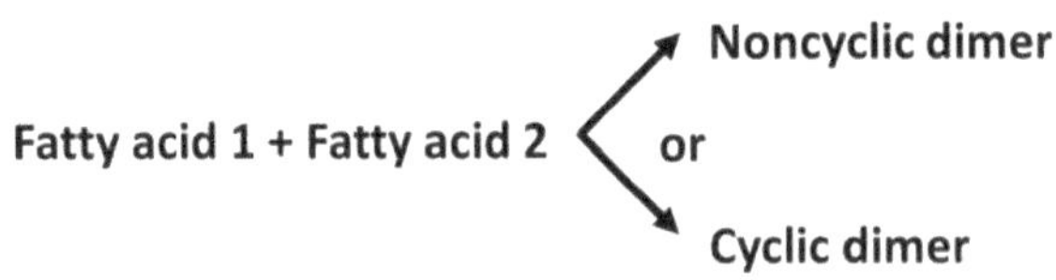


**Fig. 9.17** Polymerization process for frying oils.

temperature. The formation of cyclic monomers and polymers increases with the amount of linolenic acid and is not detectable until the linolenic acid content exceeds about 20%. During deep-fat frying with soybean oil, in specific, tricyclic dimers and bicyclic dimers of linoleate, as well as cyclic monomers, are produced (Kochhar, 2016).

## Non-polar triglyceride dimers

The mechanisms of the non-polar dimers' formation (products bearing C—C linkages without extra oxygen in the molecule) (Fig. 9.12) involve the oxidation pathway, since the major compounds identified may be generated through radical reactions from the allyl radicals. In the presence of conjugated double bonds, thermal dimers originate after a Diels-Alder reaction (Velasco et al., 2009; Zhang et al., 2012; Khor et al., 2019; Fig. 9.12). Thus, a tetrasubstituted-cyclohexene structure is formed by the reaction between the double bond of a molecule, acting as dienophile, and the conjugated

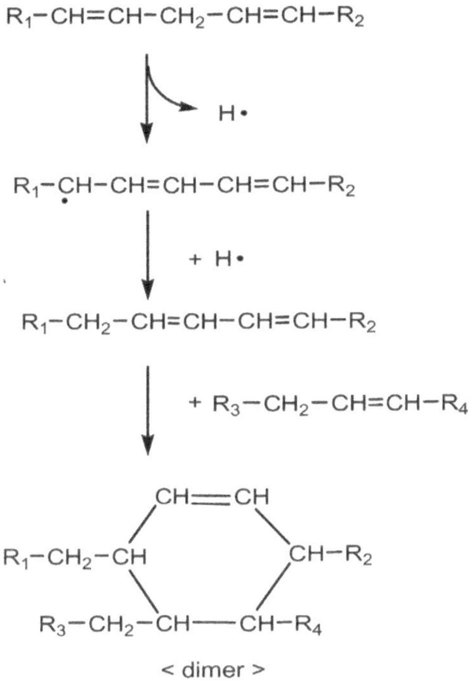


**Fig. 9.18** Formation of a cyclic compound from linoleic acid by the Diels-Alder reaction during deep-fat frying (Kochhar, 2016).

diene of another one. Different dimers result depending on the conditions (i.e., mono-, bi-, and tricyclic dimers at >250°C or dehydrodimers at lower temperatures). The non-polar dimers are probably the most relevant groups of new compounds formed during frying; however, the presence of non-polar oligomers in used frying fats has not been reported.

## Isomeric triglycerides

The formation of *trans* fatty acyl groups above 200°C in the absence of air has been studied and the structure of the main compounds has been identified. Such compounds may be present in minor amounts in fresh frying oils as they can be possibly formed during the refining process of deodorization, or the hydrogenation of unsaturated oils. However, the hydrogenation process is nowadays limited in practice, while the formation of those compounds can be avoided under proper deodorization conditions. The formation of *trans* isomers during the frying process is affected by the frying conditions; the reported levels of *trans* oleic and linoleic isomers in used frying oils are very low (at the level of mg/kg oil).

The formation of cyclic fatty acids in oils heated at high temperatures has also been studied, and their structure and formation mechanisms have been defined (Velasco et al., 2009; Zhang et al., 2012; Khor et al., 2019). Thus, the main cyclic compounds originating during frying are cyclohexenyl and cyclopentenyl rings from linolenic acid and a wide variety of cyclopentenyl, cyclopentyl, and cyclohexyl compounds, as well as bicyclic fatty acids from linoleic acid (Fig. 9.18). Additionally, saturated rings have been detected in small amounts that may generate from oleic acid. The levels of cyclic fatty acids in used frying oils reported are so low as 0.01 to 0.66 wt%. These induce toxic effects in rats but in much higher concentrations than those found in used frying fats.

Cyclic compounds are not formed to a significant extent until the oil temperature reaches between 200°C and 300°C. During deep-fat frying with soybean oil, in specific, tricyclic dimers and bicyclic dimers of linoleate, as well as cyclic monomers, are produced (Choe and Min, 2007).

Thus, the non-polar isomeric *trans* and cyclic fatty acyl groups are produced in very low amounts; however, there is interest in other compounds produced by *trans* isomerization and cyclization of the new compounds formed (i.e., oxygenated compounds formed through thermal oxidation and non-polar dimers formed through thermal reactions).

### 9.3.3 Hydrolysis

When a food is fried in heated oil, its water evaporates forming bubbles that gradually subside. Hydrolysis of triglycerides takes place during frying due to the action of water and steam, resulting in monoacylglycerols and diacylglycerols, as intermediate products, and finally in free fatty acids (FFA) and glycerol (Velasco et al., 2009; Zhang et al., 2012; Fig. 9.19).

The extent of hydrolysis depends on the frying temperature, the amount of water and steam, the interface area between the oil and the aqueous phase, and/or the presence of a surfactant (i.e., sodium hydroxide and alkalis used for fryer cleaning). The water will hydrolyze oil more quickly than steam, while the duration of frying may not be important. Thermal hydrolysis takes place mainly within the oil phase rather than the water-oil interface. It increases at higher contact between the oil and the aqueous phase of food while it slows down upon replacing the frying oil with fresh oil. Hydrolysis of short-chain and unsaturated fatty acids (which are more soluble in water) is more preferable to long-chain and saturated fatty acids. The reaction is favored at high temperatures and when frying foods with high water content, and its equilibrium is

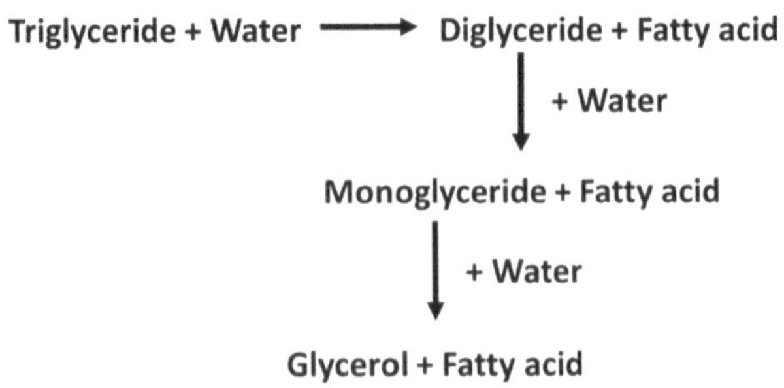


**Fig. 9.19** Hydrolysis process for frying oils.

shifted to hydrolysis products due to the evaporation of glycerol (at >150°C), thus the presence of FFAs and/or low-molecular-weight acidic products from oil oxidation accelerates the reaction. The formation of FFAs may also originate from technological problems (i.e., decrease of smoke point, formation of extra volatile and flavor compounds, decrease of interfacial tension). In used frying oils, diglycerides and FFAs have been found in levels up to 0.5% with the maximum suggested FFA% content being 0.05% to 0.08%. In well-controlled operations, the degradation products can be reduced even in the case of foods with high water content. FFAs and their oxidized compounds produce off-flavors and make the oil less acceptable for further frying. Thus, FFAs content is used as an indicator for the degradation level of oil due to hydrolysis in controlling the fry life of the oil.

## 9.4 Changes in oil and food during frying

### 9.4.1 Frying/frying cycles

During deep-fat frying heat and mass transfer phenomena take place, as analyzed previously. As frying continues, the food acquires a golden-brown color and absorbs oil creating a crisp texture and a deep-fried flavor. However, due to the high heat and moisture, the quality of oil gradually deteriorates. The life cycle of frying oil can be distinguished into five stages. In the first stage, the fresh oil provides only a slight browning, while the food remains undercooked with an imperceptible deep-fried flavor. The second stage is the optimum phase of the cycle; during this, a fully cooked food with a desirable golden-brown color and optimal deep-fried flavor can be achieved while the oil has undergone only slight deterioration. The used frying oil can be conditioned through the addition of fresh oil in the fryer at a certain rate. During the third stage, although the oil continues to deteriorate, it is still acceptable while the resulting fried foods exhibit a darker brown color and slight off-flavors. By the fourth stage, the oil has deteriorated even further, and foods have a dark brown color and moderate-to-strong off-flavors, while oil foaming prevents uniform cooking. Finally, the fifth cycle is characterized by severe oil degradation and foaming that produces a not fully cooked food with unacceptable fried flavor. Unless frying conditions are adjusted to maintain the oil at the second phase of the cycle, the oil will continue to deteriorate and will have to be discarded (Warner, 2004).

### 9.4.2 General physical and chemical changes in the frying oil

The primary purpose of frying is to produce a fried product with desirable sensorial properties: golden-brown color, and distinctive deep-fried flavor and taste. These properties can be obtained during the optimum stage of the frying cycle where chemical changes remain limited. However, with continued frying, oil unsaturation and the level of essential fatty acids (linoleic and linolenic) decreases concurrently and various decomposition products (FFA, carbonyl compounds, volatile products, and nonvolatile monomeric and polymeric compounds with high molecular weight and polarity) accumulate to levels that produce off-flavors and potentially toxic effects. The extent of the deteriorative reactions and changes as well as the amounts of the compounds formed and their chemical structures depend on many factors, including oil and food type, frying conditions, and oxygen availability, and may make the oil no longer suitable for frying.

The factors affecting the degradation of frying oil are the unsaturation level of its fatty acids, the temperature, the oxygen absorption, the metals in the food and the oil, and the type of food as shown in Table 9.5. Fresh oils (not oxidized) of high quality and stability, and good unsaturation level (low PUFA and metals levels) must be selected to achieve optimal oil fry life and fried foods with favorable sensorial and nutritional characteristics. The conditions of the frying process can be adjusted (i.e., not too high temperature or prolonged time and minimum aeration). Also, metal chelators, antifoaming, and antioxidant additives can be used. The type of food being fried affects oil decomposition. For example, fat-containing foods increase the fatty acids content of oils, while breaded and battered foods degrade oil more quickly than nonbreaded foods. Food particles accumulating in the oil also deteriorate oil more quickly; therefore, oil filtering will help remove these particles along with other oxidation products and can help extend the fry life of the oil. Frying protocols of intermittent or continuous frying affect fry life as well (Warner, 2004; Asokapandian et al., 2020).

The physical changes in oils that occur during frying mainly include increased viscosity, color darkening, and increased foaming and smoking as the frying time continues. The chemical compounds formed due to the thermal oxidation and polymerization reactions are considered responsible for these changes and as they increase, the flavor quality of the fried foods is also affected. The relationship between

**Table 9.5** Quality limits for fresh frying oils (Rossell, 2001a,b).

| Quality characteristics | Limits |
|---|---|
| Peroxide value (mEq/kg) | 0.4 max |
| p-Anisidine value | 5 units max |
| Linolenic acid (%) | 3.0% max |
| Lauric acid | 5% max |
| *Trans* fatty acids | 20% max |
| Iodine value | 130 units max (<100 units) |
| Color (Lovibond) | 2.0 max (red units, 5.25 in. cell) |
| Free fatty acid (%) | 0.1% max |
| Flavor | Bland |
| Flash point | 315°C |
| Smoke point | 200°C |
| Moisture | 0.05% max |
| Soap (sodium oleate) | 25 mg/kg max |
| Copper | 0.02 mg/kg max |
| Iron | 0.1 mg/kg max |

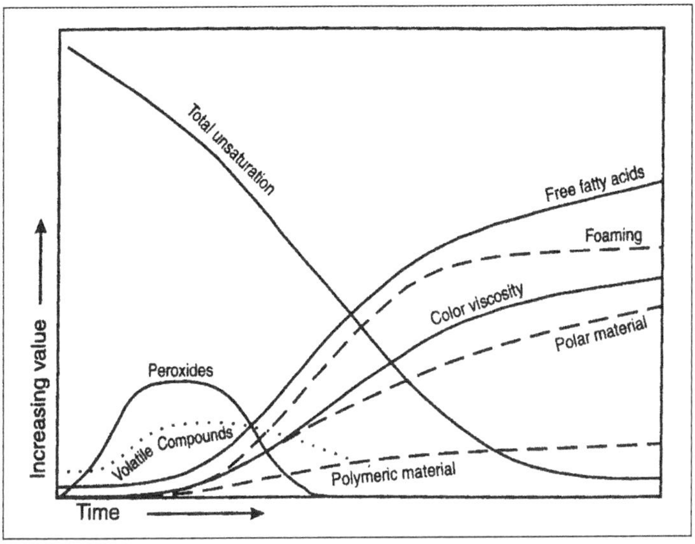


**Fig. 9.20** Physical and chemical changes of oil during deep-fat frying (Kochhar, 2016).

physical and chemical changes in the oil is usually predictable and helps at the discharge of used oils, particularly in batch operations (i.e., restaurants). The main physical and chemical changes of oil during deep-fat frying are presented in Fig. 9.20.

As far as the products of oil deterioration during deep-fat frying (volatile and nonvolatile decomposition products) are concerned, the following must be noted. Volatile compounds primarily affect the flavor of the food, whereas the nonvolatile compounds affect the fry life and the shelf life of packaged fried foods. Off-flavors and odors (i.e., grassy, fruity, plastic, or waxy odor) can be produced if the frying oil is allowed to deteriorate. Nonvolatile polymeric compounds (triacylglycerols, oxidized triacylglycerol derivatives, and cyclic polymers), at low levels, may not affect the flavor of the fried food if it is consumed immediately after frying. Oil decomposition products not only adversely affect the stability of the frying oil but may have a potentially adverse effect on the safety and flavor of the fried products. In batch frying operations due to repeated heating and cooling cycles and low turnover rates, the oil deteriorates more rapidly than in continuous frying. In the industrial frying processes, problems with fried products can be eliminated, if good operating protocols are followed by monitoring the frying conditions and using a high oil turnover rate. Oil deterioration should be controlled through parameters such as temperature and time, exposure to oxygen, continuous or batch frying, oil filtration, and oil turnover, and depending on the extent of the degradation reactions the oil should be discharged (Warner, 2004).

In small-scale frying operations (i.e., restaurants), oil discard is necessary at excessive foaming, smoking, or darkening of the frying medium. In addition, abused oil develops off-odors, such as acrid and burnt, which are transferred to the fried food as well. At the industrial level, specific methods exist for the quantitative determination of the chemical and physical changes in the frying oil.

### 9.4.3 *Interactions between the food and the frying oil*

Although fried foods are popular, mainly due to their appealing sensory characteristics, current nutrition guidelines recommend a decrease of the fat intake in the diet and draw attention to the possible health effects that arise from oil degradation during frying. For this purpose, the complex changes contributing to the characteristics of the fried food, due to the interaction between the frying medium and the food components, are examined (Fig. 9.21).

#### 9.4.3.1 *Major changes in the food*

##### Water loss and oil uptake

As already mentioned, deep-frying involves two interrelated phenomena, namely water evaporation and oil absorption. Key parameters for oil uptake, in specific, are the damaged areas on the food surface and the areas of weakness in the cellular structure of the food as well as the pores developed by food dehydration. Oil is more concentrated in the crust of the fried food and the uptake ratio (UR) is used to express

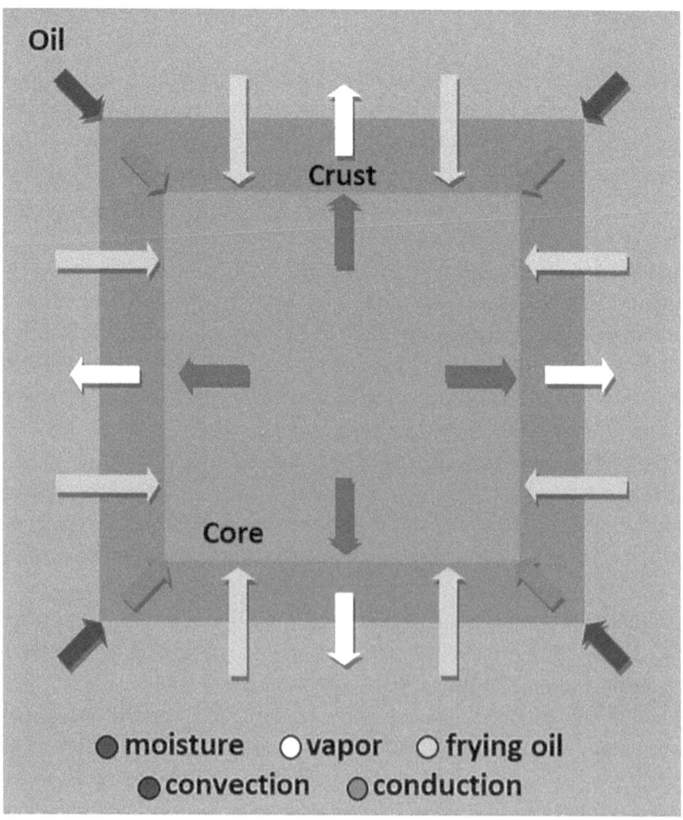


**Fig. 9.21** Major changes in the fried food.

the weight ratio between oil uptake and moisture loss (Velasco et al., 2009; Asokapandian et al., 2020).

Various factors significantly affect oil adsorption, such as the type and composition of the food and its properties (i.e., shape, size, surface-to-weight ratio, and porosity), the type of oil, as well as the temperature and the time of frying, and the pre-frying treatment of the food. For example, potatoes absorb more oil than does meat. For a typical non-lipid food, such as potatoes, the oil uptake is lower as the specific gravity increases while the larger the surface-to-weight ratio, the higher the oil absorption. Higher temperatures are also related to shorter frying times and lower oil absorption on the food surface. The influence of the type of fat or oil used is lower than that of the food and in general, oil absorption depends more on oil quality than on oil type. Low levels of oil degradation and polar compounds content, provide fried foods with a thin and crispy crust, golden color, and a delicious flavor while when degradation increases, the products can become unacceptable because of a high oil content absorbed and a darkened and thick crust.

Thus, the identification of the variables that affect oil content in fried foods, and the understanding of the physical processes involved, will help optimize the frying process and provide foods of good quality and oil with a longer fry life. However, due to consumer preferences for low-fat foods, efforts to decrease oil uptake have been made either through pre-drying or coating of the food (Velasco et al., 2009; Asokapandian et al., 2020).

## Color-texture-flavor

Foods are fried to provide desirable color, texture, and flavor. These positive attributes are obtained as a result of the chemical and physical reactions that occur during frying. Thus, fried foods will become crisp and golden-brown, and obtain a characteristic deep-fried flavor.

Texture is an important quality attribute of fried foods and is the response to the structural changes in the product during thermal processing and cooking. The porosity of the food crust during frying is considered a major contributor to the surface roughness, which in turn affects oil uptake. The texture of the fried foods may be influenced by the type of the frying oil, the oil temperature, the dry matter content of the uncooked product as well as the fat content of the final product (Mittal, 2009; Asokapandian et al., 2020).

Color is one of the most significant characteristics of foods, and particularly for fried products (i.e., potato chips). It is the consequence of the non-enzymatic browning associated with the Maillard reaction during frying. The color of the fried products is affected by the process variables such as the type of oil used, the frying temperature and time, and the pretreatment of the raw materials (i.e., lower frying temperatures result in lighter colored chips). The color of the fried products can be controlled through the blanching of vegetables, while vacuum frying can reduce the effect of higher temperatures on food color. Deep-fried flavor can be developed in products processed under the optimum frying conditions as related to the frying oil quality (Mittal, 2009; Velasco et al., 2009; Asokapandian et al., 2020).

Food shrinkage may initially occur during frying due to the rapid evaporation of moisture, or puffing due to vapor expansion inside the food.

Although in the early stages of frying many of the changes occurring are generally necessary to provide the desirable characteristics in fried foods, frying can negatively affect the properties of the fried food. At high temperatures, the oil changes significantly, and the degradation compounds produced by the deteriorative reactions affect the functional, sensory, and nutritional quality of the food. Therefore, oil deterioration should be monitored to maintain fried foods of high quality and prolong the fry life of the oil.

The shelf life of the final fried foods can be prolonged using antioxidants in the frying oils, and particularly natural antioxidants, such as polyphenols, and packaging under modified atmosphere (MAP) (Wu et al., 2019).

### *9.4.3.2 Migration of food components to the frying oil*

The quality of the fried foods is greatly affected by the quality of the frying oil, but the food can also influence the quality of the oil. Minor food components may leach into the frying oil modifying its physical and chemical properties, and thus its quality. Characteristic lipid food component reported to migrate into the oil are phospholipids that contribute to foaming, lipid-soluble vitamins, and trace metals that inhibit or accelerate oil oxidation, cholesterol from fatty animal foods, pigments and Maillard browning products that contribute to oil darkening, phenolic compounds that increase oil stability, and volatile compounds from flavored foods that contribute to special off-flavors. Particles from breaded or battered foods may also migrate into the frying oil resulting in burnt, off-flavors, and decreased oil quality (Velasco et al., 2009).

### *9.4.3.3 Chemical interactions between the frying oil and the food*

The main reactions that occur during frying are Maillard reactions between food proteins and carbohydrates contributing to the golden browning of the fried foods. Maillard reactions are of great interest because their products can migrate to the frying medium and affect the oxidative stability of the oil, but they are also associated with the formation of acrylamide 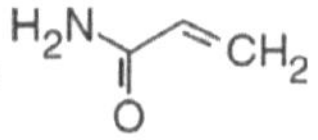

that is considered a potentially toxic substance. Non-enzymatic browning occurs upon the reaction of lipid oxidation products with amines, amino acids, and proteins. Reactions are reported between lipid radicals and secondary lipid oxidation products such as aldehydes, epoxides, and ketones with reactive groups of amino acid residues. Also, the interaction between oxidized lipids and proteins involved in the non-enzymatic browning leads to the formation of compounds of low molecular weight and high molecular weight polymers. However, during frying, the mechanisms of lipid oxidation are interrelated with the Maillard reaction as oil oxidation seems to have a key role in the browning developed during frying, and are associated with the interfacial activity of the lipid oxidation compounds (Matthäus, 2009; Velasco et al., 2009).

### 9.4.4 Nutritional and healthiness evaluation of fried foods

Fried foods are consumed by people of all ages and are either prepared in restaurants and food services and consumed directly, or are produced on industrial scale and distributed as packaged products with a long shelf life. Deep-fat frying is one of the conventional and popular food preparation methods, especially in the snack industry. French fries are widely consumed throughout the world and potato chips are a popular snack product due to their pleasant characteristics of taste, texture, and mouthfeel. However, deep-fat fried foods present a high-fat content and increased calories compared to non-fried foods. The daily dietary recommendations for fat intake to ensure adequate total energy, are set to a minimum of 15% to 20% (maximum 30%–35%) for healthy adults. The daily energy intake of saturated fatty acids should be in the range of 5% to 10%, and for trans fatty acids (industrial and from natural food) less than 1%.

The major concern due to the consumption of fried foods is the risk of obesity which is associated with cardiovascular health issues. Moreover, the type of fat and its content is directly correlated to human health. However, the consumption of fried foods is conventionally considered unhealthy mainly because frying impairs the chemical properties of food or oil, increases the overall content of *trans* fats, and decreases that of unsaturated fats. Many of these changes may be controlled by frying temperature and time, oil quality as well as products' moisture content and composition/structure. To address this, many investigations focusing on oil uptake reduction, suggest blanching, pre-drying, and coatings of foods before frying (Krokida et al., 2001; Varela and Fiszman, 2011; Kurek et al., 2017; Liberty et al., 2019). Also, novel alternative frying technologies (microwave-assisted vacuum frying (MVF), pulse-spouted microwave vacuum frying and power ultrasound, and microwave-assisted vacuum frying (USMVF) are proposed to provide products with lower oil content and superior textural characteristics compared to conventional frying (Zhanga et al., 2020).

During deep-fat frying, substances, such as fatty acids (TFAs) and acrylamide, produced are a major threat to public health. It is therefore highly imperative to identify and assess their effects on human health and also find ways to reduce oil uptake and restrict the production of such compounds without comprising the taste of fried foods. Industrially produced TFAs pose severe effects on human health as their consumption is associated with coronary heart disease (CHD) (Booker and Mann, 2008). Acrylamide is a chemical process contaminant classified as a probable carcinogen for humans (IARC) and is considered to be a human health concern (JECFA). It is formed in potato products during industrial processing, retail, catering, and home preparation. Guidelines, with specific recommendations for the mitigation of acrylamide produced in potato-based products in the food sector, are provided by FDA, while a benchmark level of 500 μg/kg in French fries has been established by the EU Regulation 2017/2158. Furthermore, mitigation strategies are proposed, most of which are targeted to the industrial sector, to ensure the application of well-controlled, supervised and standardized processing protocols. However, the contribution of the home-prepared fried foods to acrylamide intake is also considerable (Matthäus, 2009; Vinci et al., 2012; Mesias et al., 2021).

## 9.5 Industrial frying—Quality and safety management

### 9.5.1 *Fryers*

Industrial frying is one of the major segments in the frying industry. Fried foods are prepared in many countries and marketed in various forms. The most popular of these products are the salty snacks consisting of a wide variety of products with potato chips being the most highly consumed.

The industrial fryers used are of two main types: (i) batch fryers and (ii) continuous fryers. Vacuum fryers are less common and are primarily used to fry products that need to maintain their natural colors, such as sliced fruits and certain vegetables. Deep fryers are distinguished by size in: (1) small countertop fryers, (2) floor-model restaurant or foodservice fryers, (3) industrial batch or kettle fryers, and (4) continuous industrial fryers.

Fryers may be equipped with a filter for the filtration of the used oil and the removal of burnt food particles, while industrial fryers may be supplied with a CIP system for their automatic cleaning.

### 9.5.2 *Frying operation and oil filtration—Oil turnover*

The frying process should be monitored through oil deterioration and the formation of oil decomposition products as these can considerably extend or shorten the fry life of the oil. The type and the formation rate of the decomposition products vary with the food being fried, the oil/fat used, the fryer design, and the operating conditions. Food scraps should be frequently removed as they can become overcooked or char imparting a burnt flavor to the oil. Usually, scraps can fall into a cold zone (at the bottom of the fryer below the heater) and are often removed by filtration before they cause damage to the oil.

The filtration is accomplished in batch-frying operation by skimming the oil regularly or passing it through an external filter. On an industrial scale, a continuous filter unit with automatic discharge or two batch-type filtration units working in unison are used. In the case of breaded or coated products which are more friable, it is useful to shake or blow loose breadcrumbs or batter from the coated products before frying to avoid excessive deposits in the filter and consequently oil deterioration. However, the oil may be contaminated by substances migrating from the food fried, such as food fats, emulsifiers, trace metals, food scraps, free fatty acids, alkaline-reacting material, potato "whiteners" or other components. The regeneration of used oil through special filters is required to remove oil decomposition products and undesirable substances and thus extent its fry life (Rossell, 2001a; Gupta, 2005).

Filtration of the oil is considered an important task for the industry, particularly in breaded or coated foods. The selection of the filter system, as well as the carrying out of the filtration, are critical for the economic use of the frying oil and the optimization of the quality of the fried food. Filtration of the used fat/oil can be done by an external or a built-in fryer filter with automatic discharge in continuous operation, or in batch mode using a pump to drain or fill the fryer. Filtration, in respect to the filter type used, can be divided into: (i) Absolute filtration with filter papers or cloths which can remove particles down to about 5 μm in diameter and restrict the development of polar compounds and burnt flavors, (ii) Depth filtration (passive filtration) based on physical adsorption to remove particles down to 1 μm and microemulsified water, and (iii) Active filtration in which various adsorbent materials, chemically reacting with the decomposed compounds of the oil, are used; however, undesirable soluble components may be formed that adversely affect the oil (i.e., the FFAs of used oil react with some adsorbent materials forming soap). Filtration can significantly improve the fry life for the oil, but the adsorbents must be fully removed from the oil before food is fried. The use of commercial products that remove FFAs through a chemical reaction with an alkali, is not recommended as the remaining soap residues can reduce oil stability. The oil from the fryer must be routinely filtered to maintain fried food quality (Rossell, 2001a).

#### *9.5.2.1 Oil turnover*

As the frying continues, the fried food gradually absorbs oil. For this reason, it is necessary to add fresh oil to maintain the oil quantity in the fryer at the optimal levels. The amount of fresh oil added is related to the turnover rate and may be expressed as a ratio: Oil turnover = (weight of oil in the fryer)/(average weight of fresh oil added per hour) or, alternatively, as the time taken to replace 100% of the original oil charge. This is the theoretical time required by the fried food to pick up all the oil from the fryer that is filled up to the level suggested by the manufacturer and is expressed in hours.

In practice, assuming an 8-h turnover does not mean that all of the original oil will be replaced within 8 h; in each addition, the initial oil in the fryer becomes diluted by the fresh oil, and its concentration is gradually reduced through oil absorption by the food. For continuous production, a normal 5 to 12 h turnover is recommended, and this should be allowed by the fryer design. A steady-state is usually achieved in the fryer after a few oil turnovers; that means a state where the blending of fresh and used oil reaches a constant rate, and oil take-up by the food being fried becomes uniform. In batch frying, longer turnover times are likely, and depending on the frying conditions, if turnover times of 20 h or more are used, the oil should be discharged, and fully replaced with fresh oil upon completion (Rossell, 2001a,b; Gupta, 2005).

#### *9.5.2.2 Good operation practices for the fryer*

The fryers should be designed and constructed in every part (pipes, valves, etc.) with suitable materials and operated according to the equipment instructions. The frying temperature should be adjusted and the recommended cooking temperatures should not be exceeded. If solid fats are used, they may be allowed to melt before frying to avoid overheating. Other practices should also be applied such as removing the excess water or coating/batter from the food before frying, salting after frying, covering the fryer, etc. Finally, the filter system should be properly operated and the programs of sanitation and maintenance of the fryer should be assured.

#### 9.5.2.3 Fryer sanitation

Fryers may be equipped with a cleaning-in-place (CIP) system installed by the fryer manufacturer. The CIP systems are used for the daily sanitation of the fryers, especially in the case where different food products are fried in the plant (meat, poultry, or fish). A CIP sanitation program involves the fryer pan, piping, etc., and includes cleaning by caustic solution or sanitizing chemicals, followed by a diluted acid solution, rinsing with water, and heat drying. In kettle fryers, manual scrubbing for sanitation is applied, whereas a CIP in continuous fryers, respectively. Fryer sanitation is essential for maintaining oil quality and prolong its shelf life; used oil or crumbs residues can enhance oil degradation. The oil from the fryer must be routinely cleaned to remove traces of polymerized oil, which will promote the degradation of the subsequent oil batches. The fry life of the oil is further improved by timely and proper sanitation of the frying system (Gupta, 2005).

### 9.5.3 Monitoring of oil quality during frying

#### 9.5.3.1 Quality of fresh oil

The type of oil should be selected based on its quality characteristics and be used for as long as the desired flavor of the fried foods and subsequently, the shelf life of the fried products are obtained. Because the frying oils degrade under continuous use, they must be very low in detrimental minor components (i.e., FFAs, trace metals, etc.) and present high oxidative stability (resistance to breakdown, high content in natural antioxidant substances). Thus, the frying oil should be low in saturated fatty acids, ultra-high (75%–85%) in oleic acid, low in linoleic, and very low in linolenic acid (<0.2%). The working life of the oil can be further enhanced by the presence of potent, natural, non-volatile antioxidant components, such as Δ5-avenasterol, α-tocopherol, oryzanol, rosemary antioxidants, mixed tocopherols, sesamolin, and polyphenols. Olive oil combines these properties since it is a stable oil rich in oleic acids and endogenous antioxidant substances (β-sitosterol, polyphenols), containing zero TFAs, and is therefore proposed as a frying oil as well (Kochhar, 2001).

The quality control criteria and limits for a fresh (unused) frying oil are presented in Table 9.5. The fresh oil should be of high quality, stored under nitrogen to avoid oxidation, and used under proper conditions.

#### 9.5.3.2 Monitoring the quality of oil in the fryer

The oil quality and its deterioration should be monitored during frying and the used oil should be discharged when the quality limits are exceeded. In restaurants, the discharge of used oil is empirically decided through the observation of its color, viscosity, foaming, and odor. On an industrial scale, oil discharge is controlled by the changes in oil's quality parameters when reaching the point of not satisfying the recommended limits for use. The oil quality may be enhanced by operating the fryer at its designed capacity and for long hours at a time. In general, oil quality oil is better maintained in a continuous fryer compared to a batch fryer due to the lower oil turnover times.

The analyses/measurement for monitoring the quality parameters of the frying oils are presented in Table 9.6. Various test devices or sensors are also available to check the oil quality in a fryer (Table 9.7), most of which measure the oil decomposition products as polar compounds.

Four well-known rapid tests include the Food Oil Sensor (FOS) (Northern Instruments, Lino Lakes, MN), which measures the dielectric constant in the frying fat in relation to the fresh oil; the RAU-Test, which is a colorimetric test-kit that contains redox indicators reacting with the total amount of oxidized compounds; Fritest (E Merck, Darmstadt, Germany), which is a calorimetric test-kit sensitive to carbonyl compounds; and the Spot Test, which assays FFAs to indicate hydrolytic degradation. Correlation with polar compounds is higher for FOS compared to the RAU-Test, Fritest, and Spot test.

**Table 9.6** Quality parameters of frying fat/oils.

| Chemical indices | Physical properties | Sensory evaluation |
|---|---|---|
| Free fatty acids (FFA%)<br>Peroxide value (PV)<br>Thiobarbituric acid test (TBA)<br>p-Anisidine value (p-AV)<br>Totox (total oxidation value) = 2 (PV) + p-AV<br>Iodine value (IV)<br>Total polar materials (TPM%)<br>Nonvolatile/volatile decomposition products (NVDP, VDP) | Refractive index<br>Dielectric constant<br>Viscosity<br>Color | Odor<br>Taste<br>Flavor |

Based on Gupta, M.K., 2005. Frying oils. In: Shahidi, F. (Ed.), Bailey's Industrial Oil and Fat Products, vol. 4. John Wiley & Sons (Chapter 1).

**Table 9.7** Measurements/analyses for frying fat/oil monitoring.

| Tests based on chemical measurements | |
|---|---|
| **Type** | **Correlation with** |
| – Fritest Test | – Carbonyl compounds |
| – Oxifrit Test, RAU Test | – Oxidized compounds |
| – Alkaline contaminant materials | – Food-oil components |
| – Low Range Shortening Monitor (LRSM) | – Free fatty acids or oxides |
| – PCT 120 (polar compound tester) | – Polar degradation compounds |
| – Mlinar, Spot Test | – Free acids |

**Table 9.8** Recommended limits for the frying fat/oil characteristics.

| **Recommended limits for used frying fats/oils** | |
|---|---|
| Max FFA: | 2.5% |
| Min linolenic acid content: | 2% |
| Max oxidized components insoluble in petroleum ether: | 0.7% |
| Min smoke point: | 150°C |
| Max polar materials content: | 25%–27% |
| Max triglyceride polymers and dimers: | 10% |
| Max triglyceride dimers: | 6% |
| Max peroxide value (PV) (for good quality fat): | 10–30 meq/kg (after frying: 80 meq/kg) |
| Max carbonyl index (p-AnV): | 50 meq/kg |
| Max totox value: | 10 |
| Max conjugated dienes (E1%, 232 nm) (for good quality fat): | 0.2–0.4 (after frying: 4.0) |
| Max conjugated trienes (E1%, 268 nm) (for good quality fat): | 0.15–0.20 (after frying: 2.0) |
| Max ΔE268 nm (for good quality fat): | 0.10 (after frying: 0.15) |
| Soap content: | No difference after frying |
| Trans fatty acids: | No difference after frying |

In Table 9.8, the recommended limits of various parameters for the discharge of used oil are presented based on the bibliography. Regarding the polar materials' level, the USFDA has not yet imposed any regulation but most countries in Europe have adopted some form of regulations to control the frying oil quality, considering that the oil reaches its end-point when its polar materials' content exceeds a certain level ranging from 12% to 24% (Gupta, 2005).

### 9.5.4 Management of frying processing—Quality and safety assurance

#### 9.5.4.1 Healthy, palatable, and safe fried foods

Nowadays the production of healthy food, as well as food products with desirable sensory characteristics is very important. Fried snack foods have become an internal part of the eating habits for the majority of the world's population. Although they display several positive quality features (appearance, taste, consumer acceptability), it must be ensured that they do not pose safety hazards.

Apart from altering the eating quality of foods, frying simultaneously offers a preservative effect due to microorganisms' destruction and water activity reduction in foods. Therefore, frying is considered a significant process for the quality of fried foods and is characterized as a control point (CP).

However, frying can also negatively affect frying oils and fried food. Deep frying, which occurs at high temperatures (~180°C) frequently involves fat deterioration due to chemical (oxidation, polymerization, hydrolysis, color formation) and physical (odor/flavor development) reactions that produce degradation compounds, harmful to human health (such as TFAs, highly oxidized or polymerized constituents of fatty acids and acrylamide) that affect the functional, sensory, and nutritional quality of oils and fried food. In the early stages of frying, many of these changes are generally required to provide the desirable characteristics typical of fried food but if the oil is allowed to deteriorate high enough, it will result in foods of poor quality and safety and shorter fry life.

#### *9.5.4.2 HACCP requirements for food safety*

The protection of human health is a matter of high importance and is internationally governed by Codex Alimentarius. HACCP (Hazard analysis and critical control point) is a highly specialized system for food safety and is considered the most effective technological method for hazards' elimination and products' safety assurance. HACCP approaches food safety focusing on the critical control points (CCPs) of food processing. According to the HACCP principles, the possible hazards (microbiological, chemical, physical) are identified, preventative measures are taken, and the critical control points (CCPs), which are the most crucial for controlling products' safety, are determined. For each CCP, critical limits for the controlling parameters and monitoring methods are established, as well as corrective actions in case these limits are exceeded. Finally, the procedures and controls should be established and verified through a documentation system (Mortimore and Wallace, 1994; Tzia and Tsiapouris, 1996; Quaglia and Bucarelli, 2001; Vorria et al., 2004). As far as food safety is concerned, the implementation of HACCP and the control of the critical for the safety stages (CCPs) is obligatory in food processes according to legislation (Reg. EC 178/2002, Reg. EC 852/2004).

#### *9.5.4.3 Integrated hygiene/efficient management of frying systems—Economic use of frying oil/fat*

Various methods and indices have been developed for the quality control of frying oils and sensors are also commercially available for the online monitoring of oil's oxidative state. In order to produce safe fried foods, the frying fat should be controlled and renewed before the oxidative indices exceed the specification limits. Moreover, filtration of the used fat is proposed as a solution both for reducing oil consumption and extending its shelf life. Most of the filter systems use passive filters; however, active filters using porous materials to absorb undesirable constituents from the used fat also exist. The purification and/or regeneration of the frying fats or oils is effective through filtration which removes highly oxidized/polymerized fat components and other impurities from the used fat.

The integrated hygiene management of the frying systems that includes the filtration of used oil/fat (Fig. 9.22), should be adopted, and guidelines and practices for the

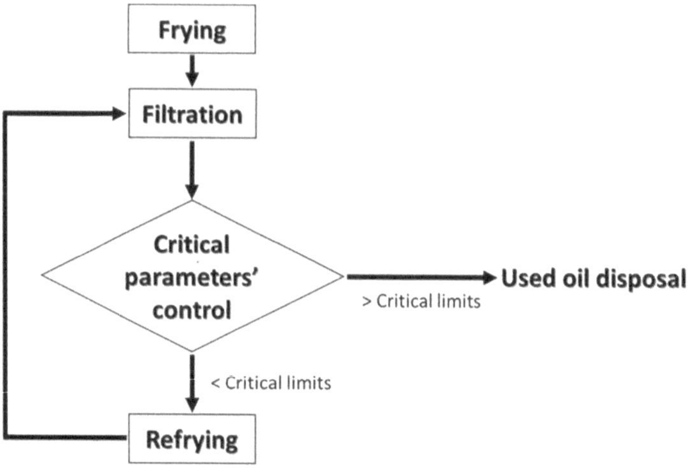


**Fig. 9.22** A fryer-filtration system operation.

operation of the fryer should be organized. At the same time, the introduction of a filtration system contributes to the economic use of frying oil. The online monitoring of oil deterioration by suitable sensors allows its timely regeneration. Therefore, good practices for oil filtration, concerning both the time intervals and the hygiene actions applied, should also be adopted.

The economic use of frying oil/fat is particularly important for the food sector (food industries and food services, restaurants, fast-foods, hotels, etc.) to minimize fat quantity demands and be able to reuse fat after its purification. Filtration of used frying fats extends fry life (by 24%–50%); thus saving cost, reducing fat/oil consumption, improving hygiene and food quality (less degraded oil absorbed, not burnt flavor), reducing odors' development in the production area (or the kitchen in food services), protecting the environment, reducing labor costs, and lowering the health/safety risks. The filtration system for the regeneration of used fat/oil should be appropriately selected and operated to achieve the above-mentioned benefits and avoid possible problems (emulsification, soap formation).

### *9.5.4.4 Frying process as CCP and CP*

Based on the above, frying is a significant step of the fried foods processing with a crucial impact on the quality (CP) and safety (CCP) of the final products (Vorria et al., 2004). Various factors related to the frying medium, the food, or the process may affect food frying, some of which are designed (size of food) while others may enhance fat/oil oxidation and should be controlled during frying (Table 9.9). The processing conditions for the production of high-quality snacks with low-fat content should be designed. The proper selection of the frying conditions may result in a significant reduction of the oil uptake and favor the development of the required sensory characteristics, such as crispiness, crunchiness, flavor, palatability, etc. (Table 9.9).

**Table 9.9** Factors affecting frying oil and fried food degradation.

| Oil/food | Process |
|---|---|
| *Fat/oil characteristics* | *Frying process* |
| – Quality indices of fat/oil | – Frying temperature |
| – Unsaturated fatty acids content of fat/oil (optimal oil fry life) | – Frying time (duration) |
| | – Alternating frying (frying-cooling-refrying) |
| – Metal ion content of frying fat/oil | – Sufficient fat/oil turnover |
| *Food characteristics* | – Fat/oil heating without foods |
| – Size of food slices | – Surface/volume rate of fat/oil |
| – Protein content of food | – Fat/oil acidity increase |
| – Metal ion content | – Oil filtering and fryer cleaning |
| – Breaded/battered foods | – Minimizing oxygen absorption |
| – Fat-containing foods | – Frying equipment maintenance |
| *Use of additives* | – Mode of frying (continuous vs batch) |
| – Use of antifoaming agent, metal chelator | |
| – Use of antioxidants | |

The extent of fat deterioration may be evaluated through various quality parameters. The most critical of these are considered to be *p*-Anisidine value, Total Polar Materials as well as UV absorbance values indicating the oxidative state and the shelf life of the fat, respectively (Table 9.6). Moreover, the continuous monitoring of fat quality can also be achieved by rapid analyses using test kits or appropriate sensors (Table 9.7).

Quality specifications for frying fats/oils have not yet been established; however, quality parameters, which oils should satisfy in order to be suitable for frying, have been suggested in various countries (Table 9.8). Therefore, the frying fats/oils are selected and their specifications are determined based on these values. The critical limits (CLs), which render fat/oil improper for refrying based on the maximum acceptable parameter values, can be determined and the monitoring actions through standard analyses or rapid tests can be designed.

The inclusion of a filtration system in continuous (food industries) or semi-continuous (food services) fryers entails the safe design and operation of the frying-filtration-refrying process (Fig. 9.22). Thus, the integrated hygienic and efficient management of the fryer-filtration system is proposed so that HACCP requirements are fulfilled and the advantages due to economic use of the frying fat/oil are exploited. Guidelines for the integrated management of the frying systems, including filtration of used fat/oil, are presented in Table 9.10. In order to produce safe fried foods, frying should be monitored, and the frying fat should be appropriately renewed at a certain deteriorative state. The purification and/or regeneration of the used frying fat is considered a complementary operation during frying to decrease fat consumption, and provide healthier fried foods.

**Table 9.10** Guidelines for the integrated management of frying systems including filtration of used fat/oil.

| Process step CCP/CP | Selected parameters/process design | Monitoring of critical parameters | Control of critical limits | Actions |
|---|---|---|---|---|
| Frying | Type of frying fat/oil<br>Specifications of fat/oil<br>Type and size of food<br>Frying temperature/time<br>Renewal rate of fat/oil<br>Type of frying (shallow/contact or deep frying) | **Continuous monitoring or sampling and control**<br>– Standard analyses (Table 9.6)<br>– Rapid tests (Table 9.7)<br>Validation of the results from measurements<br>– Continuous monitoring by sensors (Table 9.7)<br>– Calibration of sensors/ measuring equipment | – Recommended limits (Table 9.8)<br>– Correlated measures with recommended limits (Table 9.8) | **I. Values of critical parameters ≤ critical limits**<br>– Stop the frying process<br>– Cleaning of sampling tubes/probes<br>– Pumping the water layer and food particles from the bottom (cool zone) of the fryer and discharge<br>– Discharge the fryer and drain the frying fat/oil to the filter<br>– Screening of fat/oil and separation of burnt food particles<br>– Cleaning of the fryer: use of suitable cleaning agents by spraying, mechanical cleaning by brushing, and drying<br>– Cooling the fryer<br>**II. Values of critical parameters ≥ critical limits**<br>– Discharge of the used frying fat/oil<br>**Hygiene practices—GMPs**<br>– Cleaning of equipment and the surrounding area<br>– Cleaning of sampling tubes/probes after sampling<br>– Check their cleanliness |

| | | | | |
|---|---|---|---|---|
| Fat regeneration/ filtration and refrying | **Filtration system (continuous frying systems in food industries)**<br>– Selection of suitable method/ filter<br>– Use of validated filtration system<br>– Material of the filter approved as safe for food processes<br>– Construction materials of the filtration equipment (approved as safe for contact with food)<br>– Specifications of filter material:<br>Data for soap formation by using specific filter materialsData for trans fatty acids formation using clays<br>**Filtration services (external companies) (semicontinuous fryers in food services, fast-foods, etc.)**<br>– Reliability of the company offering the filtration services | **Filtration system**<br>– Measure the fat/ oil temperature<br>– Control the filtration process<br>– Control the operation of the filtration system (pumps, block, etc.)<br>– Examination of filtration efficiency (according to standard analyses or rapid tests and sensory evaluation<br>– Examination for active filtration defects (emulsification, soap, smoking, off-odors)<br>– Evaluation of oil quantity saving<br>**Filtration services (external companies)**<br>– Auditing the filtration service | **Internal limits**<br>Limits of the industry in respect to the filtration intervals (approximating the specifications of the fresh/unused fat/oil)<br>**Filtration intervals**<br>Recommended by the filter company<br>**Fat/oil temperature and contamination**<br>Metal ions from the fried food | **Efficiency of filtration (values of critical parameters ≤ corresponding internal limits)**<br>– Stop the filtration process<br>– Filling the fryer in the cool zone with the regenerated fat/oil refrying<br>– Cleaning of the filtration system according to the instructions of the construction company<br>– Discharge the concentrated (full absorbed) filter and replace filter<br>– Maintenance of the filtration system<br>– Recording<br>**Hygiene practices—GMPs**<br>– Cleaning of equipment and the surrounding area |

## 9.6 Conclusions

Frying is a popular food preparation process used both in small-scale food services and on an industrial scale as deep-fat frying. Heat and mass transfer phenomena and the oil/fat deterioration reactions (thermal oxidation, polymerization, and hydrolysis) involved in the frying process should be well understood as they contribute to changes in the final fried foods affecting both their quality (color, texture, flavor, and oil content) and safety (trans fatty acids, acrylamide). The food industry should manage the quality and safety of the products by selecting suitable oil/fats, designing the frying conditions (related both to the food and the process), and establishing an integrated hygiene/efficient management of the frying systems. The above parameters provide the potential to produce packaged fried food products with prolonged shelf life and introduce new technologies (edible coating, pre-drying) for the development of healthy (reduced oil content), palatable, and safe fried products.

## References


Achir, N., Vitrac, O., Trystram, G., 2009. Heat and mass transfer during frying. In: Sahin, S., Sumnu, S.G. (Eds.), Advances in Deep-Fat Frying of Foods. CRC Press LLC, Taylor & Francis Group, Boca Raton, FL, pp. 5–32 (Chapter 2).

Akoh, C.C., 2017. Food Lipids: Chemistry, Nutrition, and Biotechnology. Taylor & Francis Group, LLC, Boca Raton, FL.

Alvis, A., Vélez, C., Rada-Mendoza, M., Villamiel, M., Villada, H.S., 2009. Heat transfer coefficient during deep-fat frying. Food Control 20, 321–325.

Angelo, A.J.S., Vercellotti, J., Jacks, T., Legendre, M., 1996. Lipid oxidation in foods. Crit. Rev. Food Sci. Nutr. 36 (3), 175–224.

Arslan, M., Xiaobo, Z., Shi, J., Rakha, A., Hu, X., Zareef, M., Zhai, X., Basheer, S., 2018. Oil uptake by potato chips or french fries: a review. Eur. J. Lipid Sci. Technol. 120, 1800058.

Asokapandian, S., Swamy, G.J., Hajjul, H., 2020. Deep fat frying of foods: a critical review on process and product parameters. Crit. Rev. Food Sci. Nutr. 60 (20), 3400–3413. https://doi.org/10.1080/10408398.2019.1688761.

Baumann, B., Escher, E., 1995. Mass and heat potato slices transfer during deep-fat frying I. Rate of drying and oil uptake. LWT Food Sci. Technol. 28, 395–403.

Blumenthal, M.M., Stier, R.F., 1991. Optimization of deep-fat frying operations. Trends Food Sci. Technol. 2, 144–148.

Booker, C.S., Mann, J.I., 2008. Trans fatty acids and cardiovascular health: translation of the evidence base. Nutr. Metab. Cardiovasc. Dis. 18, 448–456.

Bouchon, P., Pyle, D.L., 2005. Modeling oil absorption during post-frying cooling: I: model development. Food Bioprod. Process. 83, 253–260.

Choe, E., Min, D.B., 2006. Mechanisms and factors for edible oil oxidation. Compr. Rev. Food Sci. Food Saf. 5, 169–186.

Choe, E., Min, D.B., 2007. Chemistry of deep-fat frying oils. J. Food Sci. 72 (5), 77–86. https://doi.org/10.1111/j.1750-3841.2007.00352.x.

Dana, D., Saguy, I.S., 2006. Review: mechanism of oil uptake during deep-fat frying and the surfactant effect-theory and myth. Adv. Colloid Interf. Sci. 128–130, 267–272.

Datta, A.K., 2007. Porous media approaches to studying simultaneous heat and mass transfer in food processes. I: problem formulations. J. Food Eng. 80, 80–95.

Erdogdu, F., Dejmek, P., 2010. Determination of heat transfer coefficient during high pressure frying of potatoes. J. Food Eng. 96, 528–532.
Farid, M.M., Chen, X.D., 1998. The analysis of heat and mass transfer during frying of food using a moving boundary solution procedure. Heat Mass Transf. 34, 69–77.
Farkas, B.E., Hubbard, L.J., 2000. Analysis of convective heat transfer during immersion frying. Dry. Technol. 18 (6), 1269–1285. https://doi.org/10.1080/07373930008917776.
Farkas, B.E., Singh, R.P., Rurnsey, T.R., 1996. Modeling heat and mass transfer in immersion frying. I, model development. J. Food Eng. 29, 211–226.
Frankel, E.N., 1980. Lipid oxidation. Prog. Lipid Res. 19, 1–22.
Ghaitaranpour, A., Mohebbi, M., Koocheki, A., Ngadi, M.O., 2020. An agent-based coupled heat and water transfer model for air frying of doughnut as a heterogeneous multiscale porous material. Innov. Food Sci. Emerg. Technol. 61, 102335. https://doi.org/10.1016/j.ifset.2020.102335.
Gordon, M.H., 2001. The development of oxidative rancidity in foods. In: Pokorny, J., Yanishlieva, N., Gordon, M. (Eds.), Antioxidants in Food, Practical Applications. Woodhead Publishing Ltd. and CRC Press LLC, Boca Raton, FL, pp. 7–21 (Chapter 2).
Gupta, M.K., 2004. Selection of frying oil. In: Gupta, M.K., Warner, K., White, P.J. (Eds.), Frying Technology and Practices. AOCS Press, Champaign, IL (Chapter 3).
Gupta, M.K., 2005. Frying oils. In: Shahidi, F. (Ed.), Bailey's Industrial Oil and Fat Products. vol. 4. John Wiley & Sons (Chapter 1).
Gupta, M.K., Warner, K., White, P.J., 2004. Frying Technology and Practices. AOCS Press, Champaign, IL.
Khor, Y.P., Hew, K.S., Abas, F., Lai, O.M., Cheong, L.Z., Nehdi, I.A., Sbihi, H.M., Gewik, M. M., Tan, C.P., 2019. Oxidation and polymerization of triacylglycerols: in-depth investigations towards the impact of heating profiles. Foods 8, 475. https://doi.org/10.3390/foods8100475.
Kochhar, S.P., 2001. The composition of frying oils. In: Rossell, J.B. (Ed.), Frying, Improving Quality. Woodhead Publishing Limited and CRC Press LLC, Boca Raton, FL (Chapter 6).
Kochhar, P., 2016. Thermal stability of fats for high temperature applications (Ch. 5). In: Sanders, T.A.B. (Ed.), Functional Dietary Lipids: Food Formulation, Consumer Issues and Innovation for Health. Elsevier Ltd, Cambridge, UK.
Krokida, M.K., Oreopoulou, V., Maroulis, Z.B., Marinos-Kouris, D., 2001. Effect of predrying on quality of french fries. J. Food Eng. 49, 347–354.
Kurek, M., Scetar, M., Galic, K., 2017. Edible coatings minimize fat uptake in deep fat fried products: a review. Food Hydrocoll. 71, 225–235.
Liberty, J.T., Dehghannya, J., Ngadi, M.O., 2019. Effective strategies for reduction of oil content in deep-fat fried foods: a review. Trends Food Sci. Technol. 92, 172–183.
Mallikarjunan, P.K., Ngadi, M.O., Chinnan, M.S., 2010. Principles of deep-fat frying. In: Breaded Fried Foods. CRC Press LLC, Taylor & Francis Group, Boca Raton, FL, pp. 7–31 (Chapter 2).
Matthäus, B., 2009. Acrylamide formation during frying. In: Sahin, S., Sumnu, S.G. (Eds.), Advances in Deep-Fat Frying of Foods. Taylor & Francis Group, LLC, Boca Raton, FL, pp. 143–168 (Chapter 7).
Mesias, M., Delgado-Andrade, C., Holgado, F., González-Mulero, L., Morales, F.J., 2021. Effect of consumer's decisions on acrylamide exposure during the preparation of French fries. Part 1: frying conditions. Food Chem. Toxicol. 147, 111857.
Mittal, G.S., 2009. Physical properties of fried products. In: Sahin, S., Sumnu, S.G. (Eds.), Advances in Deep-Fat Frying of Foods. Taylor & Francis Group, LLC, Boca Raton, FL, pp. 115–142 (Chapter 6).

Moreira, R.G., 2007. Deep-fat frying. In: WIT Transactions on State of the Art in Science and Engineering. vol. 13. WIT Press, pp. 209–237, https://doi.org/10.2495/978-1-85,312-932-2/07. www.witpress.com. (Chapter 7) ISSN 1755-8336.
Mortimore, S., Wallace, C., 1994. HACCP: A Practical Approach. Chapman & Hall, London.
Mousavi Kalajahi, S.E., Asefi, N., 2014. Modeling heat and mass transfer during deep-fat frying: a review. J. Middle East Appl. Sci. Technol. 14 (3), 391–394. 2305-0225.
O'Brien, R.D., 2004. Frying shortenings. In: O'Brien, R.D. (Ed.), Fats and Oils: Formulating and Processing for Application, second ed. CRC Press LLC, Boca Raton, FL (Chapter 7).
Quaglia, G.B., Bucarelli, F.M., 2001. Effective process control in frying. In: Rossell, J.B. (Ed.), Frying, Improving Quality. Woodhead Publishing Limited and CRC Press LLC, Boca Raton, FL (Chapter 11).
Rao, M.A., Rizvi, S.S.H., Datta, A.K., 2005. Engineering Properties of Foods. CRC Press LLC, Taylor & Francis Group, Boca Raton, FL.
Razali, I., Badri, M., 2003. Oil absorption, polymer and polar compounds formation during deep-fat frying of french fries in vegetable oils. Palm Oil Dev. 38, 11–15.
Rossell, J.B., 2001a. Factors affecting the quality of frying oils and fats. In: Rossell, J.B. (Ed.), Frying, Improving Quality. Woodhead Publishing Limited and CRC Press LLC, Boca Raton, FL (Chapter 7).
Rossell, J.B., 2001b. Frying, Improving Quality. Woodhead Publishing Limited and CRC Press LLC, Boca Raton, FL.
Safari, A., Salamat, R., Baik, O.-D., 2018. A review on heat and mass transfer coefficients during deep-fat frying: determination methods and influencing factors. J. Food Eng. 230, 114–123.
Sahin, S., Sumnu, S.G., 2009. Advances in Deep-Fat Frying of Foods. CRC Press LLC, Taylor & Francis Group, Boca Raton, FL.
Saravacos, G.D., Maroulis, Z.B., 2001. Transport Properties of Foods. Marcel Dekker, New York.
Shahidi, F., 2005. Bailey's Industrial Oil and Fat Products. John Wiley & Sons, Inc, Hoboken, NJ.
Shahidi, F., Zhong, Y., 2010. Lipid oxidation and improving the oxidative stability. Chem. Soc. Rev. 39, 4067–4079. https://doi.org/10.1039/b922183m.
Singh, S.V., Verma, A.K., 2016. Mathematical modeling in foods: review. In: Meghwal, M., Goyal, M.R. (Eds.), Food Engineering: Emerging Issues, Modeling, and Applications. Apple Academic Press, New York, NY, USA, pp. 75–110.
Tzia, C., Tsiapouris, A., 1996. Application of the Hazard Analysis Critical Control Point (HACCP) System in the Food Industry. Papasotiriou, Athens (in Greek).
Varela, P., Fiszman, S.M., 2011. Hydrocolloids in fried foods. A review. Food Hydrocoll. 25, 1801–1812.
Velasco, J., Marmesat, S., Dobarganes, M.C., 2009. Chemistry of frying. In: Sahin, S., Sumnu, S.G. (Eds.), Advances in Deep-Fat Frying of Foods. Taylor & Francis Group, LLC, Boca Raton, FL, pp. 33–52 (Chapter 3).
Vinci, R.M., Mestdagh, F., De Meulenaer, B., 2012. Acrylamide formation in fried potato products—present and future, a critical review on mitigation strategies. Food Chem. 133, 1138–1154.
Vitrac, O., Dufour, D., Trystram, G., Raoult-Wack, A.-L., 2002. Characterization of heat and mass transfer during deep-fat frying and its effect on cassava chip quality. J. Food Eng. 53, 161–176.
Vorria, E., Giannou, V., Tzia, C., 2004. Hazard analysis and critical control point of frying—safety assurance of fried foods. Eur. J. Lipid Sci. Technol. 106, 759–765.

Warner, K., 2004. Chemical and physical reactions in oil during frying. In: Gupta, M.K., Warner, K., White, P.J. (Eds.), Frying Technology and Practices. AOCS Press, Champaign, IL (Chapter 2).

Wu, G., Chang, C., Hong, C., Zhang, H., Huang, J., Jin, Q., Wang, X., 2019. Phenolic compounds as stabilizers of oils and antioxidative mechanisms under frying conditions: a comprehensive review. Trends Food Sci. Technol. 92, 33–45.

Yamsaengsung, R., Rungsee, C., Prasertsit, K., 2008. Simulation of the heat and mass transfer processes during the vacuum frying of potato chips. Songklanakarin J. Sci. Technol. 30 (1), 109–115.

Zhang, Q., Saleh, A.S.M., Chen, J., Shen, Q., 2012. Chemical alterations taken place during deep-fat frying based on certain reaction products: a review. Chem. Phys. Lipids 165, 662–681.

Zhanga, X., Zhanga, M., Adhikari, B., 2020. Recent developments in frying technologies applied to fresh foods. Trends Food Sci. Technol. 98, 68–81.

# Different types of fryers for the food industry

# 10

*Chi-Ching Lee and Marin Neio Demirci*
Faculty of Engineering and Natural Sciences, Department of Food Engineering, Istanbul Sabahattin Zaim University, Istanbul, Turkey

## 10.1 Introduction

Frying is an age-old processing technique, with a pan or a pot being its essential tool. The amount of oil or fat used, level of heat, frying time, and even the style of stirring, have set the foundation for many different frying methods. This in turn has given way to developing different fryers, starting with pans of different shapes and sizes, and continuing with a wide selection of electrical automated equipment. While some fryers mimic existing practices, like deep fryers or automated stir fryers, others offer alternative approaches, like air, vacuum, or pressure fryers. Thus, this chapter starts by introducing traditional frying methods and their currently available fryer types and will continue by focusing on alternative fryer options.

In addition to an overview of conventional and alternative fryer types, the chapter will elaborate on them based on different production scales. Domestic kitchen fryers are meant for light use and they should not take up much kitchen space; whereas, commercial fryers are designed for more intense use in the foodservice sector for serving customers and are thus bigger in size and more sturdy. Domestic and commercial frying are commonly batch processes, meaning that a specific amount of food is fried at once. For example, pan-frying is a batch process in which food is loaded into the pan, fried, and removed from the pan when ready. Afterward, another batch of food could be fried. In addition to commercial batch fryers, this chapter will explore their currently available continuous versions as well, where food goes continuously in from one end of the fryer and exits from the other end, fried. On the other hand, industrial batch or continuous fryers are meant for large-scale production of fried products, not for direct consumption, but to be packaged and distributed, which might end up as finished products in retail stores or intermediate products in food services.

Compared to one another, fryers have their advantages and disadvantages, such as price, size, frying speed, operating complexity, production capacity, end-product taste, texture, and heath, which become decisive in the most suitable fryer selection and oftentimes also drive fryer development and innovation. Thus, different fryer advantages and disadvantages will be compared in more detail under fryer selection, followed by insights into fryer development in food industries and current fryer innovation.

High-Temperature Processing of Food Products. https://doi.org/10.1016/B978-0-12-818618-3.00003-3

## 10.2 The conventional type of fryers

The terms "shallow" and "deep" are often used to categorize frying methods, which indicate the amount of oil used during these processes. During deep-frying, food could be completely submerged into the oil. While in comparison, shallow frying, also referred to as pan-frying, uses significantly smaller amounts of oil, either to just grease the utensil surface or leave a visible layer of oil or fat (The Culinary Institute of America, 2011; Myhrvold et al., 2011; Davidson, 2006).

### *10.2.1 Pan fryer*

Shallow or pan-frying are often used as general terms, which include frying methods like pan-frying, stir-frying, and sautéing. Although in one context the terms pan- and shallow frying might be used interchangeably, in another, they might be differentiated by the amount of oil or fat used (The Culinary Institute of America, 2011; Myhrvold et al., 2011; Davidson, 2006). Regardless, this does not affect the type of fryers used. For the sake of clarity, this book will use "pan fryers" as the umbrella term for the fryers under this category although some modern alternatives might not look like a pan at all!

According to the Culinary Institute of America, pan-frying is "A cooking method in which items are cooked in deep fat in a skillet; this generally involves more fat than sautéing or stir-frying but less than deep-frying" (The Culinary Institute of America, 2011). Sautéing and stir-frying are both quick cooking methods using small amounts of fat. Unlike sautéing, stir-frying includes tossing using a traditional wok or, nowadays, a stir-fry pan alternative. Wok is a bowl-like frying pan, which enables tossing of food during stir-frying with minimal spilling (The Culinary Institute of America, 2011; Myhrvold et al., 2011; Davidson, 2006).

Pans come in different shapes, sizes, and materials. Materials used for pans include stainless steel, aluminum, copper, cast iron, and carbon steel. Pans can also be coated with different materials, including Teflon, ceramic, and porcelain enamel. According to the amount and type of food being cooked and the frying method used, pans differ in diameter, bottom thickness, and shape, giving rise to different types of pans, like skillets, frying, roasting, stir-fry, sauce, and sauté pans, and woks (Drysdale, 2009; Zeb, 2019).

Pans can be used with electrical or gas induction heating. In the case of induction heating, pans should contain or be made out of ferromagnetic material to heat up. An induction stove heats only the part of the pan in contact with the stove (Drysdale, 2009). Since the wok is meant to be used in flames, heating up the sides as well, induction woks need a deeply curved induction stove where the wok would fit completely inside, which would enable the wok to heat from both the bottom and the sides. An example of an induction heated wok pan stove can be seen in Fig. 10.3 (Shanghai Chuanglv Catering Equipment Co., Ltd, 2020).

Griddles could be used to replace pans in a commercial setting (Fig. 10.1). A griddle is a heavy, metal surface, which has heaters directly under it. Cooking is done

**Fig. 10.1** Flat commercial griddle.
Courtesy of Ottoman Mutfak, Istanbul, Turkey.

directly on the griddle after greasing the surface with a little bit of oil. Since griddles do not have sides like pans do, it is easier to flip food and they do not hold water near the frying food, allowing it to be properly seared. Griddles have a large surface area, which enables them to cook a large amount of food compared to a pan. Therefore, they are mainly used in the foodservice sector, but smaller ones exist for domestic use as well (The Culinary Institute of America, 2011; Myhrvold et al., 2011; Drysdale, 2009).

Besides pans, automated electrical batch and continuous fryers meant to imitate pan-frying cooking methods have also been developed. As mentioned above, stir-frying and sautéing require less oil, higher heat, and shorter cooking times compared to pan-frying. While all of the techniques require stirring to some extent, stir-frying requires constant tossing of food. Thus, in order to mimic these practices, automated pan fryers should enable mixing at different speeds or even tossing of the food, in addition to having adjustable heating temperatures and oil amounts. It is important that these fryers are able to maintain a high temperature after food is added to the frying vessel and that mixing only causes minimal breakage of the food.

Different ways in which automated pan fryers can mix food are by tumbling, paddling, tossing, or with a horizontal stirring screw. Tumbling type fryers include a hollow ellipsoidal frying vessel that can be tilted to a suitable angle. The amount of oil and food used can be adjusted by the operator who adds them from the opening in front of the tank. The vessel is heated, rotates, and has protruding elements attached to its inner wall to enable uniform frying through even mixing. According to the frying vessel's capacity, tumbling type fryers can be used for domestic, commercial, or even industrial purposes. Since the frying vessel's temperature and rotating speed can be adjusted, the fryer can be used for various pan-frying techniques. Since some of these fryers are specially designed to accommodate the extreme condition of stir-frying, they are called stir fryers (ANKO Food Machine Co., LTD, 2020; Ding-Han

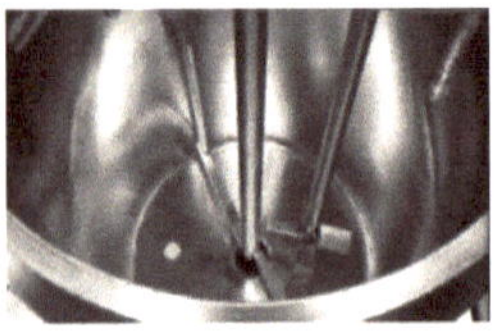

**Fig. 10.2** Batch pan fryer with a central stirring blade.
Courtesy of GEA Group, Düsseldorf, Germany.

Machinery Co., Ltd., 2020a; German Pool Group Co. Ltd, 2020; DANENG (S) Energy-Saving Technology Pte Ltd, 2020).

In the case of a batch pan fryer in which mixing is done through paddling (Fig. 10.2), the frying vessel lays horizontally. The food and oil are added to the heated frying vessel from the opening at the top by the operator. The fryer in Fig. 10.2 has a stationary frying vessel with central blades to stir the food (GEA Group, 2020).

Some of the batch pan fryers are designed especially to accurately mimic the tossing of the stir-frying process to ensure minimal breaking of food (Figs. 10.3 and 10.4). The wok pan in Fig. 10.3 is induction heated and rotates. The pan is heated, oil and food are added, after which the stirrer is lowered, which tosses the food as both the pan and stirrer rotate (IKC Asia-Pacific Pte Ltd, 2020).

The stir fryer in Fig. 10.4 has a horizontal rotating frying vessel with a tossing stirrer on one side, which picks food up from its one side and, with a circular movement, puts it down on the other side. Since it is an open vessel and the stirrer is only at one side, the operator could also easily intervene in the process by mixing or scraping any stuck food with a spatula. This minimizes food breaking during tossing and ensures uniform frying (IKC Asia-Pacific Pte Ltd, 2020).

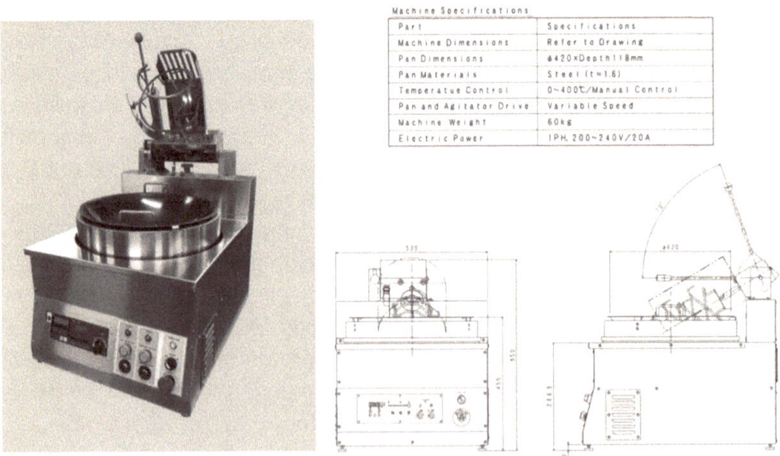

Machine Specifications

| Part | Specifications |
|---|---|
| Machine Dimensions | Refer to Drawing |
| Pan Dimensions | ϕ420×Depth118mm |
| Pan Materials | Steel (t=1.6) |
| Temperatue Control | 0~400℃/Manual Control |
| Pan and Agitator Drive | Variable Speed |
| Machine Weight | 60kg |
| Electric Power | 1PH, 200~240V/20A |

**Fig. 10.3** Batch stir fryer with the stirrer raised up *(left)* and its schematic drawing with the stirrer lowered into the wok pan *(right)*.
Courtesy of IKC Asia-Pacific Pte Ltd., Singapore.

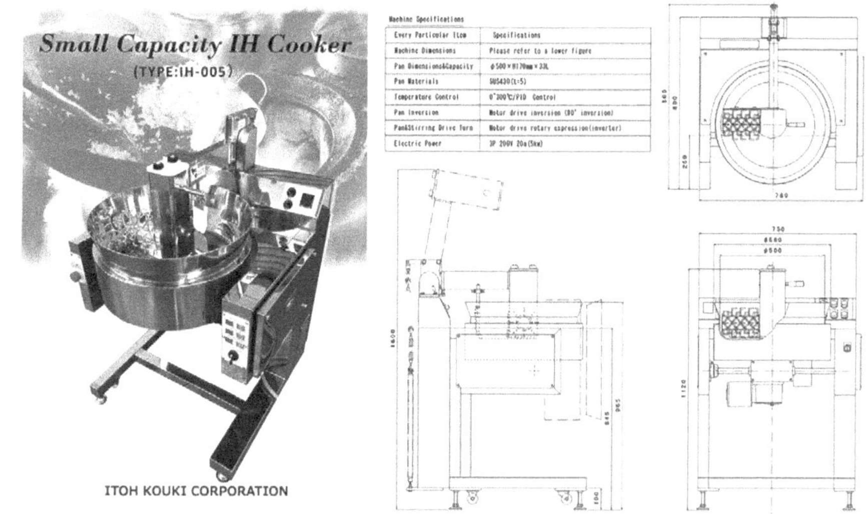


**Fig. 10.4** Batch stir fryer with a tossing stirrer *(left)*, its schematic drawing from above *(center)* and the side, with a side view of the stirrer *(right)*.
Courtesy of IKC Asia-Pacific Pte Ltd., Singapore.

In addition to commercial batch pan fryers, industrial-scale versions are also available. An example would be a fryer with a horizontal cylindrical frying vessel with a stirring screw in the middle (Fig. 10.5). The stirring screw allows sudden forward and reverses movements, which enables it to mimic the stir-frying tossing motion. It is important that food do not stick to the frying vessel and accumulate there, which would result in uneven frying. Thus, these types of frying vessels should also include special built-in scrapers. Furthermore, in some cases, it is possible to modify the size of the frying vessel to enable frying smaller batches. Some of these types of fryer options can even be adjusted for vacuum stir-frying (Blentech Corporation, 2020).

Some batch pan fryers can be adjusted to enable continuous industrial-scale frying. Like the previous batch fryer, the continuous fryer contains a screw in the frying vessel

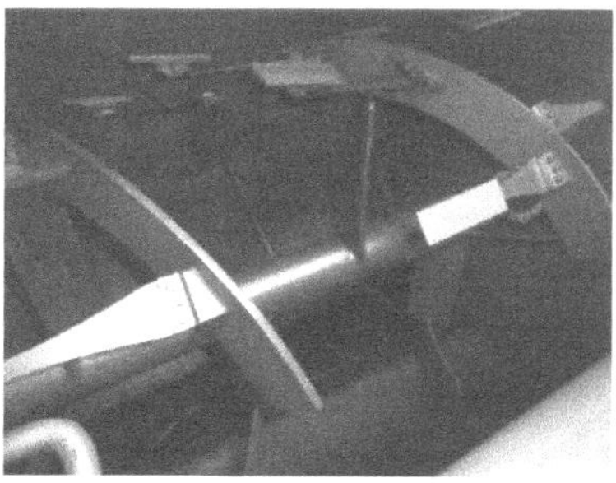

**Fig. 10.5** Close-up of the stirring screw of an industrial-scale batch or continuous pan fryer.
Courtesy of Equipment Exchange, Pennsylvania, United States.

to enable horizontal stirring, with food inlet at one end and the outlet at the other. Depending on the model, these kinds of fryers can also be accommodated to allow pan and deep-frying and blanching or boiling (Blentech Corportion, 2020).

## 10.2.2 Deep fryer

Compared to the variety of pan-frying methods, deep-frying is straightforward. It is important that there is enough oil to cover the product so that it can be submerged in the oil, turned easily in the oil if it is floating, or pushed under the oil. Thus, a deep fryer should be able to heat a large amount of oil and maintain its temperature. However, to enable cost-efficient production, the oil should be maximally reused. Thus, it is important to consider aspects like oil circulation, filtration, and usage duration, to prevent oil decomposition or harmful compounds from forming (Drysdale, 2009; Mallikarjunan et al., 2009; Zeb, 2019).

Domestic deep fryers are usually countertop batch versions, whereas commercial ones could be either countertop or floor-type (Fig. 10.6). Both domestic and commercial batch fryers have strainers, which hold the food. Together with the strainer, food is lowered into the oil and, when ready, it is lifted out. Automatic deep fryers include an elevator system, which lifts the strainer out by itself. The strainer is then left to strain the oil from the product as can be seen in Fig. 10.6 (Ottoman Dış Ticaret ve Mutfak Gereçleri Pazarlama LTD. ŞTİ, 2020).

**Fig. 10.6** The floor-type commercial batch deep fryer. Courtesy of Ottoman Mutfak, Istanbul, Turkey.

**Fig. 10.7** Standalone commercial batch deep fryer unit.
Courtesy of Ottoman Mutfak, Istanbul, Turkey.

Some commercial deep fryer units might contain a separate area to store already fried products, as can be seen on the left of Fig. 10.7. The perforated pan allows extra oil to drain, while heating lamps above keep the products warm. In addition, some commercial deep fryers have different frying programs, with varying frying time and temperature for different products, which makes it possible to get uniform quality products with less operator expertise (Ottoman Dış Ticaret ve Mutfak Gereçleri Pazarlama LTD. ŞTİ, 2020).

It is important to remove crumbs from fryer oil to prevent them from degrading and diminishing oil quality and usage time (Erickson, 2007). Although common in industrial fryers, even commercial fryers may have an oil filtration system, which works by opening a valve at the bottom of the frying oil vessel, allowing the oil and crumbs to flow through a perforated metal plate containing a filter that catches the crumbs, to a container underneath it. After all the oil has been removed from the frying vessel, the filtered oil is pumped back up through a pipe connected with the frying vessel. This system makes it possible to safely clean the frying vessel and filter the oil without spills and taking up too much time and space in the kitchen (Ottoman Dış Ticaret ve Mutfak Gereçleri Pazarlama LTD. ŞTİ, 2020).

Industrial-scale batch fryers are similar to domestic and commercial deep fryers because they include a frying vessel holding the oil and a removable strainer (Fig. 10.8). However, in accordance with the scale of production, the frying vessel is bigger and the fryer includes special mechanisms to lift the heavy strainer (Economode Food Equipment (India) Pvt. Ltd., 2020). In some cases, batch fryers

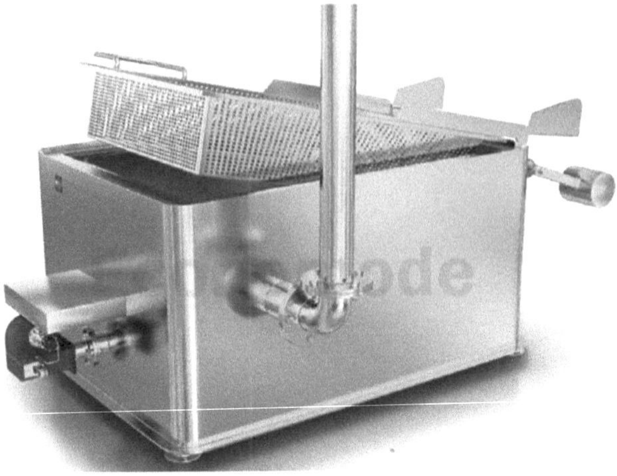

**Fig. 10.8** Rectangular industrial batch deep fryers.
Courtesy of Economode Food Equipment India Pvt. Ltd., Maharashtra, India.

are preferred to continuous deep fryers because they give specific characteristics to the product (PotatoPro, 2020).

In comparison to conventional batch deep fryers, wider batch deep fryers are also available to fry bigger products, like eggplants, fish, or doughnuts. For example, their width enables them to fry multiple doughnuts at once. A flat strainer plate is used to place the doughnuts in the oil and remove them. Since doughnuts float on oil, they need to be turned around to evenly fry both sides. In the case of batch fryers, the operator has to turn the doughnuts manually (Ottoman Dış Ticaret ve Mutfak Gereçleri Pazarlama LTD. ŞTİ, 2020). Continuous doughnut fryers include a long pool of oil, in which the doughnuts float. The pool has either a constant oil current or metal blades, which force the products to move forward. After one side has been properly fried, a wide metal blade automatically turns the doughnuts around to be fried from the other side (Heim Gerätebau GmbH and Co., 2020).

Continuous deep fryers are typically used for large-scale production but smaller commercial versions have also been developed (Ding-Han Machinery Co., Ltd., 2020b). Continuous fryers often include a conveyor belt to move the products along with the oil and in some cases two conveyor belts on top of each other to keep floating products submerged in oil. To get the same quality of product throughout the frying process, the fryer should enable precise side-to-side and end-to-end oil temperature control to avoid any unintended temperature changes. Continuous fryers are usually equipped with systems to remove food crumbs, which either float or sink to the bottom of the oil. Depending on the products, built-in or separate oil filtrations systems can be used to remove impurities enabling the reuse of oil (Fig. 10.9). Since some oil is absorbed by the products during frying, continuous fryers should have systems in place to measure oil level and add oil when required (Erickson, 2007; Sumnu and Sahin, 2008; Mallikarjunan et al., 2009).

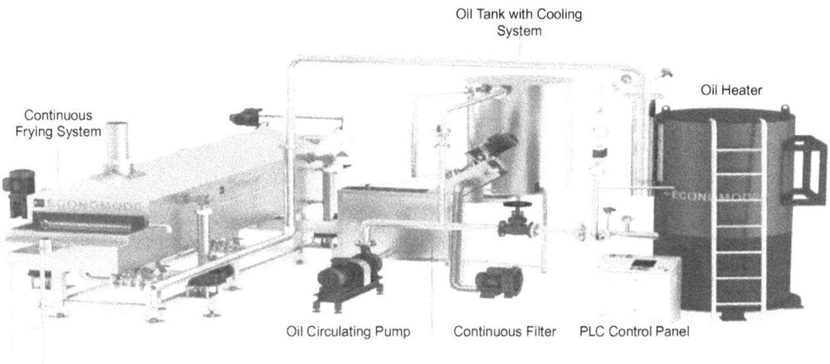


**Fig. 10.9** Continuous industrial deep fryer.
Courtesy of Economode Food Equipment India Pvt. Ltd., Maharashtra, India.

## 10.3 Alternative types of fryers

In the development of frying technology, conventional fryers including pan fryers and deep fryers have been used for centuries. Nevertheless, both frying systems suffer disadvantages, such as high oil uptake and acrylamide formation of final fried products due to high frying temperatures and the large amount of frying oil used. Because health concerns among modern consumers are growing, oil content and acrylamide reduction are the main issues in the improvement of frying technology. Thus, alternative types of fryers such as pressure fryers, vacuum fryers, air fryers, and microwave fryers have been developed to provide a higher quality of fried foods than conventional fryers.

### *10.3.1 Pressure fryer*

The pressure fryer is one of the most common and popular commercial fryers in the fast food industry, especially for fried chicken products. It is well-known that a higher quality of fried product is produced by pressure frying equipment than by an open deep fryer under atmospheric pressure. The pressure fryer was initially designed for fried chicken in fast-food chain restaurants to obtain tender and crispy meat texture within a short time. Because of the demands on fast-food restaurants, the fried chicken is required to have a rapid cooking process and short service time. Some frozen par-fried foods and prepared foods with specific recipes are recommended to cook at a specific pressure. Even though the pressure fryers provide a limited application for large-scale production, the unique flavor and taste after pressure frying are still highly desired by consumers (Mallikarjunan et al., 1997; Mohan Rao et al., 1995).

During the pressure frying process, the moisture inside the foods is released to increase inner pressure in the frying vessel. However, the application of nitrogen gas is used to reach desirable high pressure inside the fryer instead of the pressure generated from the steam from food moisture. Commercial breaded chicken breast nuggets were fried at two pressures (163 and 184 kPa) and three frying temperatures (150°C, 175°C, and 190°C) with two sources of pressure, that is, generated from nitrogen gas and steam caused by moisture of the fried products. The results showed that the frying under nitrogen gas offered the higher quality of fried products in comparison with the pressure source from steam due to the increase of moisture retention, juiciness, and texture preference in the final fried products (Innawong et al., 2006).

The texture of fried products is also influenced by any surface coating on the foods. For the breaded fried chicken nuggets, the whey protein isolate-coated samples frying under a steam environment were crisper comparing with nitrogen gas; whereas uncoated and methyl cellulose-coated samples using nitrogen gas as the pressure source were crispier than using steam. It indicated that the film-forming property of coating materials on fried products is significantly affected by the different pressure sources during the frying process (Ballard and Mallikarjunan, 2006; Sumnu and Sahin, 2008). On the other hand, chicken nuggets were fried at 175°C and 163 kPa for 4 min with a modified restaurant-type pressure fryer under nitrogen gas and steam as pressure sources. Consumer assessment for the fried breaded chicken nuggets was conducted by untrained human subjects. In this case, no significant preferences between these samples were exhibited in terms of oiliness, juiciness, flavor, and crispness (Ballard et al., 2007).

The pressure fryer can be constructed as a batch type or a continuous type with an electric or gas heating system of many different designs. When starting the pressure frying process with a sealed lid, the raw materials are heated and moisture released from the frying materials to become steam, which fills the air space of the frying chamber, thus increasing the inner pressure. The moisture comes out from fried materials continuously until the inner air holds water steam at the maximum level, reaching complete air saturation.

Because moisture content can be maintained within the food materials under the saturated water vapor environment, the final fried products are noticeably juicy, tender, and crispy. In addition, some volatile aroma compounds of foods can be much better preserved within fried products under the high-pressure frying process. However, moisture retention may cause frying oil rancidity rapidly because oils and fats undergo oxidative reactions partially or completely after exposure to moisture. The oil degradation results in the adverse formation of unfavorable compounds, which can affect consumer preferences and purchase decision-making (Mallikarjunan et al., 2009).

There are two types of batch pressure fryer, countertop type and floor type, in restaurants or catering industries. For countertop fryers, the heat capacity ranges from 7.5 to 13 kW. The oil capacity of fryer vessel can be 7 to 15 L with a single basket and gas or electrical heating system with a temperature monitor to heat the oil up to 210°C and cut off the heating source automatically to allow oil temperature cool down to 95°C within a few minutes in the frying vessel. During the pressure frying process, the operating pressure can reach approximately 105 kPa. These countertop fryers are used very commonly for small-scale production in restaurants and pilot plants because the operation is convenient, with a single pressure setting, simple temperature monitoring, and automatic heating turn-off (Gupta, 2004).

Floor-type fryers are also widely used for small- or middle-scale frying production, as they have larger frying vessels than countertop fryers. The heat capacity of these fryers ranges from 25 to 60 kW. The oil capacity of the fry container can be 20 to 40 L. These fryers are also normally equipped with electrical or gas heating systems that enable rapid heat recovery. When the demand for fried products is low, for example, between meal times or at night, the fryer can be set on standby mode. It is very important to allow the frying oil temperature to drop quickly and then let it recover gradually for gaining desirable flavor, texture, and appearance of the final fried product, especially for fried meat, fish, and chicken products. In order to prevent foodborne illness outbreaks and obtain the desired quality of fried products, the oil temperature must be controlled and monitored precisely and prudently (Gupta, 2004).

The first prototype of a pressure fryer was developed in 1917 by The United States Department of Agriculture (USDA) after the first patent of a pressure cooker was granted in 1902. However, it was not widely used until 1935. The first commercial pressure cooker was produced and released onto the market in 1939. Afterwards, it was modified into a pressure fryer by a fried chicken producer to provide high-quality fried products in their restaurants. The first model of pressure fryer was equipped with a jiggle top vent at a single pressure setting of 15 psig (lbs/in$^2$). However, these appliances with this depressurization system could cause operational safety issues. The modern design of pressure release system is a spring-loaded system equipped with a safety alert indicator. This mechanism of pressure controlling not only makes it easy to adjust the inner pressure, but is also safe to depressurize when the frying process is complete.

A floor-type stainless steel pressure fryer with 11 kW thermal power electrical heating system manufactured by Ottoman Mutfak Ltd. (Istanbul, Turkey) is shown in Fig. 10.10 (left). Its 45 L tank reservoir with 25 L oil capacity allows the frying of 30 chicken drumsticks or 50 to 60 chicken wings in one batch frying process. This pressure fryer is equipped with a digital display controller with 10 different programs, a sliding cover system, and an automatic micro stainless steel oil filter. For safety control, there are three mechanical and one electric safety control with audible and visual warning systems. Moreover, it contains an automatic pressure relief system in the event of a pressure jam. Another, automatic model of pressure fryer, similar to the fryer mentioned above but more advanced, and manufactured by the same company, is shown in Fig. 10.10 (right). The significant highlight of the later advanced model is the highly automatic and larger capacity of frying. It contains a four-layer tray basket to increase frying capacity up to 50 chicken drumsticks for one batch frying process. Furthermore, automatic sealed frying vessel and pressurization systems make frying operation more efficient and effective in comparison with the previous model.

### *10.3.2 Vacuum fryers*

Vacuum frying technology was first developed in the late 1960s, when it was used to produce potato chips. Comparing with the traditional frying process, the products produced by vacuum frying are much more desirable. Since the 1980s, vacuum frying technology has developed rapidly and is widely used, especially in East and Southeast

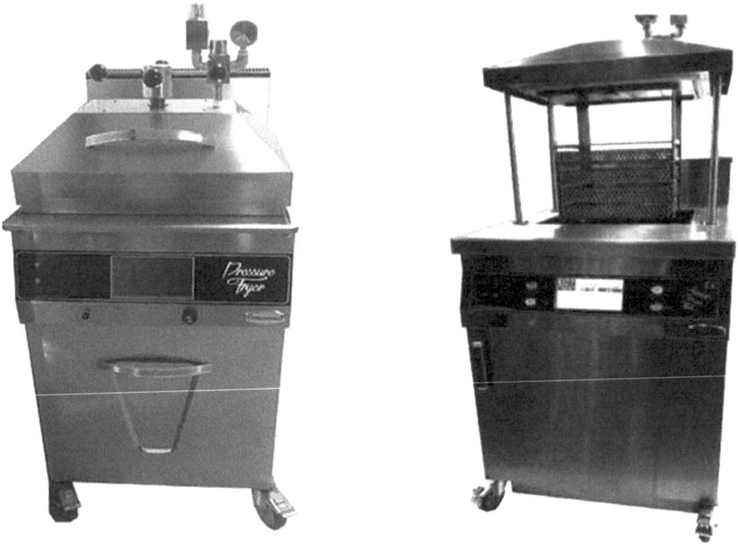

**Fig. 10.10** Two modern models of floor-type pressure fryers: manual-loaded *(left)* and automatic-loaded *(right)*.
Courtesy of Ottoman Mutfak, Istanbul, Turkey.

Asia including Taiwan, Philippines, Indonesia, China, and Thailand. Slices of fruits and vegetables cooked under the vacuum frying process have become very popular fried products. Commonly vacuum-fried vegetables include carrots, okra, sweet potato, potato, mushroom, and so on. The common vacuum-fried fruits include pineapple, mango, guava, banana, apple, and so on. In addition, some companies have established a vacuum frying process for fish and shellfish. Nevertheless, these vacuum fryers are mostly applied for small-scale production (Moreira, 2014).

Vacuum fried products are usually prepared in small chunks or slices, and fried under much lower pressure than the atmospheric pressure of 101.325 kPa; the operating pressure of vacuum fryers is preferably below 6.65 kPa to significantly decrease the boiling point of water and fats in foods so that lower frying temperatures can be applied (Moreira et al., 1999; Dueik and Bouchon, 2011). The acrylamide formation and fat absorption of the fried products in vacuum fryers are also reported to be reduced significantly in comparison with the conventional types of atmospheric fryers (Granda et al., 2004).

Because the pressure is low during the frying process, the moisture and oil absorption of fried products is also reduced. Therefore, the advantages of vacuum fryers are not only to provide oil content reduction for fried products but also to maintain the oil quality under low temperatures. Furthermore, the fried foods are protected from quality loss in terms of flavors, natural color, and texture, because of the low frying temperature and the decrease in oxidation of fried products (Shyu et al., 1998; Garayo and Moreira, 2002). The latest innovations in vacuum fryers have produced models that can operate at a frying temperature below 90°C, thus avoiding the waste oil problem produced from the higher frying temperature of approximately 130°C in the previous generation of vacuum fryers (Dueik and Bouchon, 2011). Using electricity as an energy source is not recommended as a heating system in vacuum fryers because of their low-temperature applications.

A vacuum fryer contains five major components: (a) a vacuum pump; (b) a frying container; (c) a heating element; (d) a condenser; and (e) an oil-removal system. The vacuum pump is to evacuate the sealed frying container to reach the required pressure for the frying process and remove non-condensable gases. The frying container is a sealed vessel including a basket, a heating system, and a lid with a pressure gauge. A heating element can be of tubular or finned type, comprising nickel-chromium resistance wire that is spot-welded to a terminal pin. An electric or gas heating element is usually used in household or restaurant fryers with small capacity. A water or steam heating pump is used for continuous industrial fryers for large-scale production. A condenser is either a water-cooled or a refrigerated condenser to cool down the high-temperature vapor coming from the frying container during the frying process. The steam must be removed before reaching the vacuum pump to protect the pump from any vapor that could cause damage to the vacuum system by cavitation in the pump. The refrigerated condenser is designed to provide higher efficiency condensing than a water-cooled condenser. An oil-removal system is applied with the centrifugation method to spin the product-loaded basket at a high-speed rotation by a centrifuge motor connecting to a lift shaft, which can also adjust the basket position inside the frying container.

When heating the oil in the frying container to the required temperature, the raw material is placed in the basket but it is still suspended above the heated oil by controlling a lift shaft or rod. The frying container is sealed with a lid and the product-loaded basket is not sunk into the hot oil until the frying container reaches desirable pressure at approximately 1.33 kPa for initiating the flying process. According to the requirements of different fried products and frying oils, monitoring the level of pressure, moisture, time, and temperature accurately is a crucial step. When all of the parameters reach a favorable level, the product is lifted from the hot oil, and oil removal method is applied by centrifugation at the required speed and time. Finally, the fried product is removed from the frying container by opening the lid with pressurization. In some cases, the final fried product is also centrifuged by separate equipment or just suspended inside the frying vessel to get rid of the excessive oil on the surface of products. Finally, the vacuum-fried products are placed on absorbent materials, cooled down, and packed in suitable modified atmosphere packaging with nitrogen.

Vacuum fryers can be categorized into industrial-, pilot-, and laboratory-scale, based on their size. Because of process requirements, there are also batch and continuous types of frying equipment. The batch type of fryer is used for testing in laboratories or for small-scale production in pilot and processing plants. Conversely, the continuous type of fryer is used for large-scale production in food industries, mostly for the production of fruit and vegetable crisps in recent years. A variety of vacuum frying equipment can be designed and constructed according to the customer's requirements. A variety of vacuum fryers have been patented for different industrial purposes (Andrés-Bello et al., 2011).

A vacuum fryer with a 20L camber capacity was used for frying slices of Chinese purple yam. The frozen yam slices were fried with vacuum pressure 90kPa at 100°C for 15min. Then the yam crisps were centrifuged at 450rpm under the same vacuum pressure in the frying camber for 5 min to drain the frying oil from the surface of products. Finally, the fried yam crisps were removed from the fryer, cooled down, and packed in foil-film bags (Fang et al., 2011).

Gros Michel banana was fried in a well-designed vacuum fryer including a liquid ring vacuum pump, a condenser, a frying vessel, and a centrifuge. The frying vessel, containing a gas-heating system, was built with stainless steel composed of a 300mm height, a 400mm diameter, and a 6mm thick wall. The lid of the frying vessel consisted of 8mm thick stainless steel. The stainless steel condenser comprised a 9425mm long tube of 19mm internal diameter coiled inside the chamber with water-cooled circulation. After the bananas were peeled and cut into 4mm thick and 25 to 30mm cross-section diameter slices, they were fried in 13.7L of soybean oil under vacuum pressure 8.0kPa and at 100, 110, and 120°C for 20min. After frying, the products were centrifuged at 450rpm for 5min to get rid of the excess oil from the surface of banana slices to decrease oil absorption (Yamsaengsung et al., 2011).

GASTROVAC vacuum frying equipment was designed and constructed under patent at the Polytechnic University of Valencia (Martínez-Monzó et al., 2004). It was used for frying sea bream fillets. This equipment contained a membrane vacuum pump, a pressure chamber with an inner basket, and a heating system with a temperature detector. The fillets were placed in 2L sunflower oil and the ratio of fillet and oil ratio was approximately 0.07 w/w. Frying procedures were comprised of initial heating the oil to target temperature 30min before frying, a depressurization before the fillet-loaded basket was submerged under the oil, the oil immersion of the basket while the temperature and vacuum reached the required values, a frying period of 10min, and opening the lid after pressurization. The samples were fried at 90°C with 15kPa, 100°C with 20kPa, and 110°C with 25kPa, respectively. However, the fried fillets were dried with paper towels after taking out from the frying chamber because this equipment had no oil removal mechanism (Da Silva and Moreira, 2008; Andrés-Bello et al., 2010).

There is a variety of different commercial vacuum fryers designed and constructed by many companies all over the world. A Taiwanese company, I-TUNG machinery industry, designs and produces three types of commercial batch vacuum fryers: Pilot-type, TWIN-5, and TWIN-100. These stainless steel fryers are designed for frying 25 to 100kg potato, sweet potato, and taro crisps per hour and 10 to 50kg fruit/vegetable crisps under 1.33kPa of vacuum pressure (I-Tung Machinery Industry Co., Ltd, 2020). There is a cryogenic vacuum fryer designed to dry-fried products rapidly in a short time after the frying process to lower the oil content in the final product. It contains an electrical heating system and 50kg capacity for each batch. The food range of the vacuum fryer includes fruits, vegetables, meats, and fish products (America TCA Machinery Inc, 2019).

### 10.3.3 Air fryers

Air frying, on the other hand, is the current modern alternative to deep-frying, using a small amount of oil and hot air in the closed compartment of an automated fryer. Due to health-related issues, like the amount of fat or oil, or levels of acrylamide in deep-fried products, air fryers were developed as alternatives to deep fryers (Zaghi et al., 2019). Unlike deep fryers, where oil does the cooking, air fryers use hot air, like convection ovens. However, unlike cooking in convection ovens, air frying is done in a small, compact, closed compartment, which heats up faster, gets hotter, gives off less heat to the surroundings, and moves air faster, resulting in a quicker and crispier final product (Paster, 2019; Pitre, 2019).

Air fryers are equipped with heaters and fans designed to circulate hot air evenly at high speeds through the products, providing a high level of heat transfer and, with a little bit of added oil, give a crispy product similar to deep-frying. Depending on the fryer model, heaters might be separate from the frying compartment, leaving only hot air for cooking, or directly heat the frying compartment together with the hot airflow. Hot air enters the frying compartment either from below, from one side (Fig. 10.11), or from both below and sides and is pulled upward by the fan, thus evenly heating the frying compartment (Zeb, 2019; Ubert Gastrotechnik GMBH, 2020). Additionally, some air fryers are also designed to first cook the products with steam and then with hot air (Ubert Gastrotechnik GMBH, 2020; Posner, 2020).

Currently, air fryers are more commonly used in domestic and commercial settings. Although domestic air fryers might have different styles, like paddle, basket, and countertop oven types, the common commercial air fryer style is the basket type (Fig. 10.11). Paddle-type air fryers include a stirrer in the middle of the frying pan, which moves in a circular motion throughout the frying process and eliminates food build-up, enabling the products to be evenly fried. In this case, the airflow is directed so that the hot air will circulate inside the pan (Peterson and Shaw, 2020; Review GiG, 2020).

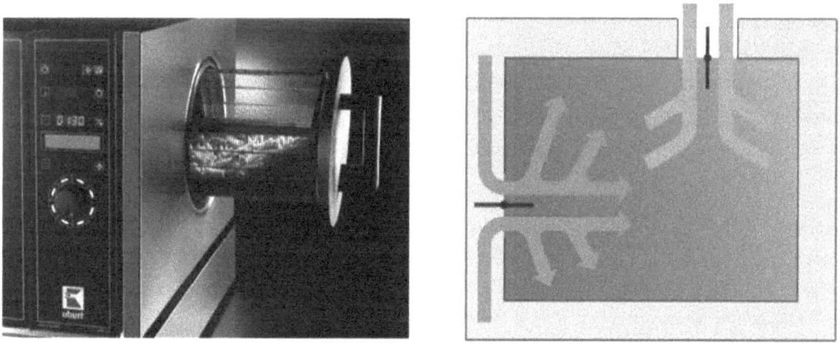

**Fig. 10.11** Commercial air fryer with a cylindrical basket *(left)* and its airflow during air frying *(right)*.
Courtesy of Ubert Gastrotechnik GMBH, Raesfeld, Germany.

The basket-type air fryer includes a basket where the food items are placed, which is then inserted in the fryer. One side of the basket typically works as a door to close the frying compartment. Unlike the paddle type, where food is automatically moved, some basket-type air fryers might require taking the basket out and manually shaking it for even frying. The basket is perforated either from the bottom or from both bottom and sides, to allow excess oil drainage and airflow. Either the food and fat or oil is mixed beforehand or the frying basket is lightly greased with oil. Depending on the fryer model, the amounts of necessary oil or fat vary.

To get even product quality throughout the batch, which is especially important for commercial settings, food should be evenly distributed in the basket before frying. Some commercial models include a cylindrical basket, which could be completely filled and rotates during the frying process, ensuring even frying (Fig. 10.11). At the end of the process, the frying basket is removed and emptied by the operator manually or the products are poured out automatically by the air fryer to a tray beside it (Ubert Gastrotechnik GMBH, 2020; Quik 'n'Crispy Greasless Fryer, 2020). Although the frying process might take longer compared to typical deep-frying of French fries, it does offer a potentially healthier alternative (Zaghi et al., 2019; Zeb, 2019; Santos et al., 2017).

### 10.3.4 Microwave fryer

The microwave application for food production has developed gradually since 1990 and is considered to be an innovative method to improve food quality in a variety of processes including drying, thawing, baking, and frying. The use of microwaves in the food industry provides three major advantages: high energy efficiency, high food quality, and short processing time.

The multiphysics phenomenon of microwave heating can be described as heat transfer between food materials by electromagnetic radiation. When food materials absorb the microwave energy, it can be used to generate heat based on dipolar and ionic mechanisms. The common dipolar molecules in foods are water molecules resulting in dielectric heating. When the oscillation of water molecules in the electromagnetic field occurs, the polarized dipolar water molecules tend to realign with the direction of the field. Because of the high frequency of electromagnetic field provided by microwaves, the realignment and rotation of dipolar water molecules occurs at a million times per second, causing them to collide with other molecules with high internal friction. The motion and kinetic energy generated is tremendous, leading to the heating of food materials.

On the other hand, the ionic conduction mechanism also plays a critical role in microwave heating due to the oscillations and pattern formation of ionic migration in foods. Positive ions move on the direction of the electromagnetic field and negative ions move opposite to the direction of the field. The high-frequency shifting of the field direction brings about the high frequency of ion locomotion oscillating back and forth to generate heat in food materials by the collision and internal friction of molecules (Datta and Davidson, 2000; Sumnu and Sahin, 2008; Chandrasekaran et al., 2013). Modern microwave ovens usually offer oscillation of electromagnetic waves at a frequency of 2.45 GHz. Therefore, microwave heating and its heat transfer can be significantly influenced by the dielectric properties of the food materials and the depth of penetration.

Microwave frying is a new type of application in food production. In addition to the heat of its own frying oil, it can heat up and cook the fried materials from the outside to the inside, and also the energy from the microwave can be used as a second source of heating to quickly cook the fried materials from the inside to the outside because of the dipolar and ionic mechanisms of water molecules. This leads to a short frying time. In contrast with microwave frying, conventional open deep-frying provides a single heat source only from the hot oil to fried foods through the food surface by heat conduction. As a result, the application of microwave fryers is a great alternative to conventional fryers.

The evaporation rate of microwave frying is higher than conventional frying because the oscillation of electromagnetic waves increases moisture loss from fried products. The pressure on the food surface increases quickly and then pressure declines after moisture loss from the surface. Even though the surface pressure is only around 1.0 kPa, it affects the evaporation significantly under convective heat transfer (Datta, 2007). Microwave frying creates a pressure-driven flow to enhance water vaporization, resulting in higher oil absorption in comparison with conventional deep-frying during the same processing period (Feng and Tang, 1998). However, there is less processing time in microwave frying than in conventional frying, which may result in lower moisture loss and less oil absorption due to short frying time. Although the microwave fryer may cause high moisture loss in fried products, this disadvantage can be overlooked due to less frying time. (Oztop et al., 2007).

The domestic microwave oven can be used as a microwave fryer. The microwave frying of potato slices was studied under three power levels of 400 W, 550 W, and 700 W, over 2.0, 2.5, and 3.0 min of frying time. Three types of oil were used including sunflower, corn, and hazelnut oil. Microwave frying was operated in a commercial microwave oven to heat up 0.4 L oil in a glass container to 170°C before seven pieces of potato were fried in hot oil at the desired power level for a specified frying time (Oztop et al., 2007). A microwave fryer was used to fry frozen fish nuggets at 180°C for 3.5 min to compare with conventional deep-fat frying at the same temperature for 5 min. The microwave deep-fat fryer was manufactured by Chin Ying Fa Mechanical Industry Co., Ltd., Taiwan. It had $397 \times 397 \times 360$ mm interior dimensions and $610 \times 510 \times 630$ mm exterior dimensions. It was operated at 2500 W for microwave input and electric power of 2400 W for heating oil. However, the crust qualities of fish nuggets indicated no significant difference between the methods of conventional deep-fat frying and microwave frying (Chen et al., 2009).

The main components of the recently developed microwave fryer are a microwave generator, a frying tray, an oil sprayer, a heating layer, a partition plate, and a cover plate. Raw materials are placed evenly in the frying tray and then this tray is put into the heating layer. The fried oil is sprayed on the raw material surface with the sprayer. The electrical power supply of the microwave generator can provide approximately 4000 v for the magnetron, which is a high-powered vacuum tube, to produce self-excited microwaves by high-frequency oscillation. There is a rotatable agitator with metal fans in the frying chamber to reflect and distribute microwave energy homogeneously for frying products evenly (Tang, 2013).

Most microwave fryers are modified by equipping a microwave generator on conventional deep fat fryers. A deep-frying apparatus is applied with microwave energy from separate magnetrons to a pair of ceramic vessels placed side by side. The magnetron is not in an energized state until the lid of the frying vessel is completely closed to push down a plunger switch (Thomas, 1986). However, the microwave fryer has developed rapidly to produce a user-friendly design that is simple and easy to control. A microwave fryer contains a magnetron, an oil vessel, a detachable frying container, a lifting mechanism to move the frying container and the oil vessel, a temperature monitor, a device protection system, and an oil amount detector (Andoh et al., 1996). A microwave generator device is equipped in a frying chamber in a closed container. During the frying process, the container rotates around with an inclined axis of 20 to 45 degrees at a speed of 1 to 100 rpm (Vandenbussche, 2009).

A combined batch microwave/frying appliance was invented in Taiwan (Cheng and Peng, 2006). It contains a frying device, a microwave generator, and a supporting component. The frying device includes a housing containing an oil groove in the upper end with an upward opening. The Mesh is installed in the oil groove so that fried food items can be placed there. Besides, the gas stove uses a number of heat tubes to heat oil in order to provide heat. The microwave generator comprising the magnetron and the waveguide is equipped in the top lid of the frying device. When raw materials are fried in hot oil in the oil groove, the microwave also directly heats the materials from the top. The products can be fried evenly on the food surface and also internally to avoid the risk of burnt food surface and undercooked interior of final products. Because of providing short processing time, crispy and crunchy fried products can be obtained. The supporting component is located between the microwave generator in the lid and the container of the frying device; these components including two telescopic pneumatic rods, which support the body of the microwave generator in the lid when the frying device is in the open position.

There is a continuous microwave hybrid fryer developed by Tsung Hsing Food Machinery in Taiwan (Tsung Hsing Food Machinery Co., Ltd, 2020). A microwave launch cavity and a microwave transmitter are equipped above the middle section of the oil-frying tank. The microwave, as the second heating source, can achieve the processing optimization of the fried products by configuring the microwave emission power. There is a microwave anti-leakage design at the inlet and outlet to prevent operators from any microwave exposure. A clean-in-place automatic cleaning system is equipped in the frying tank, the heating pipes, and the inner frying hood. A pressure protection system is installed in the oil circulation tube to prevent the pump from idling or oil-free ignition. There are other features including overheating protection in exhaust gas, an automatic temperature controller, and an emergency stop system.

## 10.4 Fryer selection

To select the most appropriate fryer for the desired final products is very important in the frying industry. In order to attract the purchasing preference of customers, producers need to improve their final fried products for all sensory attributes including appearance, texture, aroma, flavor, and mouth-feel. Furthermore, in order to lower the manufacturing cost, producers also need to understand the factors that affect the frying process. However, it may occur to some manufacturers unfamiliar with industrial frying that just one single type of fryer can be used for frying a variety of foods. Actually, the characteristics of raw materials, their sensory qualities, the frying operation, and processing designs all have to be considered when deciding on a frying system. In Fig. 10.12, the frying line for Indian snacks of moong dal is a typical industrial continuous fryer. It includes feeder, heater, oil circulation pump, oil tank, continuous fryer, cooling conveyor, and flavor applicator.

Many factors should be considered in choosing an appropriate fryer for the specific requirements of a food business (Gupta, 2004; Gupta et al., 2004); these include:

- The physical and chemical property of food materials
- The desired sensory attributes in the final fried product
- Productive capacity
- Oil turnover rate
- Oil filtration
- Heat load
- Heating system
- Pollution control
- Fryer maintenance and sanitation

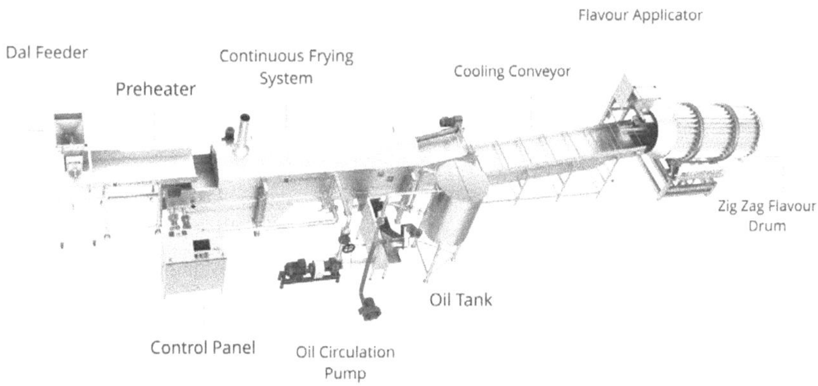


**Fig. 10.12** Moong dal frying line.
Courtesy of Economode Food Equipment India Pvt. Ltd., Maharashtra, India.

### 10.4.1 The physical and chemical properties of food materials

Different types of fried products made from different raw materials have different physical and chemical properties. Batter and breaded coating products have a more complicated process design than uncoated products such as meatballs, potato fries, and vegetable chips. The design of conveyor belts in fryers should be based on the density of fried products. The buoyancy of fried products results in floating, sinking, or submerging in hot oil. Some food materials even change their buoyancy during the frying process. In addition, the chemical components of food materials also influence the fryer design. When the raw materials have high water content, the dehydration and emission of the fryer have to be considered.

### 10.4.2 The desired sensory attributes in the final fried product

The sensory attributes include color, shape, size, texture, flavor, and aroma of final products. The frying method and the process plan affect these attributes significantly. A variety of optical analyses can be used for color measurement of fried foods (Lee et al., 2020). The color of vacuum-fried mango slices was lighter than atmospheric frying (Da Silva and Moreira, 2008). Mango chips with atmospheric frying had less carotenoid retention compared to vacuum frying (Nunes and Moreira, 2009). The special feed belt and product-resting zone also need to be designed based on product characteristics.

### 10.4.3 Productive capacity

The productive capacity represents the maximum output that can be achieved by a processing line or equipment. The desirable productivity of the food industry affects the selection of fryer types. The fryer capacity is usually expressed as a certain figure in weight or volume per hour or per minute, such as in lb/h or kg/min. The floor type or industrial continuous type of fryer has higher productive capacity in comparison with the countertop type or batch type of fryer. The area of the processing plant or operation zone determines the physical dimensions of the fryer, the conveyor length and width, and the loading rate. In addition, there are other factors such as operation time of the fryer, the required frying time for a high quality of final products, and the facility of storage and distribution. All of them are taken into account to choose an appropriate productive capacity in food manufacturing.

### 10.4.4 Oil turnover rate

The definition of oil turnover rate is the time (in hours) required for the total oil absorption of fried products from a whole oil content during the entire frying process. The theoretical oil turnover rate can be calculated as:

$$\begin{matrix}\text{Theoretical} \\ \text{oil turnover rate(h)}\end{matrix} = \frac{\text{The weight of total oil capacity in frying process(kg of lb)}}{\text{The weight of oil absorption by fried materials per hour(kg or lb/h)}}$$

The above calculation exhibits a 100% capacity frying operation. Nevertheless, fryer utilization is reduced to 80%–90% in reality. So the actual oil turnover rate would be calculated as:

$$\text{Actual oil turnover rate (h)} = \text{Theoretical oil turnover rate (h)}/0.8 \text{ or } 0.9$$

Nowadays, a fryer that requires large volumes of oils costs enormously to operate. A low oil volume in a fryer guarantees a low oil turnover rate. When the oil turnover rate decreases, the deterioration and rancidity of fried oil can be reduced to improve the end-product quality. The oil turnover rate of the fryer should be sustained in the 5 to 12 h range for the maintenance of fried oil quality (Dunford, 2003).

### 10.4.5 Oil filtration

During the frying process, food debris is produced and then remains within the fried oil resulting in oil degradation and causing undesirable flavor in the final fried products. Therefore, the purpose of the oil filtration system is to remove fines from fried oil periodically to maintain oil quality at a certain level. The combination of a centrifugal and a motorized filter is commonly applied to remove and treat approximately 5% of fried oil. The selection of the type of oil filter depends on the amount, the size, and the texture of the food particles required to be removed. For example, battered and breaded products usually deposit huge amounts of debris in oil so that bottom-dredging needs to be applied. Furthermore, there should not be a significant quantity of oil loss. Besides, the oil quality should not deteriorate throughout the oil filtering process, similar to the downsides of paper filters.

### 10.4.6 Heat load

Heat load requirement depends on frying various products. Potato chips and corn chips typically require more heat energy than other coated fried products in the frying system. The mass and the energy of heat transfer have to be calculated accordingly, including all heat transfer and loss in the frying chamber, pipes, tubes, exhausts, and filters. The important factors of the delta-T ($\Delta$T) along a whole fryer and temperature recovery time have to be considered for heat load calculation. $\Delta$T is defined as the difference in the gradient of the temperature difference between the feed of raw materials (inlet) and the discharge of final fried products (outlet).

### 10.4.7 Heating system

There are two types of oil-heating systems: direct heating systems and indirect heating systems. A direct heating system uses electrical heating elements or burner tubes to heat up the frying oil in the frying vessel directly. Because the heating elements are submerged in and contact the frying oil directly, the heat energy is transferred into the fried oil completely through the currents of natural convection. A direct heating system is commonly applied for small-scale production in batch frying in the restaurant or household. An indirect oil-heating system is normally applied by large industrial continuous frying plants. The oil is heated in a heat exchanger with an external thermal fluid like a broiler or a heater by electricity or gas. The design of these heat exchangers is commonly of cross-counter flow type and these external thermal fluids can be chlorinated hydrocarbons or water vapor (Wu et al., 2013).

### 10.4.8 Pollution control

All types of fryers, especially continuous industrial fryers, can release certain (sometimes large) amounts of odor, steam, gases, and volatile chemical compounds to the surroundings. According to the local regulations and policies in different cities and countries, scrubber systems are required to be equipped on the fryer compulsorily. A scrubber system uses liquid to remove undesirable pollutants from a gas stream to achieve the purpose of environmental pollution prevention.

### 10.4.9 Fryer maintenance and sanitation

A comprehensive management program for fryer operation should be established to improve the fryer efficiency and minimize any errors and mistakes during the frying process. The start-up and shut-down of frying equipment do more damage to devices than the period of processing. The hardness of water used in food processing plants also affects the effectiveness of cleaning and sanitation (Lee et al., 2015). Heat control for oil loading and circulation needs to be well-programmed to avoid the fryer operating in the idle mode. Austenitic stainless steel should be used for the construction materials on the contact surface instead of copper and brass, because these enhance oil rancidity. Sanitation procedures should include a CIP (clean-in-place) system following the general provisions such as good manufacturing practice (GMP) and hazard analysis and critical control points (HACCP). Pathogens can attach to different kinds of organic and inorganic materials on the surface to develop biofilms; this not only increases the risk of foodborne outbreaks but also reduces the apparatus efficiency. Chlorine solution is widely used in facility sanitation (Lee et al., 2016). All welding joints should be intact, not rough, and all pipes and tubes should be cleaned regularly.

## 10.5 Industrial fryer

Frying technology has improved and evolved enormously for a century and the conventional frying methods such as deep-fat frying in the kitchen are no longer sufficient for the current demands of consumers. The requirements of a fryer in the modern food business include efficient process, rapid production, simple and safe operation, high-quality products, and low cost. Most consumers desire low-oil content, high quality, and convenient fried products to achieve a healthy lifestyle. The frying oils and solid fats used in frying procedures may cause adverse effects such as a high amounts of trans fats and saturated fatty acids. Alternatives to frying oil such as the oleogel can be used during frying processes (Demirci et al., 2020). The industrial frying operation has become a large-scale process and it can produce thousands of kilograms of products hourly. Unlike a fryer used in restaurants or food services, the process of industrial fryers is highly complicated to allow all manufacturing steps following precise requirements automatically or sometimes semi-automatically.

Any kind of food materials can be fried, including coated or uncoated forms. There are various industrially produced fried products in the market, as listed below:

- Plain or seasoned chips: potato chips, corn chips, tortilla chips, vegetable chips, and fruit chips.
- Fried meat, poultry, and seafood products: chicken-fried steaks, batter-coated chicken, batter-coated fish fillet, and batter-coated shrimp.
- Fried nuts: seasoned or sweet peanut, pecan, cashew, pistachio, walnut, hazelnut, and so on.
- Par-fried products: French fries, cheese, vegetables, chicken/fish nuggets or patties.
- Extruded corn products.
- Fabricated products.

## 10.5.1 Types of industrial fryers

In the frying industry, there are two types of fryers: the batch type and the continuous type. Batch fryers are used for small-scale production in the restaurant, catering, and other food services. Continuous fryers are applied in food manufacturing and processing plants for mass production.

### 10.5.1.1 The batch fryer

The batch fryer was described in the previous sections when the frying systems in the restaurant and food services industries were discussed. Countertop frying kettles (Figs. 10.3 and 10.4) are commonly used. The commercial deep fryer can be classified into four categories: flat-bottom fryers (Fig. 10.1), tube-type fryers (Fig. 10.6), open-pot fryers (Fig. 10.7), and commercial pressure fryers (Fig. 10.10).

Flat-bottom fryers are used for frying dough such as Spanish churro and Middle Eastern falafel. The fried dough products can float in the top of fried oil freely to maintain the desired shape of the final products. Tube-type fryers are used for bread-coated products such as onion rings and cheese sticks because these fryers contain large cold zones to prevent all remaining food particles from deteriorating fried oil in order to maintain the oil quality. Open-pot fryers are usually applied for French fries and hash browns because their V-shaped frying vessels can decrease the usage of oil. Commercial pressure fryers are used for frying bread-coated products like fried chicken because high pressure enhances the oil heating system and maintains the fried chicken quality (KaTom Restaurant Supply, Inc, 2020).

### 10.5.1.2 The continuous fryer

The continuous fryer is operated in fried snacks processing plants rather than in restaurants or catering businesses because of the high volume of production required. There are various designs to fulfill different requirements from food manufacturers including the designs of a flat straight frying vessel, a U-shaped frying pan, and an automatic multi-zone fryer.

**Fig. 10.13** Indirect heated fryer with the electrical heat exchanger.
Courtesy of GEA Group, Düsseldorf, Germany.

Raw materials are fed into one end of the fryer (inlet) and fried products are transported to the opposite end by conveyor systems. The fried oil is heated in a straight or U-shaped vessel under frying temperature control with a direct or indirect heating system (Fig. 10.13). An automatic multi-zone fryer is widely used because of the decent maintenance of fried oil quality. The oil is heated indirectly by an external heat exchanger and is then driven into different designated oil sections in the frying vessel (Fig. 10.14). However, this design of oil heating and capacity may retain fried oil at high temperature constantly and enhance oil turnover time, leading to increased oil oxidation and rapid deterioration.

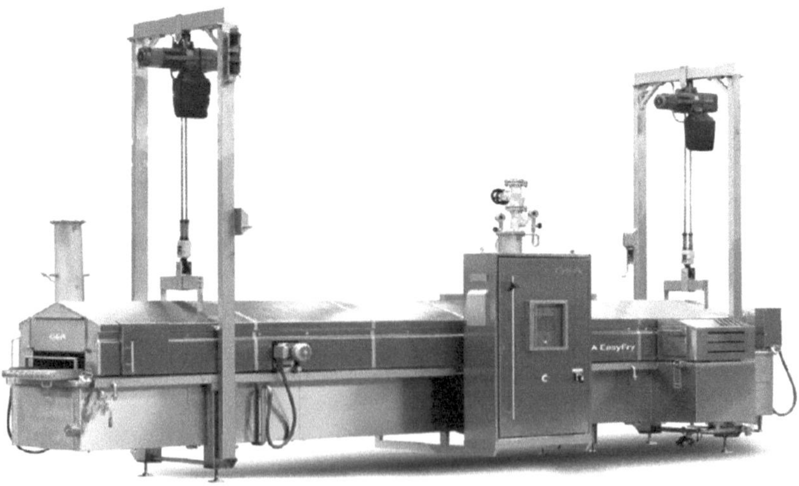

**Fig. 10.14** Multi-zone fryer.
Courtesy of GEA Group, Düsseldorf, Germany.

## 10.5.2 Components of an industrial frying process

An entire industrial frying process consists of a variety of devices and mechanisms. A semi-automatic production line of potato chips is a combination of batch and continuous processing arrangements (Fig. 10.15). It includes feeder with hopper, shaker, control panel heater, oil filter, oil tank, continuous fryer, inspector shaker, and flavor applicator. It is an economical operating line with optimal automation in which the major processes are automated, except for potato grading, peeling, and slicing.

Even though the fryer certainly plays the most crucial role in a whole frying system, many other peripheral and integral apparatuses and devices also integrate to facilitate and complete production. In Fig. 10.16, a fully automatic frying line of potato chips is shown. Its components are as listed below:

- Hopper with feeder: potatoes supplied at a controlled feed rate
- Grader: the removal of undesirable shape and size potatoes by grading and sorting
- De-stoner
- Peeler: potato peeling
- Inspection conveyor: visual inspection taking place
- Single potato feed conveyor
- Slicer: single potato slicing
- PLC control panel: a programmable logic controller (PLC) is used to control manufacturing processes by an industrial digital computer
- Washing station: potato slices washed before blanching
- Transfer conveyor
- Blanching system: scalding potato slices in steam or boiling water in a short time to inactivate enzymes within the potatoes and clean the surface of slices
- Air-sweep system: After blanching, nozzles release a large volume, high pressure burst of compressed air to dewater the potato slices.
- De-watering conveyor with air knife
- Vibro-shaker: extra water removal
- Continuous frying system: this has a variety of designs based on specific requirements, such as submerged conveyors and paddle wheels for potato chip fryers
- Oil receiving and storage system: oil pump recirculating and piping
- Oil heating system: direct and indirect heating system

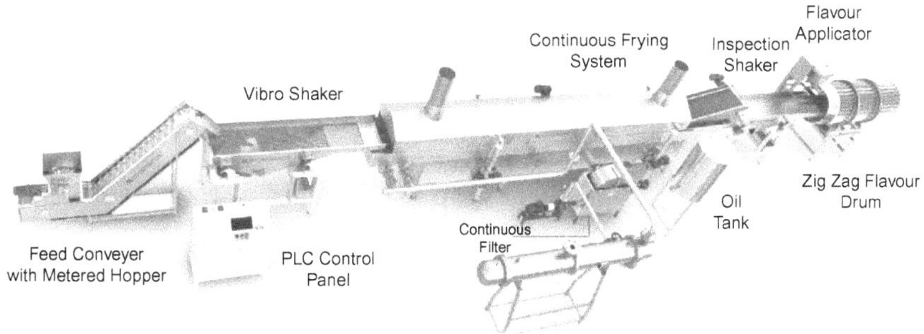


**Fig. 10.15** A semi-automatic potato chips line.
Courtesy of Economode Food Equipment India Pvt. Ltd., Maharashtra, India.

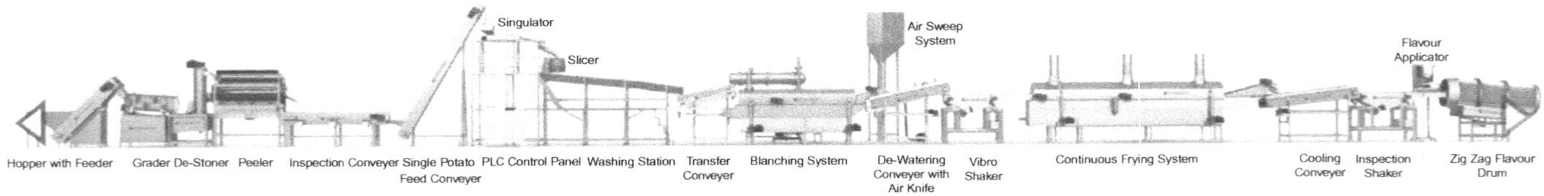


**Fig. 10.16** A fully automatic potato chips line.
Courtesy of Economode Food Equipment India Pvt. Ltd., Maharashtra, India.

- Oil temperature and level monitors
- Oil filtration system: specifically designed conveyor may be equipped to dredge the bottom of fried oil for the removal of precipitated bulk particles or sludge, especially for coated products
- Cooling conveyor or equilibrator for cooling
- Inspection shaker: fried chips color monitoring with optical sorters
- Flavor applicator: salt and seasoning adding
- Zig-zag flavor drum: flavor mixing
- Extruder: fried products extrusion
- Filling/packaging machine: modified atmosphere packaging system applied with a nitrogen flush
- Metal detectors
- Fire-restraint controller
- Process alarm for fault detection and diagnosis in each process unit

## 10.6 Current innovations in fryers

Food producers in the frying industry make efforts to improve frying systems so that ideal processing conditions can be achieved, such as the life extension of fried oil, highly efficient manufacture, high-volume production, low-cost operation, low-energy requirement, and favorable fried products.

There are different advantages and disadvantages for the different frying systems. In Table 10.1, there are comparisons between different fryers in relation to various frying factors. Researchers and manufacturers seek to develop innovative frying technologies to invent new types of fryers that contain advantages from different fryers. For example, the microwave-assisted frying technology was reviewed in this chapter to indicate that this innovative technology can enhance energy-usage efficiency and the quality of the fried product (Ekezie et al., 2017). The microwave-assisted vacuum fryer was studied for the oil-uptake reduction and the quality improvement of fried potato chips. Microwave-assisted vacuum frying decreased oil absorption of fried potato chips in comparison with vacuum frying only. Furthermore, the microwave application also improved the color and crispness attributes of fried potato chips significantly. Under the observation of scanning electron microscopy, the integrity of the cell structure of fried chips was well-preserved after microwave-assisted vacuum frying (Su et al., 2016).

Ultrasound and microwave have also been studied in relation to a vacuum frying system. Purple-fleshed sweet potato was fried in a vacuum fryer with the assistance of ultrasound and microwave technology. The results exhibited that the combination of these technologies not only enhanced the moisture diffusivity and evaporation rate but also reduced oil absorption of fried potatoes compared to vacuum frying without the application of ultrasound and microwave. An improvement of the crispness and color in fried samples was also observed. The higher the power level of ultrasound and microwave is, the better is the quality of fried products produced (Su et al., 2018). The same combination of ultrasound and microwave technology was used in a vacuum

**Table 10.1** A comparison of different types of fryers.

| Fryer | Frying temperature | Frying pressure | Frying time | Oil oxidation | Common fried product | Oil uptake of fried products | Fried product characteristics | Production capacity | Energy efficiency | Operational costs |
|---|---|---|---|---|---|---|---|---|---|---|
| Pan fryer | High 180–200°C | Atmospheric pressure at 101 kPa | Long | High | All types of foods | High | Oxidized easily, may cause acrylamide formation | Low | Low | Low |
| Deep fryer | High 180–200°C | Atmospheric pressure at 101 kPa | Long | High | All types of foods, snacks | High | Oily, may cause acrylamide formation, high oxidation | High | Low | Medium |
| Pressure fryer | Medium 110–130°C | Additional 35 to 105 kPa pressure given | Short | High | Fried chicken, breaded shrimp/fish, vegetables, rice, pudding, custard | Low | The retention of intact flavor in the product | Medium | Medium | Medium |
| Vacuum fryer | Low below 90°C | Low below 7 kPa | Short | Low | Fruit and vegetable chips | Low | Low oxidation | Low | Medium | High |
| Air fryer | High heated air up to 200°C | Atmospheric pressure or additional pressure given | Short | Low | All types of foods | Low | High preservation of flavor and aroma | Low | Medium | Medium |
| Microwave fryer | Medium 160–180°C | Atmospheric pressure at 101 kPa | Short | Low | All types of foods | Variable | The highest quality | Low | High | High |

fryer for button mushroom chips. The findings also showed that this frying-enhanced technology preserved the nutrition and quality of fried mushroom chips (Devi et al., 2018; Devi et al., 2019).

A pressure cooker is a common kitchen appliance in many households and air fryers have become popular in recent years. Therefore, a manufacturer developed a cooker/fryer combined countertop apparatus including a housing, a heater, a fan, a pressure pot, and an air fryer pot. The heater heats external air from outside and then the fan blows the hot air into the inner space of the air fryer pot (Itzkowitz, 2020). Another development of fryers is to have alternative frying platforms. This fryer contains a switchable mechanism of operation between a pressure fryer and an atmosphere open fryer. The selectively closeable frying vessel contains a cover with a locking device by rotation of a handle when an open frying vessel is used. A pressure assistance feature is also provided to reach a required pressure for small fried-loading when a pressure frying platform is used (Eros et al., 2018).

Countertop air fryers have become widely used and are designed as an indoor appliance. However, outdoor air frying apparatuses have been demanded by customers recently because of the advantage of air frying. Hence, an outdoor cooking appliance equipped with an air frying function has been developed (He et al., 2020). This gas air fryer comprises a cooking chamber, a heat exchanger, a blower, and a frying burner. Ceramic and infrared burners have been used in recent years to obtain more heat energy from the fuel and dispense heat energy more uniformly (Bendall, 1998). The frying burner is designed to combust fuel for creating heat and then the blower is adapted to circulate the heated airstream from the heat exchanger into the cooking chamber to fry food materials. This outdoor air fryer can be used in open areas like backyards or parks. It is also applied in catering businesses and any other outdoor food services.

The cool zone has become a popular additional design in the modern fryer. A cool zone is an extra space underneath the heating element at the bottom of the frying chamber into which food debris and chunks sink to be collected. Because the oil temperature is much lower in the cool zone than in the regular frying zone, the food particles are prevented from burning and oil quality can be maintained. Cool zones are usually equipped in huge floor-mounted continuous fryers but sometimes also in small batch fryers. A tube-type fryer contains a large cool zone because the tubes are installed above the bottom of the vessel to allow oil and food particles to cool down in a large space.

## 10.7 Conclusions and perspectives

Frying is one of the most traditional methods for food preparation in human history all over the world. Different regions developed various frying techniques based on their own indigenous ingredients and food culture. The geographical territory and local climate not only affect agricultural products but also influence consumer preference for fried products. In this chapter, we discussed the major types of fryers, including two types of conventional fryers and four types of alternative fryers. The development of frying technology always depends on the demands of food manufacturers and of customers. Some important elements were mentioned made for the improvement of the frying system two decades ago (Moreira et al., 1999). Even though many mechanical engineers have invented a variety of fryers over a century, there are still some crucial factors currently required to be considered in the future. They are listed as follows:

- The quality of fried products: This includes the reduction of fat content, the prevention of nutritional loss, and the improvement of sensory attributes. Thereby, the purchase desire of customers to fried products will be increased significantly.
- Frying process efficiency: Automatic systems are less labor-intensive and decrease human error during operation. Besides, the fryer design should focus on heat capacity and heat exchanging in order to save energy and reduce fuel costs.
- Oil control: Frying oil is the most crucial ingredient in frying production. Oil receiving and storage, oil circulation, oil heating, and oil filtration need to be designed based on the requirements to maintain oil quality.
- Emission control: In recent years, environmental protection and its related regulations have gradually become emphasized. The emission control of industrial fryers is of great concern nowadays. Emission control devices are applied to industrial frying systems for the removal or reduction of pollutants. Mist collectors, scrubbers, incinerators, air filters, catalytic converters, and electrostatic precipitators are all examples of technology for emission control purposes.
- Real-time online monitors: A monitoring system should be established for the whole frying process. The sensors can measure frying oil condition and fried product quality in real-time through online reporting. Alarm systems should be applied and corrective action taken when deviations occur unexpectedly during operation. Computer-based controllers are required to be programmed well.
- Combination systems: Two or more frying systems can be merged into one system under one single operation to develop innovative fryers.

## Acknowledgments

This work was supported by several frying equipment manufactures in the fried food bussiness. They are Ottoman Mutfak in Istanbul, Turkey; Zirve İnternational in Gaziantep, Turkey; GEA Group in Düsseldorf, Germany; Ubert Gastrotechnik GMBH in Raesfeld, Germany; Equipment Exchange in Pennsylvania, United States; Economode Food Equipment India Pvt. Ltd. in Maharashtra, India; and IKC Asia-Pacific Pte Ltd. in Singapore.

## References

America TCA Machinery Inc, 2019. Vacuum fryer 50-150. Retrieved from: http://www.americatca.com/machine/2019-04-29/1696.html.

Andrés-Bello, A., García-Segovia, P., Martínez-Monzó, J., 2010. Vacuum frying process of gilthead sea bream (*Sparus aurata*) fillets. Innovative Food Sci. Emerg. Technol. 11 (4), 630–636.

Andrés-Bello, A., García-Segovia, P., Martínez-Monzó, J., 2011. Vacuum frying: an alternative to obtain high-quality dried products. Food Eng. Rev. 3 (2), 63.

Andoh, Y., Katabe, M., Suenaga, H., 1996. U.S. Patent No. 5,568,765. U.S. Patent and Trademark Office, Washington, DC.

ANKO Food Machine Co., LTD, 2020. Multiple Function Stir Fryer. Retrieved from: https://www.anko.com.tw/en/product/Multiple-Function-Stir-Fryer/sf.html.

Ballard, T.S., Mallikarjunan, P., 2006. The effect of edible coatings and pressure frying using nitrogen gas on the quality of breaded fried chicken nuggets. J. Food Sci. 71 (3), S259–S264.

Ballard, T.S., Mallikarjunan, P., Duncan, S.E., 2007. Consumer assessment of crispness of pressure fried chicken nuggets using nitrogen gas. J. Foodservice 18 (1), 1–6.

Bendall, D., 1998. Well-equipped frying equipment. Food Manage. 33 (3), 72. 74–76, 78.

Blentech Corportion, 2020. Steady-state Oil Cooking Techonology. Retrieved from: https://blentech.com/product/continutherm-cpo/.

Blentech Corporation, 2020. VersaWok Industrial Wok. Retrieved from: https://blentech.com/wp-content/uploads/2014/04/VersaWok-Aug14.pdf.

Chandrasekaran, S., Ramanathan, S., Basak, T., 2013. Microwave food processing—a review. Food Res. Int. 52 (1), 243–261.

Chen, S.D., Chen, H.H., Chao, Y.C., Lin, R.S., 2009. Effect of batter formula on qualities of deep-fat and microwave fried fish nuggets. J. Food Eng. 95 (2), 359–364.

Cheng, F.J., Peng, K.J., 2006. U.S. Patent No. 7,053,346. U.S. Patent and Trademark Office, Washington, DC.

Da Silva, P.F., Moreira, R.G., 2008. Vacuum frying of high-quality fruit and vegetable-based snacks. LWT Food Sci. Technol. 41 (10), 1758–1767.

DANENG (S) Energy-Saving Technology Pte Ltd, 2020. Commercial Double Stir Fryer. November 15. Retrieved from: https://www.daneng.com.sg/product-page/double-stir-fryer.

Datta, A.K., Davidson, P.M., 2000. Microwave and radio frequency processing. J. Food Sci. 65, 32–41.

Datta, A.K., 2007. Porous media approaches to studying simultaneous heat and mass transfer in food processes. II: property data and representative results. J. Food Eng. 80 (1), 96–110.

Davidson, A., 2006. The Oxford Companion to Food. Oxford University Press, Oxford.

Demirci, M., Lee, C.C., Çavuş, M., Çağlar, M.Y., 2020. Oleogels for food applications. In: Biopolymer-Based Formulations. Elsevier, pp. 781–811.

Devi, S., Zhang, M., Law, C.L., 2018. Effect of ultrasound and microwave assisted vacuum frying on mushroom (*Agaricus bisporus*) chips quality. Food Biosci. 25, 111–117.

Devi, S., Zhang, M., Mujumdar, A.S., 2019. Influence of ultrasound and microwave-assisted vacuum frying on quality parameters of fried product and the stability of frying oil. Dry. Technol. 39, 1–14.

Ding-Han Machinery Co., Ltd., 2020a. Stir-fry Machine. Retrieved from: https://www.ding-han.com.tw/en/category/Stir-fry-Machine/A09.html.

Ding-Han Machinery Co., Ltd., 2020b. Tabletop Electric-Heating Frying Machine. Retrieved from: https://www.ding-han.com.tw/en/product/Tabletop-Electric-Heating-Frying-Machine/DH509S.html.

Drysdale, J., 2009. Restaurant Food Service Equipment. Pearson Education, Inc., New Jersey.

Dueik, V., Bouchon, P., 2011. Development of healthy low-fat snacks: understanding the mechanisms of quality changes during atmospheric and vacuum frying. Food Rev. Int. 27 (4), 408–432.

Dunford, N.T., 2003. Deep-Fat Frying Basics for Food Services: Fryer, Oil and Frying Temperature Selection. Division of Agricultural Sciences and Natural Resources, Oklahoma State University, Oklahoma Cooperative Extension Service.

Economode Food Equipment (India) Pvt. Ltd., 2020. Product Gallery. Retrieved from: https://www.economodefood.com/product-gallery.html.

Ekezie, F.G.C., Sun, D.W., Han, Z., Cheng, J.H., 2017. Microwave-assisted food processing technologies for enhancing product quality and process efficiency: a review of recent developments. Trends Food Sci. Technol. 67, 58–69.

Erickson, M., 2007. Deep Frying - Chemistry, Nutrition, and Practical Applications. AOCS Press, Urbana.

Eros, K., Cartwright, R.W., Long, S.J., 2018. U.S. Patent Application no. 15/561,569.

Fang, Z., Wu, D., Yü, D., Ye, X., Liu, D., Chen, J., 2011. Phenolic compounds in Chinese purple yam and changes during vacuum frying. Food Chem. 128 (4), 943–948.

Feng, H., Tang, J., 1998. Microwave finish drying of diced apples in a spouted bed. J. Food Sci. 63 (4), 679–683.

Garayo, J., Moreira, R., 2002. Vacuum frying of potato chips. J. Food Eng. 55 (2), 181–191.

GEA Group, 2020. GEA Processing equipment. Retrieved from: https://www.gea.com/en/binaries/gea-food-processing-equipment-brochure_tcm11-56982.pdf.

German Pool Group Co. Ltd, 2020. Automatic IH Stir-Fryer. Retrieved from: https://www.germanpool.com/eng/products/sf_isf.php?mdl=622.

Granda, C., Moreira, R.G., Tichy, S.E., 2004. Reduction of acrylamide formation in potato chips by low-temperature vacuum frying. J. Food Sci. 69 (8), E405–E411.

Gupta, M.K., 2004. Critical elements in the selection and operation of restaurant fryers. In: Frying Technology and Practices. AOCS Publishing, pp. 115–129.

Gupta, M.K., Grant, R., Stier, R.F., 2004. Critical factors in the selection of an industrial fryer. In: Frying Technology and Practices. AOCS Press, pp. 91–109.

He, A., Murad, U., Simon, P., Giebel, M., 2020. U.S. Patent Application No. 16/848,019.

Heim Gerätebau GmbH & Co., 2020. Continuous Donut Fryers Serie USD-VSD. Retrieved from: https://www.jufeba.de/en/jufeba-2/continuous-doughnut-fryer/.

I-Tung Machinery Industry Co., Ltd, 2020. Vacuum Fruit & Vegetable Chips Equipment. Retrieved from: http://www.taiwan-agriculture.org/itung/index.html.

IKC Asia-Pacific Pte Ltd, 2020. Food and Food Processing Industries. Retrieved from: http://ikcasiapac.com/food-and-food-processing-industries/.

Innawong, B., Mallikarjunan, P., Marcy, J., Cundiff, J., 2006. Pressure conditions and quality of chicken nuggets fried under gaseous nitrogen atmosphere. J. Food Process. Preserv. 30 (2), 231–245.

Itzkowitz, B., 2020. U.S. Patent Application No. 16/743,556.

Posner, J.M., 2020. Light Fry – Cooks with Steam & Hot Air not Oil. Retrieved from: https://www.jmposner.co.uk/products/light-fry—cooks-with-steam-and-hot-air-not-oil.html#Specification.

KaTom Restaurant Supply, Inc, 2020. Types of Fryer. Retrieved from: https://www.katom.com/cat/fryer-types.html.

Lee, C.C., Chen, J., Frank, J.F., 2015. Role of cellulose and colanic acid in attachment of Shiga toxin–producing *Escherichia coli* to lettuce and spinach in different water hardness environments. J. Food Prot. 78 (8), 1461–1466.

Lee, C.C., Chen, J., Frank, J.F., 2016. Influence of extracellular cellulose and colanic acid production on the survival of Shiga toxin–producing *Escherichia coli* on spinach and lettuce after chlorine treatment. J. Food Prot. 79 (4), 666–671.

Lee, C.C., Tomas, M., Jafari, S.M., 2020. Optical analysis of nanoencapsulated food ingredients by color measurement. In: Characterization of Nanoencapsulated Food Ingredients. Academic Press, pp. 505–528.

Mallikarjunan, P., Chinnan, M.S., Balasubramaniam, V.M., 1997. Mass transfer in edible film coated chicken nuggets: influence of frying temperature and pressure. In: Narsimhan, G., Okos, M.R., Lombardo, S. (Eds.), Food Engineering. Purdue University, West Lafeyette, IN, p. 107111.

Mallikarjunan, P., Ngadi, M.O., Chinnan, M.S., 2009. Breaded Fried Foods. CRC Press.
Martínez-Monzó, J., Salvador, J., Torres, S., Sanjuán, N., García-Segovia, P., 2004. System for Vacuum Cooking of food. ES 1057342U.(A47J27/04). 10, p. 8.
Mohan Rao, V.N., Delaney, R.M., Mermelstein, N., 1995. An engineering perspective on deep-fat frying of breaded chicken pieces: engineering aspects of deep-fat frying of foods. Food Technol. 49 (4), 138–141.
Moreira, R.G., Castell-Perez, M.E., Barrufet, M., 1999. Deep Fat Frying: Fundamentals and Applications. Boom Koninklijke Uitgevers.
Moreira, R.G., 2014. Vacuum frying versus conventional frying–an overview. Eur. J. Lipid Sci. Technol. 116 (6), 723–734.
Myhrvold, N., Young, C., Bilet, M., 2011. Modernist cuisine: the art and science of cooking 2. In: Techniques and Equipment. The Cooking Lab, LLC, Bellevue.
Nunes, Y., Moreira, R.G., 2009. Effect of osmotic dehydration and vacuum-frying parameters to produce high-quality mango chips. J. Food Sci. 74 (7), E355–E362.
Ottoman Dış Ticaret ve Mutfak Gereçleri Pazarlama LTD. ŞTİ, 2020. Products. Retrieved from: https://www.ottomanmutfak.com.tr/.
Oztop, M.H., Sahin, S., Sumnu, G., 2007. Optimization of microwave frying of potato slices by using Taguchi technique. J. Food Eng. 79 (1), 83–91.
Paster, E., 2019. Epic Air Fryer Cookbook: 100 Inspired Recipes that Take Air-Frying in Deliciously Exciting New Directions. Harvard Common Press, Boston.
Peterson, W., Shaw, E., 2020. Air Fryer Cookbook for Dummies. John Wiley & Sons, New Jersey.
Pitre, U., 2019. Every Day Easy Air Fryer: 100 Recipes Bursting with Flavor. Houghton Mifflin Harcourt, Boston.
PotatoPro, 2020. Specialised Fryer Designs Improve Productivity and Quality. Retrieved from: https://www.potatopro.com/news/2019/specialised-fryer-designs-improve-productivity-and-quality.
Quik 'n'Crispy Greasless Fryer, 2020. Products. Retrieved from: https://q-n-c.com/products/products/.
Review GiG, 2020. Best Air Fryer review – Your Kitchen Special. Retrieved from: http://www.reviewgig.com/home-n-garden/best-air-fryer-review-kitchen-special/.
Santos, C., Cunha, S., Casal, S., 2017. Deep or air frying? A comparative study with different vegetable oils. Eur. J. Lipid Sci. Technol. 119, 1–14.
Shanghai Chuanglv Catering Equipment Co., Ltd, 2020. Stainless Steel Commercial Induction Wok Cooker 2 Burners Wok Cooking Equipment. Retrieved from: http://www.teppanyakigrilltable.com/sale-10769707-stainless-steel-commercial-induction-wok-cooker-2-burners-wok-cooking-equipment.html.
Shyu, S.L., Hau, L.B., Hwang, L.S., 1998. Effect of vacuum frying on the oxidative stability of oils. J. Am. Oil Chem. Soc. 75 (10), 1393–1398.
Su, Y., Zhang, M., Zhang, W., Adhikari, B., Yang, Z., 2016. Application of novel microwave-assisted vacuum frying to reduce the oil uptake and improve the quality of potato chips. LWT 73, 490–497.
Su, Y., Zhang, M., Bhandari, B., Zhang, W., 2018. Enhancement of water removing and the quality of fried purple-fleshed sweet potato in the vacuum frying by combined power ultrasound and microwave technology. Ultrason. Sonochem. 44, 368–379.
Sumnu, S.G., Sahin, S. (Eds.), 2008. Advances in Deep-Fat Frying of Foods. CRC Press, pp. 293–299.
Tang, C.H., 2013. China Patent No. CN2013203181226U. China Patent and Trademark Office, Hu Nan.
The Culinary Institute of America, 2011. The Professional Chef. John Wiley & Sons, Inc., New Jersey.

Thomas, P.W., 1986. U.S. Patent No. 4,580,024. U.S. Patent and Trademark Office, Washington, DC.

Tsung Hsing Food Machinery Co., Ltd, 2020. Continuous Microwave Hybrid Fryer. Retrieved from: https://www.tsunghsing.com.tw/en/product/continuous_microwave_hybrid_fryer.html.

Ubert Gastrotechnik GMBH, 2020. Foodservices Equipment. Retrieved from: https://www.ubert.com/en/equipment/rofry/rofry-units.html.

Vandenbussche, F., 2009. F.R. Patent No. 2918249A1. French Patent and Trademark Office, Paris.

Wu, H., Tassou, S.A., Karayiannis, T.G., Jouhara, H., 2013. Analysis and simulation of continuous food frying processes. Appl. Therm. Eng. 53 (2), 332–339.

Yamsaengsung, R., Ariyapuchai, T., Prasertsit, K., 2011. Effects of vacuum frying on structural changes of bananas. J. Food Eng. 106 (4), 298–305.

Zaghi, A., Barbalho, S., Guigeur, E., Otoboni, A., 2019. Frying process: from conventional to air frying technology. Food Rev. Int. 35, 763–777.

Zeb, A., 2019. Food Frying: Chemistry, Biochemistry, and Safety. John Wiley & Sons Ltd., Oxford.

# Index

Note: Page numbers followed by *f* indicate figures and *t* indicate tables.

## G

## P



## Q



## R



## S

FSC
www.fsc.org
MIX
Papier | Fördert
gute Waldnutzung
FSC® C083411

Kolibri 360 GmbH
Ferdinand-Jühlke-Straße 7
99095 Erfurt, Deutschland
produktsicherheit@kolibri360.de